COLOUR AND CLARITY OF NATURAL WATERS
Science and Management of Optical Water Quality

COLOUR AND CLARITY OF NATURAL WATERS
Science and Management of Optical Water Quality

R. J. DAVIES-COLLEY
W. N. VANT and D. G. SMITH
all at the NIWA-Ecosystems, National Institute of Water and Atmosphere,
Hamilton, New Zealand

THE BLACKBURN PRESS

Reprint of First Edition, Copyright 1993

Colour and Clarity of Natural Waters
Science and Management of Optical Water Quality

ISBN-10: 1-930665-71-7
ISBN-13: 978-1-930665-71-2

Library of Congress Control Number: 2003103130

THE BLACKBURN PRESS
P. O. Box 287
Caldwell, New Jersey   07006 U.S.A.
973-228-7077
www.BlackburnPress.com

Table of contents

Plates 1 to 14   are between pages 160 and 161

Foreword

There is one area of water quality where human beings can detect degradative change even without the assistance of scientific measurement, and that is optical water quality. We can see for ourselves when a previously clean river is turned brown by effluent, or turbid by soil particles from an eroding catchment. And while our initial response may be primarily aesthetic — humankind has, after all, found beauty in limpid streams and clear blue seas since time immemorial — our concern at such changes is well founded. Not only do they mar the visual quality of our lakes, rivers and seashores, and lessen their attractiveness for water-based recreation, but they also have serious implications for the health of the aquatic ecosystem. The plants which form the base of the whole food chain cannot photosynthesize, and the fish and other fauna cannot seek their food, if deprived of sunlight. The murkier the water becomes, the shallower the surface layer within which solar heat is trapped, so that stratification is intensified, deep circulation is impeded, and anoxic layers grow.

Given that impairment of the optical quality of our surface waters is of common occurrence, and is so obvious when it occurs, it is puzzling that, by comparison with concerns about chemical and microbial quality, it has received so little attention from the wastewater industry and environmental managers. Part of the problem might perhaps have been the perceived difficulty in establishing clear-cut and physically meaningful, but ecologically and psychologically relevant, criteria in terms of which to assess optical water quality.

Fortunately, however, to arrive at such criteria has now become much easier. It has been made possible by the establishment, mainly over the last 50 years, of a sound body of physical theory and observation concerning the nature of the underwater light field. This has arisen more from studies in marine, than in freshwater, ecosystems, but the principles of hydrologic optics are everywhere the same, and can be applied with as much confidence to a muddy lowland river as to the clearest ocean. The task of setting up criteria has also been made easier by the ready availability in recent years of photoelectric instrumentation for characterizing the underwater light field, and by the development of laboratory methods for measuring the inherent optical properties of natural waters and wastewater.

Armed with the new instruments and methods, and basing ourselves firmly within the robust theoretical framework of hydrologic optics, we can now hope to arrive at criteria for optical water quality that can be applied universally and interchangeably throughout the world. Indeed a beginning has already been made, and some proposals for the establishment of practical standards of optical water quality, for the benefit of environmental managers and industrial water users, have already appeared in the scientific literature.

The time is therefore ripe for just such a book as Rob Davies-Colley, Bill Vant and Dave Smith have written. A book to carry the message to that audience — wastewater engineers, industrial dischargers of effluent, environment protection agencies, limnologists, coastal oceanographers — who need to hear it, to grasp its implications, and to frame policies in its light. This book provides, for anyone interested in, or charged with the responsibility for, the management of the optical quality of our surface waters, a comprehensive treatment which outlines the fundamental concepts, explains what the problems are, describes the sort of measurements that need to be made, and discusses the criteria in terms of which we can assess optical water quality together with possible specific guidelines which could be applied.

The authors are well qualified to address this topic: they have been actively engaged in research into optical water quality for a number of years, and have made notable contributions to it. The valuable case studies described in the book arise directly from their own professional involvement in problems relating to colour and clarity of water bodies in New Zealand.

There already exist, of course, other books and reviews dealing with the optics of our surface waters. But these have been aimed primarily at research scientists and students, in oceanography and limnology. This book is the first to be written specifically for the environmental managers and the water-using industries. I believe it will become a standard work, and I commend it to all those in our society who have a responsibility for, or whose activities have a significant impact on, the quality of our inland or marine waters.

John T. O. Kirk
Canberra, March 1993

Preface

This book is an outgrowth of a suite of studies on the optical quality of water, and before that, on the visual appearance of waters ('visual water quality'), that have been undertaken at the Water Quality Centre since 1983. The main impetus for the work was that New Zealand water law, then as now, specifically protects the colour and clarity of waters. Under the Resource Management Act (1991), discharges cannot be permitted if they will give rise to 'Any conspicuous change in the colour or visual clarity' of waters. Initially the research focussed on colour and clarity of waters as perceived by humans, but the work was rapidly extended to light for plant photosynthesis, and more recently, to optical methods for monitoring and tracing of waters, to vision of aquatic animals, and to the penetration of biocidal ultraviolet radiation into waters.

Internationally there has been little attempt to apply the methods and concepts of aquatic optics in water quality management. There exists, still, a large communication gulf between scientists working in the field of aquatic optics (most of whom are concerned with remote sensing applications primarily in marine environments) and those applied scientists and engineers concerned with management of inland and coastal waters. The former group generally have little comprehension of, and possibly little interest in, the water quality problems of shallow inland waters such as rivers. The latter group have little understanding of the principles of aquatic optics in spite of their frequent use of such basic optical tools as the Secchi disc and the turbidimeter. Given their very different perspectives it is perhaps not surprising that there has been little interchange or cross-fertilization between these two groups of professionals. One notable exception has been the work of John Kirk (CSIRO, Canberra, Australia) whose publications show an acute awareness of the potential applications of aquatic optical concepts in water management. Kirk (1982) introduced the term 'optical water quality' which is the subject of this book. The purpose of this book, written from our perspective as water quality scientists, is to bridge the gap between aquatic optics and water quality management.

During the writing of this book, the authors have all been employed as government scientists at the Water Quality Centre. In 1986, when the project was first suggested, the

Water Quality Centre was part of the Water and Soil Division of the New Zealand Ministry of Works and Development. In 1988 the Water Quality Centre became part of the New Zealand Department of Scientific and Industrial Research (DSIR) and remained under that umbrella until the demise of DSIR in June 1992. Most of the writing of the book was carried out when the Water Quality Centre was part of DSIR, and the book was completed after the Water Quality Centre became part of the National Institute of Water and Atmospheric Research (NIWA).

This book was originally conceived of (in 1986) as a handbook for water managers on colour and clarity of waters. The scope and purpose of the book was considerably widened in 1988 and since then the book has become more of a monograph with an applied flavour, rather than a handbook. A handbook, as we understand the term, is more appropriate to fields where recipes for applying the well known are required. There is still too much that is poorly known about 'optical habitat' of aquatic organisms, or, for that matter, of the needs of human recreational users of waters, for a handbook to be produced. There was a definite need to introduce concepts and to discuss the underlying science (of 'aquatic optics') in greater detail than would be appropriate in a handbook. Nevertheless, this book has some of the attributes of a handbook, and included in these pages is much guidance on practical problems of water management involving optical water quality. Although the book focusses on natural waters, we have also considered the optical character of effluents which may affect the colour and clarity of receiving waters.

The level of treatment in this book varies somewhat with subject matter but generally presumes that the reader has a good general grounding in fundamental sciences (chemistry, biology, and especially physics) and also in some less-fundamental sciences and relevant applied sciences (geology, soils, hydrology, physical oceanography, physical geography, environmental engineering). We have assumed that readers can cope with some algebra and trigonometry and introductory-level calculus. We trust that natural scientists will not be dismayed at some social science which has crept into this book — and is entirely appropriate in such a necessarily interdisciplinary field as water quality management.

This is a jointly authored book. However, RJD-C has led the writing project from the outset and is primarily responsible for the structure and scope of the book. As authors, we have acted as our own editorial board, with RJD-C as editor-in-chief. With the multiple rounds of review and editing, the contributions to the various chapters and sections of the book have inevitably become somewhat blurred. Nevertheless, some may be interested in the primary contributions which are as follows: Chapters 1, 2 and 3 are by RJD-C; Chapter 4 was written mainly by WNV (RJD-C wrote Section 4.3); Chapter 5 has contributions from all three of us (Section 5.2 is by DGS; Section 5.3 is by RJD-C and DGS; Sections 5.4 and 5.5 are by RJD-C; Section 5.6 is by WNV); Chapter 6 is by RJD-C; Chapter 7 is by DGS.

Hamilton, New Zealand
September, 1992

RJD-C
WNV
DGS

Acknowledgements

As is often the case with a book of this kind, a large number of people have contributed (many unwittingly!) to the book's completion. It is not possible to thank everyone but special mention is due to Rob McColl (Department of Conservation, Wellington) who first suggested we prepare a handbook on colour and clarity for water managers.

This book was written while the authors were all with the Water Quality Centre in Hamilton, New Zealand. We thank all the staff of the Water Quality Centre who were involved in any way with the project (which is most). Dr Noel Burns was scientist-in-charge of the Centre for most of the duration of the writing task. Graham McBride was a helpful and encouraging co-ordinator of the programme from which this project was funded in 1991–1992, and he provided a very thorough review of the complete manuscript. Dr Kit Rutherford proffered valued advice about books and publishers from his own experience with the 'Handbook on mixing in rivers' (to be published by Wiley).

We have done most of our own review and editing but Dr Bruce Williamson reviewed Sections 5.4 and 5.5, Christine Smith and Keith Smith reviewed Section 5.5, and Rohan Wells checked Section 6.3. Christine Thompsen assisted with the index preparation and with dataprocessing matters. Mary Stokes typed much of the earlier material (Chapters 2 and 3) before we obtained our own Macintosh PCs, and Frank Bailey prepared the hand-drawn figures (Figs 2.11, 2.18, 3.16, 6.4, 6.6, 6.12, 6.14 and A1.1).

We are grateful for the patience, support and understanding of our three families.

Nomenclature

Symbol	Quantity (definition or explanation, units of measurement)
a	**Absorption coefficient**, m^{-1}. The proportion of light absorbed from a perfectly collimated light beam over a small path length, divided by the length of the path.
a_c	**Absorption coefficient of all the constituents of water**, m^{-1}.
a_f	**Absorption coefficient of a water filtrate**, m^{-1}.
a_p	**Absorption coefficient of the particulate constituents of water**, m^{-1}.
a_w	**Absorption coefficient of pure water**, m^{-1}.
A_i	**Absorption cross-section**, $m^2 \, g^{-1}$. The absorption coefficient per unit mass concentration of a constituent, i.
b	**Scattering coefficient**, m^{-1}. The proportion of light scattered from a perfectly collimated light beam over a small path length, divided by the length of the path.
b_b	**Backscattering coefficient**, m^{-1}. The proportion of light scattered from a perfectly collimated light beam, between $90°$ and $180°$ to the incident direction, over a small path length, divided by the length of the path.
b_w	**Scattering coefficient of pure water**, m^{-1}.

B **Luminance,** $lm\ m^{-2}\ sr^{-1}$. The photopic equivalent of radiance; see Appendix 1.

B_b **Background luminance,** $lm\ m^{-2}\ sr^{-1}$.

B_i **Scattering cross-section,** $m^2\ g^{-1}$. The scattering coefficient per unit mass concentration of a constituent, i.

B   **Biomass** (of phytoplankton — as chlorophyll a concentration), $mg\ m^{-3}$.

BOD   **Biochemical oxygen demand,** $g\ m^{-3}$. An index of the biochemically labile organic content of a water — the oxygen likely to be consumed in biochemical oxidation of a wastewater discharged to a natural water.

c **Beam attenuation coefficient,** m^{-1}. The proportion of light removed from a perfectly collimated light beam over a small path length by the processes of absorption and scattering, divided by the length of the path.

c   **Speed of light,** $2.9979 \times 10^8\ m\ s^{-1}$.

C **Beam attenuation cross-section,** $m^2\ g^{-1}$. The beam attenuation coefficient per unit mass concentration. Sometimes given a subscript, i, denoting constituent, i.

C_r **Apparent contrast of an object** (at range, r), dimensionless.

C_T **Threshold contrast of an object** (at extinction of its image), dimensionless.

C_i **Concentration of the ith constituent of water,** $g\ m^{-3}$.

C_0 **Inherent contrast of an object** (at zero range), dimensionless.

d **Characteristic dimension of a particle,** μm. The diameter for a sphere, and the diameter of a sphere of the same volume for a non-spherical particle.

D **Absorbance** or **optical density,** dimensionless. $-\log_{10}$(transmittance), as measured by a spectrophotometer.

D   **Ratio of diffuse to global sunlight,** dimensionless.

DOC   **Dissolved organic carbon concentration,** $g\ m^{-3}$.

E **Irradiance (general symbol)**, einsteins (moles of photons) $m^{-2} s^{-1}$, or, alternatively, $W m^{-2}$. The light flux incident per unit area of a surface. The flux may be measured in photons per second or, alternatively, in watts. See Appendix 1.

E_d **Downwelling vector irradiance**, einsteins $m^{-2} s^{-1}$, or, alternatively, $W m^{-2}$. The irradiance detected by a vector irradiance sensor facing upwards.

E_o **Scalar irradiance**, einsteins $m^{-2} s^{-1}$, or, alternatively, $W m^{-2}$. The irradiance detected by a suitably calibrated spherical irradiance sensor, $E_o = 4E_s$.

E_r **Reference irradiance**, einsteins $m^{-2} s^{-1}$, or, alternatively, $W m^{-2}$.

E_s **Spherical irradiance**, einsteins $m^{-2} s^{-1}$, or, alternatively, $W m^{-2}$. The irradiance actually detected by a spherical irradiance sensor, $E_o = 4E_s$.

E_u **Upwelling vector irradiance**, einsteins $m^{-2} s^{-1}$, or, alternatively, $W m^{-2}$. The irradiance detected by a vector irradiance sensor facing downwards.

$\vec{E}$ **Net vector irradiance** $(= E_d - E_u)$, einsteins $m^{-2} s^{-1}$, or, alternatively, $W m^{-2}$.

F **Dilution factor or concentration factor**, dimensionless.

g **Absorption coefficient of a membrane-filtered water sample**, m^{-1}. The absorption coefficient at 440 nm, g_{440}, is a useful index of the yellow substance concentration of a water. g stands for gilvin or *Gelbstoff* (Kirk 1976).

h **Planck's constant**, 6.6256×10^{-34} J s.

I **Illuminance**, $lm m^{-2} = lx$. Light as seen by the human eye. The photopic equivalent of irradiance. Appendix 1.

j **Angle to the vertical of the direct solar beam in water**, degrees, or, alternatively, rad.

k **General irradiance attenuation coefficient**, m^{-1}. The proportion of radiant flux removed over a small path in a water body, divided by the path length.

k_c **Specific irradiance attenuation coefficient for phytoplankton (as chlorophyll *a*)**, m^2 (mg chlorophyll *a*)$^{-1}$. The contribution to the total irradiance attenuation coefficient attributable to phytoplankton, divided by the phytoplankton chorophyll *a* content.

K **Illuminance attenuation coefficient**, m^{-1}. The proportion of ambient irradiance removed over a small depth interval in a water body, divided by the depth interval. Sometimes referred to as the diffuse attenuation coefficient.

K_d **Irradiance attenuation coefficient for downwelling irradiance,** m^{-1}.

K_E **Attenuation coefficient for net vector irradiance,** m^{-1}.

K_o **Irradiance attenuation coefficient for scalar irradiance,** m^{-1}.

K_u **Irradiance attenuation coefficient for upwelling irradiance,** m^{-1}.

K_{max} **Maximum luminous efficiency,** 680 lm W^{-1}. The luminance of radiant flux at 555 nm.

L **Radiance,** W m^{-2} sr^{-1}. See Appendix 1.

L_s **Radiance reflected from the surface of a water body,** W m^{-2} sr^{-1}.

L_w **Radiance emanating from within a water body,** W m^{-2} sr^{-1}.

M   **Concentration of mineral suspensoids,** g m^{-3}. Measured as non-volatile suspended solids.

m   **Mass of a particle,** g.

n   **Refractive index,** dimensionless.

NVSS   **= SS – VSS, non-volatile suspended solids,** g m^{-3}. Ash residue, an index of the concentration of mineral suspensoids.

P **Photosynthetic rate of aquatic plants,** s^{-1}. Measured as rate of oxygen production or bicarbonate uptake per unit biomass.

P_{max} **Light-saturated photosynthetic rate of aquatic plants,** s^{-1}.

POC   **Particulate organic carbon concentration,** g m^{-3}.

Q **Discharge of water,** m^3 s^{-1}.

Q_c **Attenuation efficiency factor,** dimensionless. Ratio of the attenuation cross-section of a particle to its geometric cross-section.

r **Length of light path through water,** m.

R **Irradiance reflectance coefficient ('reflectance'),** dimensionless. The ratio of upwelling to downwelling irradiance in water.

$R(0-)$ **Irradiance reflectance coefficient immediately below the water surface,** dimensionless.

R_{disc} **Illuminance reflectance of the white surface of a Secchi disc,** dimensionless.

R_I **Illuminance reflectance coefficient,** dimensionless. The ratio of upwelling to downwelling illuminance in water.

$R_r(\lambda)$ **Relative irradiance reflectance spectrum,** dimensionless.

$S(\lambda)$ **Spectral response function,** dimensionless.

s **Projected surface area of a particle,** m^2.

S **Slope parameter** (describing an exponential absorption spectrum), nm^{-1}.

S   **Area,** m^2.

SS   **Suspended solids concentration,** $g\ m^{-3}$.

T **Transmittance,** dimensionless. Ratio of transmitted to incident radiant flux.

T_c **Beam transmittance,** dimensionless. Transmittance of a perfectly collimated light beam.

T_E **Irradiance transmittance,** dimensionless. Ratio of irradiances at two different depths in a water body.

TOC   **Total organic carbon concentration,** $g\ m^{-3}$.

v_s **Settling velocity of a particle,** $m\ s^{-1}$. Fall velocity of particles as given by Stokes' law.

V **Volume,** m^3.

VSS   **Volatile suspended solids concentration,** $g\ m^{-3}$. An index of the concentration of organic suspensoids.

x, y, z	**Proportions of red, green and blue primary colours in the CIE system which specify a given colour** (i.e. match the colour when mixed), dimensionless.
$x(\lambda), y(\lambda), z(\lambda)$	**Red, green and blue tristimulus functions (colour mixing curves) in the CIE system**, dimensionless.
X, Y, Z	**Amounts (luminance) of red, green and blue primary colours in the CIE system which specify a given colour**, dimensionless.
y_{BD}, y	**Black disc visibility (horizontal direction)**, m. The range in water at which the image of a black disc, viewed horizontally, is judged to be extinguished.
z	**Depth in a water body**, m. Conventionally measured vertically downwards from the water surface.
z_c	**Maximum colonization depth** of macrophytes in a water body, m.
z_{eu}	**Euphotic depth in a water body**, m. The depth at which irradiance is reduced to 1% of that at the water surface.
z_m	**Mid-point of the euphotic zone** ($= z_{eu}/2$), m.
z_{SD}	**Secchi depth**, m. The depth at which the image of the white surface of a Secchi disc, viewed vertically through water, is judged to be extinguished.
z_{BD}	**Black disc visibility (vertical direction)**, m. The depth at which the image of a black disc, viewed vertically through water, is judged to be extinguished.
α	**Shape factor in the Stokes law equation for particle settling velocity**, dimensionless. **Coefficient in the power law relation between river discharge and visual water clarity**, dimensionless.
β	**The 'path amplification factor' (ratio of optical to geometric path) for a diffuse light field**, dimensionless. **Exponent in the power law relation between river discharge and visual water clarity**, dimensionless.
$\beta(\theta)$	**Volume scattering function** (scattering as a function of angle, θ, to the incident beam), $m^{-1}sr^{-1}$.
Γ	$= \ln(C_0/C_T)$, the log of the ratio of inherent contrast to threshold contrast of a Secchi disc, dimensionless.

ε **Scattered light collection efficiency**, dimensionless. Proportion of scattered light detected in a spectrophotometer.

E **Energy (quantum) of an electromagnetic photon**, J.

η **Absolute viscosity (dynamic viscosity)**, g m^{-1} s^{-1}.

θ **Angle of the light path to the vertical (or angle of scattering)**, degrees or, alternatively, rad.

κ **Product of Secchi depth and irradiance attenuation coefficient**, dimensionless.

λ **Wavelength of light**, nm.

Λ **Wavelength of light — different from λ**, nm. For example, a long wavelength.

μ **Average cosine** (of light rays in water), dimensionless. A measure of the degree of diffusion of the light field; Appendix 1.

μ_d **Average cosine for downwelling light**, dimensionless. A measure of the degree of diffusion of the downwelling light field.

μ_u **Average cosine for upwelling light**, dimensionless. A measure of the degree of diffusion of the upwelling light field.

ν **Frequency of electromagnetic oscillations**, s^{-1}.

ρ **Optical size parameter**, dimensionless.

ρ **Density**, g m^{-3}.

ρ_s **Density of a suspended particle**, g m^{-3}.

τ **Optical depth or attenuation depth ($K_d z$)**, dimensionless.

ϕ **Azimuthal angle of the light path**, degrees or rad.

Φ **Radiant power (light flux)**, W or, alternatively, photons s^{-1}. See Appendix 1.

Φ_a **Absorbed light flux**, W or, alternatively, photons s^{-1}.

Φ_b **Scattered light flux**, W or, alternatively, photons s^{-1}.

Φ_0 **Incident light flux,** W or, alternatively, photons s^{-1}.

Φ_r **Light flux after path length** r **through water,** W or, alternatively, photons s^{-1}.

χ **Apparent absorption coefficient,** m^{-1}. Absorption coefficient of a light-scattering water sample as measured in a spectrophotometer.

Ψ $= \ln(-1/C_T)$, the log of the ratio of inherent contrast $(= -1)$ to threshold contrast of a black disc $(= C_T$, negative), dimensionless.

ω **Solid angle,** sr.

1

Introduction

1.1 OPTICAL WATER QUALITY

The colour and clarity of water are manifestations of the behaviour of light in this optical medium. That is, the colour and clarity of a water body are determined by its optical properties. Kirk (1982) introduced the concept of *optical water quality* which is the main concern of this book. Optical water quality can be defined as follows (Kirk 1988):

> *'The extent to which the suitability of water for its functional role in the biosphere or the human environment is determined by its optical properties.'*

This is an excellent definition and one we enthusiastically adopt. There are two main categories of water use that we are concerned with: habitat for living organisms and human use. The influence of the optics of water on aquatic organisms has long been recognized and has received considerable attention, particularly by aquatic botanists and plant physiologists concerned with the capture of light energy in photosynthesis. The influence of the optical quality of water on its suitability as a habitat for sighted aquatic animals has been much less studied.

The effects of the optical character of waters on the 'human organism' have also received little research attention, and one of the purposes of this book is to swing some of the attention of aquatic scientists towards what we may call the *visual quality* of water, that is, the quality as perceived directly by the sense of sight. With this perspective the appearance of water is regarded, not merely as an element of scenery, but as a guide to its overall quality and hence suitability for a variety of uses.

1.1.1 Light in water and human use
People are essentially visual creatures. It has been estimated that, on average, about 80% of the neural processing of our sensory input is associated with the sense of vision. Some indication of the importance of vision in our everyday experience is demonstrated by the abundance of visual images and metaphors in the English language. We speak of

something being 'clear' to us and say that we 'see' an argument when we follow the logic. A person of demonstrated intelligence is described as 'bright'. An emotional factor may 'colour' our judgement.

Not surprisingly then, the appearance of water bodies is of profound importance. For the general public, it may be the only clue as to the unseen quality—particularly as regards disease-causing organisms, submerged hazards or toxic constituents. Besides appearance, only the sense of smell has anything to contribute to an observer's on-site assessment of water quality and, therefore, suitability for a particular desired use, unless sources of information external to the water, such as media reports or official warning signs, are available.

The visual colour and clarity of water are the main aspects of visual water quality or water appearance, although other 'less fundamental' aspects can be recognized (Table 1.1). These 'less fundamental' aspects, including surface phenomena such as foams and slicks, are not considered here, but we recognize their importance.

Colour is of concern, not just as a guide to water quality but because of the aesthetic response it evokes when water is an element of scenery. Visual clarity is also of aesthetic

Table 1.1. Aspects of the appearance of waters

Fundamental aspects[a]

Clarity
    Visual clarity (distance objects can be seen through water)
    Light penetration into water (for illumination or photosynthesis)
Colour
    Hue (described as blue, green, yellow etc.)
    Colour purity (ranging from neutral greys to spectral colours)
    Brightness (related to the amount of light scattered back by water)

Other aspects

Surface slicks and scums
Surface foams
Nuisance growths (macrophytes, phytoplankton, benthic algae, bacterial slimes)
Visible ingress of wastewaters
Trash and litter
Nature of the sediment bed
Appearance of beaches and margins

---

[a] These fundamental aspects are defined in Chapter 2.

significance and, as we shall see, a minimal clarity is desirable for the safe enjoyment of water by recreational users. Blue–green-hued and clear waters are always aesthetically preferable to yellow-hued waters or turbid, 'muddy' waters.

People seem to be taught to regard blue as the 'proper' colour of water at a very early age, certainly by elementary school. This expectation is probably reinforced by the use of blue paint or tiles in swimming pools. Strangely, many adults hold the erroneous notion that 'clean' water is truly colourless and only attains a blue or green hue because of sky reflection or the reflection from water-marginal or benthic vegetation—or, of course, from blue paint in swimming pools. Water that is any colour other than blue–green is therefore 'dirty'! This belief that pure water is colourless may derive from school science classes which usually speak of water as a 'colourless' liquid when in fact, it has a pale blue colour (Section 2.4.1).

1.1.2 Light and aquatic ecology
As well as colour and clarity as perceived by the human organism, we are concerned here with colour and clarity as they affect aquatic organisms. Both the quantity of light penetrating to a given depth and its quality (colour) affect the habitat of aquatic animals and the photosynthesis of aquatic plants. The profound importance of light penetration into water becomes evident when we consider that solar radiation is the driving energy source of all aquatic ecosystems, with the exception of those strange aphotic communities based on energy from the earth's volcanism. In the last decade or two there has been something of a renaissance of research interest in light as an ecological factor, possibly as a result of 'technology push' with the ready commercial availability of a new generation of light sensors of appropriate spectral and directional sensitivity. Previously most measurements of the light field in water were carried out with photocells equipped with optical filters to isolate reasonably narrow bands of wavelength. The renewed interest in light and primary production in aquatic ecosystems has extended to consideration of the quality of light for photosynthesis, again perhaps in response to technology push, in this case the availability of submersible spectroradiometers following the pioneering work of Tyler & Smith (1970).

   In considering clarity as it affects aquatic animals we are concerned with both the amount of light present to support vision (i.e. the illumination) and the sighting distance at which objects can be seen (which is itself ultimately dependent on the illumination). The maximum sighting distance (or visibility) may be expected to be of very great importance to visual predators such as fish at higher trophic levels, and also to some predatory aquatic birds, although, apparently, few studies to test this hypothesis have been reported. In considering clarity and aquatic habitat we may also be concerned with penetration into water of potentially biologically damaging photons of high energy in the ultraviolet part of the spectrum. The penetration of energy of different wavelengths in the total spectrum of solar radiation that is incident on water bodies (about half of which is in the visible range of wavelengths between about 400 and 700 nm; Kirk, 1983, p. 28) is important because it affects their thermal regime.

1.2 'AQUATIC OPTICS'—THE BEHAVIOUR OF LIGHT IN WATER

The colour and clarity of natural waters depend on the transmission of (sun)light through water; that is colour and clarity of water depend on its *optical character*. As we shall see, at a minimum only two optical quantities need to be specified to characterize a completely mixed body of water. These properties relate to the *bulk* optical processes occurring in water (as opposed to *surface* properties such as reflection) of *absorption*, or transfer of light energy to another form (ultimately heat), and *scattering*, or change in direction (but not energy) of the light photons ('particles' of light). The directional distribution of scattered light must be specified, as well as the magnitude of light scattering, in order to characterize fully the transmission of light in water.

Other optical phenomena besides absorption and scattering occur within natural waters but these are, in the main, outside the scope of this book. *Fluorescence* is one example. This process, which involves absorption of part of the energy of each photon and re-radiation of a photon of lower energy in an arbitrary direction, can sometimes be important in affecting the colour of natural waters, and indeed some techniques for measuring phytoplankton biomass make use of the fluorescence of chlorophyll *a*. Another example is *polarization* of light in which electrical oscillations perpendicular to the direction of travel are greater in a particular direction than in other directions. Polarization has little influence on the colour and clarity of waters and is not considered in this book. The interested reader is referred to Jerlov's (1976) excellent introduction to both these topics.

The subject of aquatic optics has been developed to a high degree particularly by physical oceanographers interested in such applications as the tracing of water masses and the remote sensing of water constituents (e.g. Preisendorfer 1976). Indeed, most of the scientific knowledge about the optical behaviour of waters has arisen as a result of studies in physical oceanography. The classic text in the field is Jerlov's (1976) 'Marine optics', an update (second edition) of his earlier (1968) 'optical oceanography' (the names belie the broad applicability in principle to *all* surface waters, not just marine waters). 'Marine optics' is a very useful introduction, for the aquatic scientist, albeit at a fairly advanced level. Preisendorfer's (1976) six volume 'Hydrologic Optics' is an advanced work. The collection of papers in 'Optical aspects of oceanography' (Jerlov & Steemann-Nielsen 1974) remains useful to the aquatic scientist who, having gained a good grounding in aquatic optics, wishes to pursue some of the subtopics.

Part I of Kirk's (1983) book 'Light and photosynthesis in aquatic ecosystems' provides a very useful and comprehensive introduction to the field of aquatic optics. This review was written as a foundation (as the title suggests) for a detailed consideration of aquatic photosynthesis in Part II of the same book which is intended for the biological limnologist and marine biologist. (A new edition of this very useful book is being prepared at the time of writing.)

That the field of aquatic optics is now recognized as a distinct, if derivative, specialty in its own right was demonstrated by the publication of a special issue on 'Hydrologic optics' (Spinrad 1989) in the journal *Limnology and oceanography* (Volume 34 , number 8).

1.3 SCOPE AND PURPOSE OF THE BOOK

From the outset in writing this book our intended audience has been mainly applied scientists such as ourselves who are concerned with practical problems of stewardship of natural water. We have long recognized the need for technology transfer from the scientific community to the decision makers in the complex multidisciplinary field of water management. This book is our contribution to a topic that is important in the management of natural waters.

We recognize that specialists in aquatic optics may find much of the material in these pages too elementary to be useful and the applications too broad-scope for their taste. However, we would hope that such specialists would find the book of interest because of the insights it provides into water management applications of their field of science, particularly to aquatic ecology and to concerns with recreational water use.

Although this book is not a specialist text on aquatic optics, this subject is introduced fairly comprehensively. The main purpose of the book is to introduce and review the *application* of aquatic optics to concerns with the behaviour of light in water as it affects two important categories of water use: aquatic habitat and recreation. We have written the book in the style of a scientific monograph, but it nevertheless has many of the features of a handbook or manual of practice for water resource managers. In particular, the book deals with measurement of the optical quality of water in some detail, and also gives guidelines to protect optical water quality and guidance on the control of optical-quality-degrading materials.

We expect that this book will be particularly useful to water resources scientists, engineers and managers, and to aquatic biologists and ecologists, particularly applied ecologists involved in water management. Parts of the book should also be useful to recreation specialists and physical geographers and planners, and perhaps also to those professionals concerned with water law and the drafting of water quality guidelines and standards. Finally, although the book is not written as a textbook, we expect that it should be useful for students in the above subjects, at graduate or advanced undergraduate levels.

Readers will find the review of aquatic optics herein shorter and pitched at a more elementary level than the material in Kirk (1983) or Jerlov (1976) for example. In this regard the present book may serve as an introduction to the more comprehensive treatment of aquatic optics in these two texts. Aquatic biologists should find the pointers to areas requiring further research useful, for instance the need for research on the optical habitat of sighted organisms. These scientists, and recreation specialists and geographers, will find useful material in the sections dealing with problems of colour and clarity of waters, as will water resource engineers and scientists. Most of the above groups will find useful material in this book on measurement of aquatic optical properties and colour and clarity of waters, and also on the main optical characteristics of different types of water body.

1.4 STRUCTURE OF THE BOOK

In Chapter 2 of this book we introduce the fundamental concepts of aquatic optics and examine the relationship of the different optical properties to the composition of the water and to the colour and clarity phenomena of concern. (Appendix 1 defines the relevant quantities used to measure electromagnetic radiation in 'optical' wavelengths.) The so-called *apparent optical properties* that characterize the light field in natural water are contrasted with the *inherent optical properties* that quantify the fundamental optical processes of absorption and scattering. The various light-absorbing and scattering constituents of natural water, mainly various living and non-living organic materials and suspended mineral particles, are then discussed. Suspended solid particles (suspensoids) of both organic and inorganic composition are particularly emphasized. Finally the concepts of colour and clarity of water are defined and their relationship to the apparent optical properties is explained.

Chapter 3 considers methods for measurement of the different optical variables and related water quality. (Appendix 2 contains 'recipe style' instructions for the various measurements.) Measurement of the apparent optical properties focusses on the use of irradiance sensors, whilst measurement of the inherent optical properties emphasizes readily available technology: beam transmissometers and laboratory spectrophotometers. Measurement of the two distinct aspects of water clarity, i.e. visual clarity (using standard visual targets) and light penetration (using light sensors) is then discussed. Measurement of water colour requires special instrumentation and considerable computation, but simple colour matching by eye is often useful. Measurement of the optically significant (i.e. light-attenuating) water quality constituents is reviewed. Finally we give a very brief overview of the use of optical sensors for the monitoring or remote sensing of water quality. Advanced techniques, such as the measurement of the directional structure of the light field in water, are beyond the scope of this book and, we would suggest, beyond the concern of all but the specialist in aquatic optics.

In Chapter 4 we discuss the optical characteristics of three main categories of surface water bodies: lakes, rivers and estuaries. Deep sea marine waters are not considered since there is little potential for the management of the quality of such waters. New Zealand examples are given in most instances, reflecting the very wide range of aquatic habitat in this country, which in turn, reflects a diverse physical geography. New Zealand waters can be regarded as something of a natural laboratory for studies of aquatic optics and probably most optical 'types' of water are represented in this country, from the very clear to the extremely turbid. Fig. 1.1 gives a map of New Zealand, showing the locations of water bodies mentioned in the text. Chapter 4 concludes by comparing and contrasting the optical character of the different categories of natural water bodies. Phytoplankton are commonly the dominant light-attenuating constituent of lake waters but are usually less important in rivers and estuaries.

Chapter 5 considers the management of colour and clarity of water to protect particular beneficial uses. Inevitably the focus here is on inland and coastal waters since, as mentioned above, the optical quality of the deep sea is unlikely to be manageable. The chapter commences with a review of overseas guidelines (often called 'criteria') for preventing degradation of optical water quality. Guidelines, which we believe are a significant advance over those promulgated to date for the protection of optical water quality, are developed to protect different water uses sensitive to colour and clarity, particularly aquatic recreation and aquatic habitat. Worked examples, showing how calculations are made using these guidelines, are given in Appendix 3. The solution to problems of colour and clarity lies in the control of those light-attenuating constituents causing the degradation of colour or clarity, including wastewater constituents, mineral suspensoids, and phytoplankton. Chapter 5 discusses the optical character of wastewaters in some detail and reviews control of suspended solids from soil erosion and of phytoplankton growth in lakes.

In Chapter 6 we illustrate many of the foregoing concepts, methods of measurement, characteristics of different water bodies, and management principles with discussion of six case studies. Four of the case studies deal with lake management problems: two of these discuss reponse to eutrophication control measures; the other two concern firstly, the diversion of turbid glacial waters and, secondly, the close relationship of macrophytes and optical water quality in shallow lakes. The two river case studies have to do, firstly, with the severe optical quality impact of a Kraft pulp mill effluent and, secondly, with diversion of clear water tributaries from a large river system.

168°E   170°E   172°E   174°E   176°E   178°E

N

36°S

NORTH
ISLAND

L.Omapere

L.Kaiiwi

Wairau Ck
Waitemata Hbr

Auckland
Manukau Hbr

Kaueranga R.

1 *L.Waikare*
2 *L.Hakanoa*
3 *Lake D*
4 *L.Whangape*
5 *L.Waahi*
6 *L.Rotomanuka*

Waikato R.
Raglan Hbr

Waihou R.

Tarawera R.

Hamilton
(*L.Rotorua*)

Motu R.

38°S

L.Maraetai
L.Ohakuri
L.Taupo

L.Tarawera
L.Okaro

L.Waikaremoana

Waipaoa R.

Tongariro R.
L.Tutira

Whaka-papa R.

T a s m a n

S e a

40°S

Whanganui R.

Rangitikei R.

L.Horowhenua

Waikoropupu Springs

Wellington

Motueka R.

P a c i f i c

42°S

Red Jacks Ck
L.Haupiri
Grey R.
Kapitea Ck
L.Brunner

O c e a n

Wilberforce R.
Harper R.
L.Coleridge

Christchurch

Rakaia R.

Tasman R.
L.Pukaki
L.Tekapo

L.Ellesmere

44°S

Shotover R.

0   100 km

Waitaki R.

**SOUTH
ISLAND**

Maerewhenua R.

46°S

Dunedin

Clutha R.

Oreti R.

Fig. 1.1. Map of New Zealand showing the locations of water bodies discussed in the text.

Finally in Chapter 7 we draw together the diverse threads of this inquiry and attempt to identify future directions for research in the field of optical water quality. Some of the most significant problems are at the boundary between natural science and social science because they involve the response of the 'human organism'. There are difficulties in working at disciplinary boundaries, and this may explain why there has been only very limited research on public perceptions of water quality including optical water quality, and on what might be called 'optical habitat' of waters for aquatic life. However, if these difficulties can be overcome, we predict expanding research interest and activity in the perception of suitability for use of water, particularly those uses sensitive to optical quality, and on water colour and clarity as factors in aquatic ecosystems.

2

Concepts

2.1 INTRODUCTION—WATER AS AN OPTICAL MEDIUM

The colour and clarity (i.e. transparency) of water are manifestations of the behaviour of light in this optical medium. The behaviour of light is determined by the optical properties of the water. These, in turn, depend on the constituent composition of the water. Appendix 1 gives formal definitions of quantities important for describing the light field in water.

We can introduce the concepts relevant to the transmission of light in water by reference to Fig. 2.1. This shows a collimated beam of light shining through a sample of water held in a clear container. If we were to follow the trajectories of individual light photons in this beam, some would be found to simply disappear within the water volume. The energy of these photons has been converted to another form (ultimately heat) because of the process of *absorption*. Other photons would be found to change direction abruptly and then to continue moving on their new path unchanged in energy content. These photons have been *scattered*.

The clarity of the water is related to how efficiently light is transmitted through this medium: the greater the proportion of photons transmitted unabsorbed, and undeviated in their path by scattering, the greater the clarity. Colour of water is more complicated, being related to how well photons of different energy content (corresponding to different wavelengths and therefore different 'colours') are transmitted through the water. Essentially all important aspects of the colour and clarity of waters can be explained in terms of the light-scattering and absorbing properties of water and its constituents.

The absorption of light initially involves the excitation of an electron in a molecule from its ground state energy level to a higher energy level, the difference in energy level being exactly equal to the energy of the photon which disappears in this quantum process. The details involved in the relaxation of the electron back to its ground state energy level need not concern us here. (For a fuller, but still 'introductory', account the reader is referred to Kirk (1983, pp. 42–45).)

A few photons are absorbed but then almost immediately re-radiated in any and all directions with unchanged energy content. This 'elastic' scattering process is usually

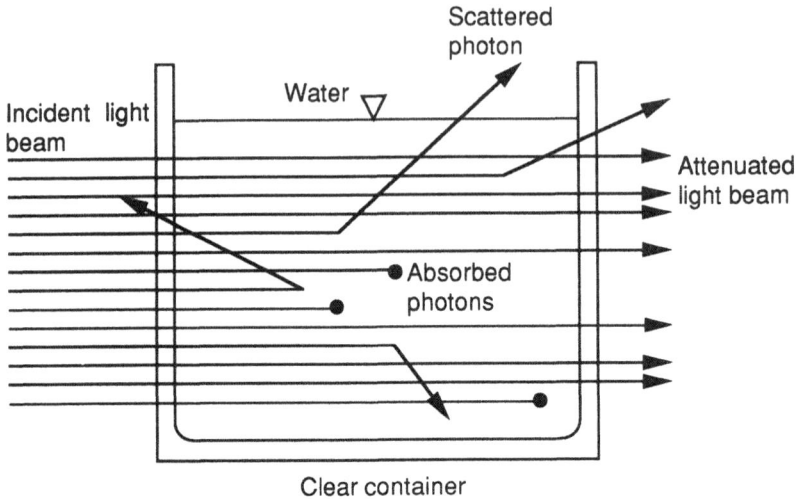

Fig. 2.1. Schematic diagram showing interaction of light photons in water by absorption and scattering.

referred to as *molecular scattering* and is generally much less important than particle scattering in waters. Other photons may lose part of their energy content on absorption but are re-radiated in all directions with lower energy content (i.e. a longer wavelength). This is 'inelastic' scattering, known as *fluorescence*. Fluorescence of natural water constituents, notably the plant pigment chlorophyll *a*, can sometimes affect water colour, but this phenomenon is generally of much less importance than absorption and scattering by particles and will not be considered further. Most of the photons that are absorbed have their energy content completely converted to non-radiant forms.

The change in photon direction involved in scattering is distinct from that in *refraction* which occurs at the boundaries of media (notably the air–water interface) because of change in refractive index (the ratio of light speed in a vacuum to that in the medium). However, this distinction becomes somewhat blurred when we are discussing water containing particulate constituents, because the surface of each suspended particle acts as a refractive index interface across which refraction occurs. Indeed, the process of refraction within particles is an important contributory mechanism to the overall phenomenon of particle scattering (Section 2.4.2). Scattering is also dependent on *diffraction*, a 'wave-like' phenomenon in which light waves bend around obstructions (particles) and create interference patterns, together with the more familiar phenomenon of *reflection* from particle surfaces. Total scattering, to a good approximation, can be taken as the sum of diffraction, reflection and refraction by particles (e.g. Kirk 1983). Overall, scattering by particles is of greater importance in natural waters than molecular scattering, although the latter process sometimes needs to be taken into account in the very clearest natural waters.

2.2 INHERENT OPTICAL PROPERTIES OF WATER

The so-called *inherent optical properties* describing the absorption and scattering processes in water can be defined by reference to Fig. 2.2. This figure shows the interaction, within a thin layer of water, of photons in a perfectly collimated, monochromatic light beam of radiant flux (or, alternatively, 'radiant power'), Φ_0.

Absorption is quantified by an *absorption coefficient* (symbol a, units m^{-1}), which can be conceptualized as the proportion of photons absorbed (= the probability of absorption of a photon) per unit length along a very short length of light path (Fig. 2.2). The light path has to be short so that all parts of the path are uniformly lit; that is the path must be short enough so that the proportion of photons absorbed is small and the radiant flux is essentially undiminished. The proportion of photons absorbed over the small path length Δr can be written as $\Delta\Phi_a/\Phi$, where the subscript, a, refers to absorption. Dividing this proportion by Δr we obtain the definition of the absorption coefficient:

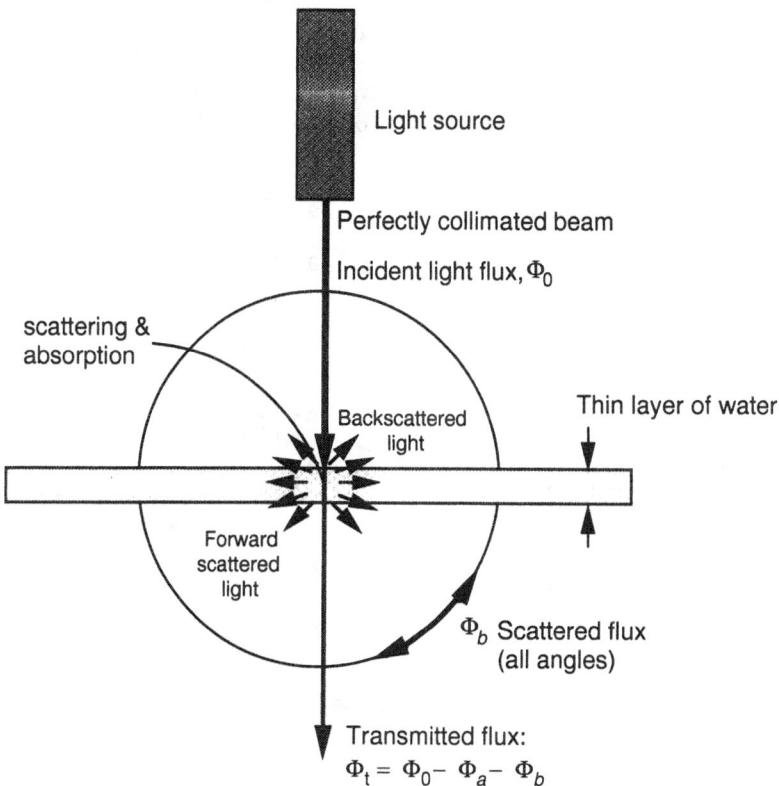

Fig. 2.2: Definition of the inherent optical properties in terms of interaction of a perfectly collimated monochromatic light beam transmitted over a very short light path in water. Φ_0 is the radiant power (light flux) of the incident beam. A fraction of the incident flux, Φ_a, is absorbed and another fraction, Φ_b, is scattered through all angles. The transmitted (attenuated) beam is of flux, $\Phi_t = \Phi_0 - \Phi_a - \Phi_b$.

$$a = \frac{\Delta\Phi_a/\Phi}{\Delta r} \qquad (2.1)$$

This coefficient is fundamental to the description of the optical character of a water.

Scattering can be quantified by a scattering coefficient (symbol b, units m^{-1}), which can be conceptualized as the proportion of photons scattered ($\Delta\Phi_b/\Phi$ = the probability of scattering of a photon) per unit length along a very small length of light path, Δr (Fig. 2.2):

$$b = \frac{\Delta\Phi_b/\Phi}{\Delta r} \qquad (2.2)$$

Note that the form of this equation is identical to that for absorption (equation (2.1)).

Scattering is a somewhat more difficult process to describe than absorption because of the need to account for the new direction taken by photons after scattering. Directional dependence of scattering is quantified by the *volume scattering function* (symbol $\beta(\theta)$, units m^{-1} sr^{-1}), a function of angle (θ in radians or degrees) of scattering with respect to the incident beam direction. The scattering coefficient, b, is the integral of the volume scattering function over all directions (e.g. Kirk 1983, p. 17).

Scattering by molecular water and its dissolved constituents is isotropic (same in all directions) because of the random orientation of these molecular scatterers. However,

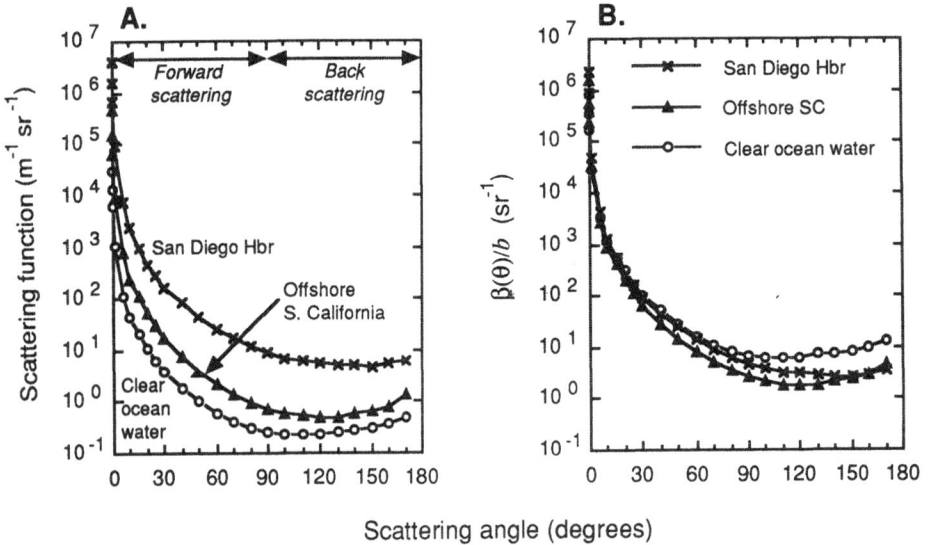

Fig. 2.3. Angular dependence of scattering in water. Plotted from the data of Petzold (1972). (A) Volume scattering functions for a clear ocean water near the Bahama Islands ($b = 0.037$ m^{-1}), for seawater in Avalon Cove, Catalina Island, off Southern California ($b = 0.28$ m^{-1}), and for a station in San Diego Harbour ($b = 1.82$ m^{-1}). Scattering in the angular range 0° to 90° is termed 'forward scattering' and 'backscattering' is in the range 90° to 180°. (B) The same scattering functions divided by the values of the scattering coefficients. The curves are very similar in shape for forward scattering but deviate slightly for backscattering. In particular, the enhanced backscattering attributable to water itself is seen for clear ocean water.

scattering by particulate constituents in natural waters is highly biased in forward directions (low values of angle θ) particularly as a result of diffraction. The shapes of the volume scattering functions of diverse natural waters are rather similar, that is, the ratio $\beta(\theta)/b$ is approximately constant (e.g. Petzold 1972). Fig. 2.3 shows this constancy in shape for some marine and estuarine waters ranging 50-fold in scattering coefficient. The only marked differences between different waters occur at high scattering angles. The volume scattering functions go through a minimum typically at angles around 120°. Overall, far fewer photons are backscattered ($90° < \theta < 180°$) than are forward scattered ($0° < \theta < 90°$). Typically the ratio of the backscattering to total scattering, b_b/b, where b_b is the backscattering coefficient, is about 1.5% (e.g. Petzold 1972). Jerlov (1976) and Kirk (1983) give more information on angular dependence of scattering by waters.

The absorption coefficient, a, scattering coefficient, b, and scattering function, $\beta(\theta)$, are three of the four so-called 'inherent' optical properties (IOPs) of water (Preisendorfer 1961). The remaining IOP is the beam attenuation coefficient, c, which quantifies total attenuation of an idealized, perfectly collimated, monochromatic, beam (e.g. a laser beam) by both absorption and scattering (Fig. 2.2).

$$c = \frac{\left(\Delta\Phi_a + \Delta\Phi_b\right)/\Phi}{\Delta r} = \frac{-\Delta\Phi/\Phi}{\Delta r} \tag{2.3}$$

Of the four IOPs, a, b, c, and $\beta(\theta)$, a and $\beta(\theta)$ are the most fundamental, since b is merely the volume integral of $\beta(\theta)$ and $c = a + b$. Together, a and $\beta(\theta)$ fully specify the optical character of a water.

The IOPs are distinguished from the apparent optical properties (AOPs, discussed below) in being dependent only on the water composition and not on the characteristics of the ambient light (Preisendorfer 1961). This feature of the IOPs is important because it permits immediate comparison of different water bodies one with another, or the same water body sampled at different times, irrespective of light conditions at the times of sampling.

Furthermore, the IOPs, again in contrast to the AOPs, are rigorously additive. For example, the scattering coefficient of the phytoplankton in a water can be added to that of suspended mineral particles, together with that of pure water, to obtain the total scattering coefficient of a water containing both constituents.

The absorption coefficient of a water containing N constituents is the sum of the absorption coefficients of the individual constituents:

$$a = a_w + a_1 + a_2 + ... + a_i + ... + a_N = a_w + \sum_{i=1}^{N} a_i \tag{2.4a}$$

where a_i is the absorption coefficient of the ith constituent and a_w is the absorption coefficient of water itself. Similarly we can write the scattering coefficient, b, as

$$b = b_w + b_1 + b_2 + ... + b_i + ... + b_N = b_w + \sum_{i=1}^{N} b_i \tag{2.4b}$$

where the subscripts have the same meaning as in equation (2.4a). From the additive principle it also follows that the partial absorption and scattering coefficients, a_i and b_i, are proportional to the concentration of the ith constituent, C_i (g m^{-3}):

$$a_i = A_i C_i \qquad (2.5a)$$

$$b_i = B_i C_i \qquad (2.5b)$$

in which the proportionality coefficients (A_i, B_i) are referred to as 'cross-sections' because they have the units of area per unit mass of the constituent, m^2 g^{-1}. A_i is the absorption cross-section and B_i is the scattering cross-section, and similarly we have a beam attenuation cross-section, $C_i = c_i / C_i$.

These equations permit the scattering and absorption of light, and the total light beam attenuation, to be partitioned into contributions from all constituents including the water itself. It is often valuable to know, for example, that 90% of the light attenuation in a water is attributable to the mineral suspended solids and only 10% to the phytoplankton biomass. If this were the case, measures designed to improve water clarity would have to focus on reducing suspended solids rather than controlling phytoplankton growth.

The additivity of the IOPs also permits rigorous calculation of the effect of mixing of water masses as in the discharge of effluent or tributary water into a river (Sections 4.3 and 5.3.6). Indeed, the IOPs are, in this respect, analogous to mass concentrations of constituents such as chloride or suspended solids.

2.3 APPARENT OPTICAL PROPERTIES OF WATER

The apparent optical properties, or AOPs, describe the optical behaviour of water bodies in a given light field (Preisendorfer 1961). For most practical purposes we are concerned with the ambient light field arising from the sun during daylight hours. Of course the spectral, and particularly the directional, characteristics of daylight are rather variable so that the apparent optical properties, while mainly dependent on the IOPs of water (and thus on water composition), will vary with time through a day or between different days. Clearly the daylight light field is quite different under high midday sun compared with sunset conditions. The daylight light field is also different under overcast skies compared with a clear sun. Jerlov (1976) and Kirk (1983) have both given discussions of the characteristics of natural light fields at different solar altitudes and in different atmospheric conditions.

The most familiar of the apparent optical properties is the Secchi disc depth, z_{SD} (m) (Preisendorfer 1986), the maximum visual range of a white or black-and-white disc viewed vertically (Section 2.5.1). The other apparent optical properties are defined in terms of measurements by light sensors in water. For the moment, and indeed, for most practical purposes, we need only consider a submersible light sensor with a flat-plate sensing element which measures the irradiance, E (see Appendix 1) in units of radiant power per unit area (W m^{-2}) or, alternatively, flux of photons per unit area (photons s^{-1} m^{-2}). The irradiance sensor can be fixed to a frame which holds it 'looking up' so that it measures the downwelling irradiance, E_d. Alternatively the sensor can 'look down' at the upwelling irradiance, E_u (Appendix 1). The difference, $E_d - E_u$, is the *net* vector irradiance, symbol $\vec{E}$.

The ratio of E_u to E_d defines the first important apparent optical property that we must consider, the *irradiance reflectance coefficient*:

$$R = E_u / E_d \qquad (2.6)$$

The reflectance coefficient is useful for estimating the scattering coefficient in a water, as we shall see later (Section 3.3). However, this quantity is also of interest in its own right since it relates to the *brightness* of the colour of a water body. A water with a high value of R will generally appear bright-coloured or 'reflective' in contrast to the dark colour of a water with a low value of R.

The irradiance in a water body declines approximately exponentially with depth as illustrated in Fig. 2.4. This decrease (or attenuation) of ambient light (irradiance) with depth is quantified by the *irradiance attenuation coefficients*:

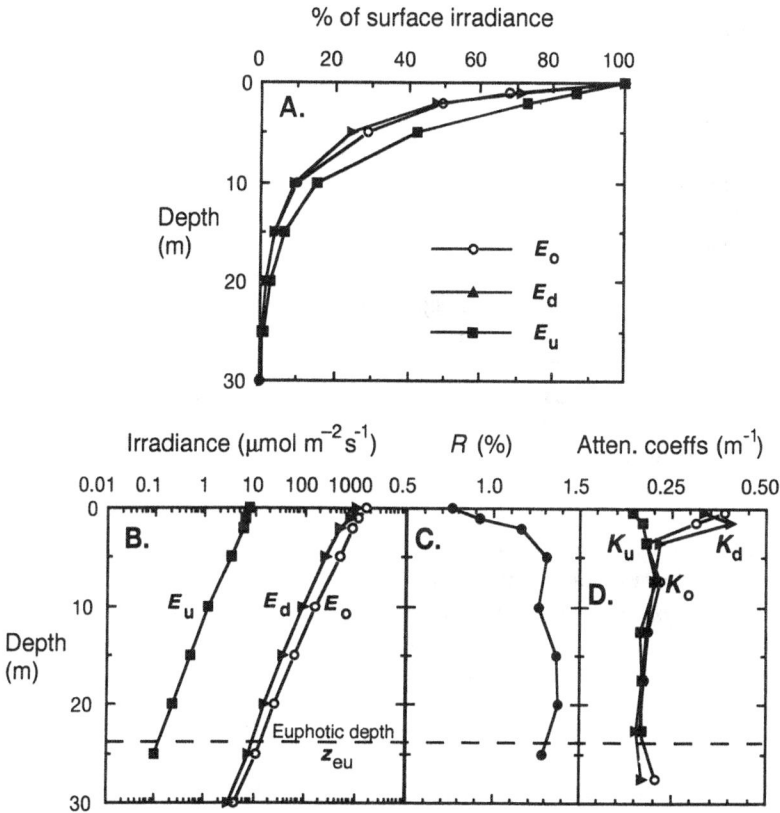

Fig. 2.4. Apparent optical properties (AOPs) as a function of depth (z) measured in Lake Waikaremoana, North Island, New Zealand, on 7 August 1984 (authors' unpublished data). Measurements were made with PAR sensors: LiCor LI 192SB sensors of vector quantum irradiance and a Biospherical scalar quantum irradiance sensor. (A) Profiles of upwelling irradiance, E_u, downwelling irradiance, E_d, and scalar irradiance, E_o (see Appendix 1), relative to the surface values of each of these quantities. (B) Profiles of E_u, E_d and E_o plotted on a logarithmic scale. (C) Profile of the reflectance coefficient, $R = E_u/E_d$. (D) Profiles of the irradiance attenuation coefficients (e.g. $K_u = $ d ln $E_u/$dz) showing the near-surface differences and the asymptotic approach to a constant value ($k = K_u = K_d = K_o$) at depth. The depth at which irradiance is 1% of that immediately below the surface of the water is indicated. This is the 'euphotic depth', an index of the maximum depth to which plants can grow.

$$K_u = -d(\ln E_u)/dz = -(dE_u/dz)/E_u \qquad (2.7a)$$

$$K_d = -d(\ln E_d)/dz = -(dE_d/dz)/E_d \qquad (2.7b)$$

$$K_E = -d(\ln \vec{E})/dz = -(d\vec{E}/dz)/\vec{E} \qquad (2.7c)$$

Note that these definitions for the irradiance attenuation coefficients are identical in form to that for the beam attenuation coefficient, c (equation (2.3)). That is, we can think of the irradiance attenuation coefficients as being the proportion of light (irradiance rather than radiance — refer Appendix 1 for definitions) attenuated per unit depth over a small depth interval in the water column. K_u is typically approximately equal to K_d and becomes exactly equal at great depth (e.g. Fig. 2.4D). Near the surface K_u and K_d diverge because the ambient light does not decrease with depth in a perfectly exponential fashion, and the deviation from an exponential function is most pronounced immediately below the water surface. In Fig. 2.4 it can be seen that the non-equality of K_u and K_d in the surface water of Lake Waikaremoana is associated with non-constancy of the reflectance coefficient R. Most of this depth dependence of the apparent optical properties arises from the fact that light of certain wavelengths, notably red light, is attenuated faster than light at other wavelengths.

So far we have considered only the irradiance measured by a light sensor with a flat diffuser element. Such a sensor is usually designed to have a cosine response. That is, the signal produced by the sensor is proportional to the cosine of the angle of incidence of a light beam impinging on its diffuser surface. A light sensor with a cosine response measures *vector irradiance*. For some purposes, however, we may wish to measure light with a sensor that is equally sensitive to all light passing through a point, irrespective of direction. For example, the photosynthetic response of algal cells is not expected to depend on light direction. Algal cells respond to so-called *scalar irradiance* (symbol E_o) which can be measured with a sensor equipped with a spherical diffuser (Appendix 1). The scalar irradiance attenuation coefficient is defined by

$$K_o = -d(\ln E_o)/dz = -(dE_o/dz)/E_o \qquad (2.8)$$

Fig. 2.4 shows that K_d and K_o are very similar in magnitude, and they are both similar to K_u and K_E except near the water surface. All these quantities become equal in magnitude at great depth (Preisendorfer 1958), as indicated by the ∞ (infinity) symbol: $K_{u\infty} = K_{d\infty} = K_{E\infty} = K_{o\infty} = k_\infty$.

The irradiance attenuation coefficients, being AOPs, are weakly dependent on the ambient light field, including direct sunlight and skylight, which is merely sunlight scattered by gas molecules and particles (aerosols) in the atmosphere. Thus the AOPs are variable with time through a day, and from day to day, as solar altitude and cloud patterns change. Nevertheless, this variation is fairly weak and it is often convenient to think of a particular water body as characterized by a particular value of K_d (Kirk 1983,

1986) and also by a given value of the reflectance, R, just as it is a useful convenience to consider the water as having a particular visual clarity.

The irradiance attenuation coefficient (K_d) is always lower in magnitude than the beam attenuation coefficient, c. This is because any scattering of the light in a collimated beam results in non-detection by the (very small) sensor, whereas in a diffuse light field scattering of light away from the sensor is compensated by scattering towards the sensor. A very useful inequality given by Preisendorfer (1958) shows that K_d is always between c and a in magnitude: that is $a < K_d < c$. Actually, K_d is typically closer to a than c, because irradiance attenuation depends more strongly and directly on absorption than on scattering. In a range of lakes we found, as expected, that irradiance attenuation correlated closely with a, whilst visual clarity (quantified by the Secchi depth) correlated with c (Fig. 2.5). Thus it is often helpful to think of irradiance attenuation as being mainly an expression of absorption while visual clarity depends mainly on beam attenuation (since, in order to form an image, light must travel in straight lines).

2.4 LIGHT-ATTENUATING CONSTITUENTS OF WATER

Natural waters contain a vast array of dissolved, colloidal and particulate constituents, most in trace quantities too low for detection except by the most sensitive analytical procedures. All water constituents can, in principle, modify light transmission, but very few are significant absorbing or scattering agents. For example, the high salt content of seawater is almost optically irrelevant, so the clearest seawaters are almost identical in appearance to the most optically pure freshwaters.

Table 2.1 lists the main categories of light-attenuating constituents of natural waters. The measurements used as indices of the concentrations of these constituents are indicated.

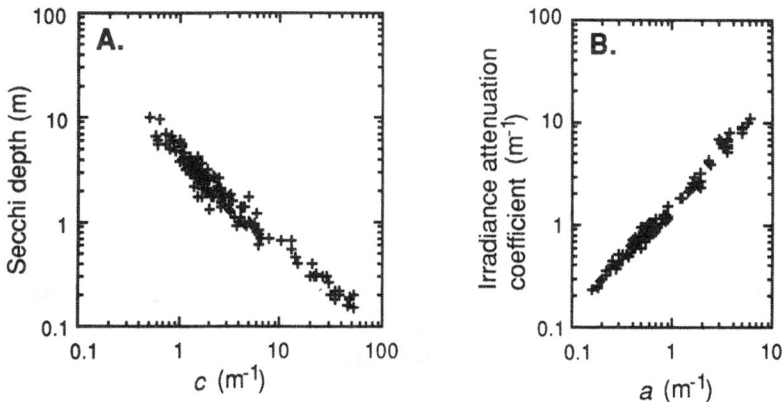

Fig. 2.5. Relationship between apparent optical properties and inherent optical properties in 12 New Zealand lakes each sampled monthly for a year. Inherent optical properties are PAR waveband (400–700 nm) averages. (A) Secchi depth versus beam attenuation coefficient, c. (B) Attenuation coefficient for downwelling irradiance, K_d, versus absorption coefficient, a. (Data from Vant & Davies-Colley (1988, unpublished).)

Besides water itself we need consider only one filter-passing ('dissolved') constituent and three particulate constituents (or categories of constituents).

Table 2.1 Categories of light-attenuating constituents in waters and wastewaters

Constituent	Optical character	Measured by (reference)
Water itself	Absorbs light, mainly red. Scatters light very weakly.	Optical properties tabulated by Smith & Baker (1981).
Dissolved constituents		
Yellow substance (= *Gelbstoff* or gilvin)	Strongly absorbs light, particularly blue light. Scattering is negligible	Absorption of membrane-filtered (0.2 μm) water sample at 440 nm (Kirk 1976).
Particulate constituents		
Total suspensoids	Responsible for almost all light scattering. Organic solids may also absorb light.	Total suspended solids (TSS) (APHA 1989).
Mineral suspensoids	Scatter light intensely but usually absorb light weakly (hydrous ferric oxides are a strongly absorbing exception).	Ash content of TSS (ignite at 400°C for 6 h).
Detritus	Spectral absorption similar to yellow substance. Also scatters light (variable optics).	Volatile SS (VSS), corrected for living algal biomass.
Phytoplankton	Absorb light strongly with strong spectral selectivity. Scatter light strongly.	Chlorophyll *a* concentration (Strickland & Parsons 1972).

2.4.1 Water itself

Water is not really the 'colourless' liquid of elementary chemistry, although it may appear so in small volumes such as a drinking glass. Pure water (and pure seawater) is pale blue in colour because of selective absorption of red light. Water absorbs light strongly in the infrared (IR) region of the spectrum and the visible red light absorption has been attributed to higher (shorter wavelength) harmonics of the IR absorption bands (Tam & Patel 1979). Absorption of light by pure water is minimal in the 400–500 nm (violet to green) range which acts as a visible 'window' (Fig. 2.6) into clear waters.

Scattering by water molecules is important only in the very clearest natural waters since scattering by particles larger than the wavelength of light is so much more efficient than molecular scattering (Jerlov 1976). Because of the random orientation of water molecules, scattering by water itself is isotropic (equal in all directions). This contrasts with the typical scattering by natural waters which is strongly peaked in the forward direction (Fig. 2.3). Even the clearest natural waters, such as the 'clear ocean water' for which $\beta(\theta)$ is given in Fig. 2.3, have volume scattering functions very different from the isotropic scattering of pure water.

As we saw in Section 2.2, most scattering by natural waters is at low angles, and because scattering at angles >90° ('backscattering') is typically only about 1.5% of total scattering, molecular scattering by water molecules can be important in comparison with particle scattering as a contributor to backscattering in clear natural waters. The total backscattering coefficient is often estimated for modelling purposes as $b_b = 0.015b + 0.5b_w$, that is, 1.5% (or some other small fraction—if known) of the total particle scattering coefficient, plus half the water scattering coefficient.

The molecular scattering by water is strongly wavelength dependent (Fig. 2.6) and Morel (1974) has shown that a power law expression characterizes the spectrum: $b_w \propto \lambda^{-4.3}$. The exponent is close to the Rayleigh law value of –4.0 that applies to molecular scattering by gases. This strong wavelength dependence means that the water scattering contribution to total backscattering $(0.5b_w)$ is most likely to be significant at blue wavelengths in the visible spectrum. The blue-dominant backscattering by water itself in the clearest natural waters, such as Crater Lake, Oregon (Plate 1), contributes significantly to their striking blue colour (Section 2.6.2).

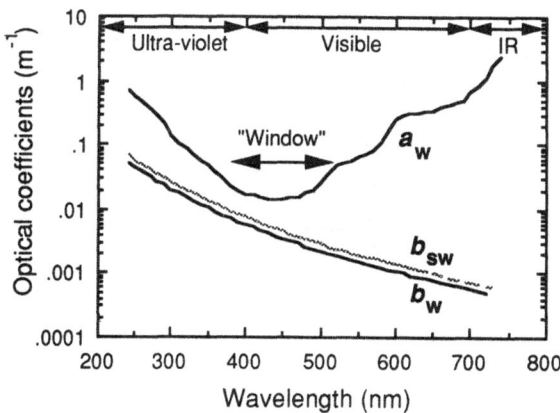

Fig. 2.6. Spectral absorption coefficient, a_w, and spectral scattering (molecular scattering) coefficients of pure water and seawater, b_w and b_{sw}. (Plotted from the data of Smith & Baker (1981, with permission.)

2.4.2 Suspensoids

There are a vast range of suspensoids in water, including grains of rock-forming minerals such as quartz, clay mineral particles derived from rock weathering, detrital organic matter, virus particles and living cells of bacteria and algae. We are concerned here mainly with relatively fine particles, usually clay or silt sized (<50 μm) since, as we shall see, larger particles, such as mineral sand, are comparatively weakly light attenuating and, in any case, tend to settle rapidly out of the water column. Kirk (1985) has given a cogent introduction to the optical effects of suspensoids in water and his paper should be consulted for more detailed information.

The three most important attributes of suspended particles in water are their *concentration*, their *composition*, and their *grain size distribution*. The light attenuation by a particular type of suspensoid is proportional to its concentration, measured as the mass of material captured by filtration of a given volume of water on a glass fibre filter (e.g. Whatman GF/C) (Section 3.6.2). This measurement does not distinguish between particles of different composition and grain size, and therefore, different optical behaviour. Thus, overall, suspended solids concentration is only weakly related to light attenuation in waters, although correlation between the two quantities may be close where particles of the same composition and size locally dominate attenuation.

The composition of suspensoids determines their refractive index and light-absorbing properties, and also their density with respect to water, which in turn affects gravitational settling velocity. The most important distinction here is between mineral particles such as quartz and kaolinite (a common clay mineral), and organic particles such as algal cells or organic floccules. Mineral particles have a higher refractive index relative to water (quartz: 1.16) than organic material (variable refractive index, but 1.04 is typical), which means that organic and mineral particles of the same size scatter light differently. Also, common mineral particles, with the exception of iron pigments, absorb light only weakly

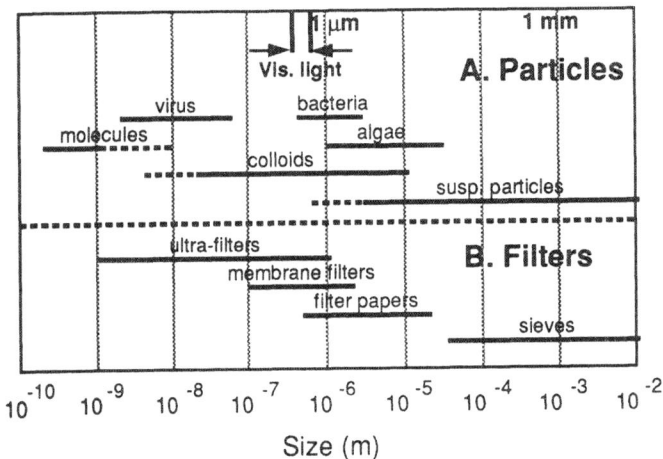

Fig. 2.7. Size spectrum of suspensoids in water and wastewater in relation to characteristic pore scale of filter media. (Redrawn after Fig. 10.16 of Stumm & Morgan (1981), with permission.)

whereas organic particles are often strongly pigmented (light absorbing). Thus mineral particles suspended in waters usually affect scattering more than absorption, and give bright-coloured, turbid waters (Plate 2). Mineral particles are also much denser than organic particles (e.g. quartz density relative to water is about 2.65 whereas density of organic particles is typically slightly greater than water) and therefore their settling velocity (e.g. Lerman 1979) is much higher and they tend to be removed from the water column by sedimentation, as is further discussed later in this section.

The characteristic grain size of suspensoids in waters covers a vast range from macro-molecules around 10^{-8} m to sand-sized particles around 10^{-3} m. Fig. 2.7 shows the size ranges of various types of particle commonly found in waters in relation to different types of filters used for separating suspensoids from waters. Particles larger than the filter pore size of a given filter are removed by simple physical straining. However, typically filters can remove some material smaller than the nominal filter pore size owing to bonding of particles with each other and with the filter material. Also, the progressive clogging of filter pores reduces their effective size as the filtration proceeds. For example, the GF/C filter that is typically used for suspended solids analysis has a pore size around 1 μm, but will typically remove many of the particles ranging in size down to about 0.1 μm, including most of the clay mineral particles commonly found in waters. Filtration of a water sample through a GF/C filter typically removes almost all of the light scattering, in spite of the fact that some particles smaller than the nominal pore size of around 1 μm may pass through the filter. The reason is that light scattering by these particles, being generally smaller than the wavelength of light, is much weaker than that by particles of larger sizes, as we now discuss.

Suspended particles will attenuate a certain fraction of the light impinging on their surface areas. This fraction is known as the *attenuation efficiency*, symbol Q_c (dimensionless). Very small particles (e.g. those not removed by a standard glass fibre or membrane filter) are simply too 'thin' to have much effect on light waves travelling near them

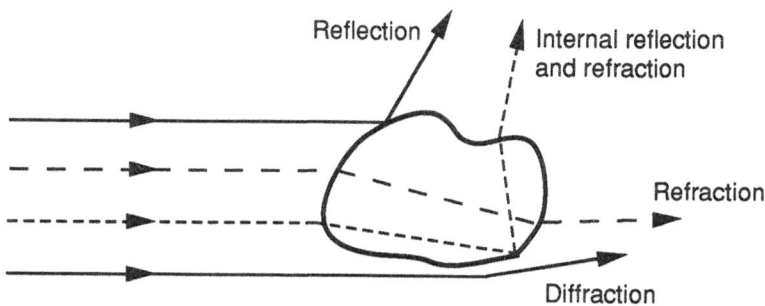

Fig. 2.8. Schematic diagram of an optically large suspended particle in an extended light beam. The particle may absorb any fraction of the light impinging on its surface. The remainder of the light impinging on the particle is scattered by the processes of reflection and refraction. An equal quantity of light passing near the particle is also scattered (by diffraction) such that the particle attenuates twice the light. actually impinging on its surface. Thus the particle may absorb up to half of the total attenuated light. (A similar diagram has been given by Kirk (1985).)

and only a small fraction of the intercepted light is attenuated (i.e. Q_c is low). By contrast, all of the light impinging on the surface of relatively large particles, say those considerably larger than the wavelength of light (i.e. diameter, $d \gg 0.55$ μm = 550 nm), is attenuated by scattering due to reflection or refraction or both, or by light absorption within the particle (Fig. 2.8). However, an equal amount of light passing round the particle will be scattered by the optical process of *diffraction*. Thus the attenuation efficiency of large particles is not unity, but is 2, because twice as much light is attenuated as directly impinges on the particle.

This counterintuitive result, which holds for all large particles, has been termed the 'extinction paradox' (van de Hulst 1957, p. 107). The apparent paradox is understood when it is remembered that the observation of attenuated light is made far from the obstructing particle. Very near the particle, where its shadow is recognizable, obviously only that light impinging on the particle is recorded as removed from the beam. However, at a distance of many particle diameters, an equal quantity of light deviates from its original path owing to diffraction, so that $Q_c = 2$. Further discussion is given by van de Hulst (1957).

More generally the attenuation efficiency varies in a complex way with the size of particles. For spheres with a refractive index similar to that of the optical medium, which includes most particles suspended in natural waters, van de Hulst (1957) gave the following expression for the attenuation efficiency factor:

$$Q_c = 2 - \frac{4}{\rho}\sin\rho + \frac{4}{\rho^2}(1 - \cos\rho) \qquad (2.9)$$

where $\rho = 2\pi d(n - 1)/\lambda$ is the optical size parameter (dimensionless), in which n is the refractive index of the particle relative to the surrounding medium, d is the sphere diameter, and λ is wavelength of light in the medium (not the reference wavelength in a vacu-

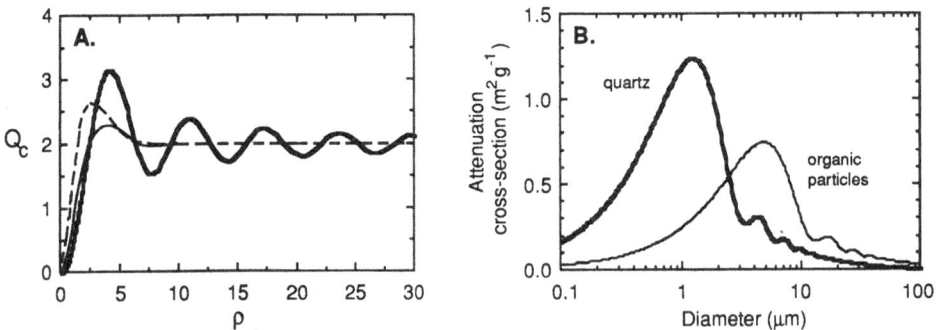

Fig. 2.9. (A) Attenuation efficiency factor (Q_c) as a function of the optical size parameter (ρ), ——— , quantitative representation of the expression for the attenuation efficiency of spheres of uniform size given by van de Hulst (1957) (equation (2.9)). The oscillations arise from light wave interference phenomena. The other two lines in (A) qualitatively indicate the smoothing out of the oscillations with dispersion of particle size (polydispersion, – – –) or with absorption (——). (B) Attenuation cross-section of a suspension of spherical particles as a function of their diameter. The points were calculated from equation (2.11) using the curve of Q_c for monodisperse particles in (A). The curves shown in (B) are for spheres of quartz (representative of mineral particles, refractive index and density relative to water of 1.161 and 2.65 respectively) and representative organic matter with a relative refractive index of 1.04.

um). This function is plotted in Fig. 2.9A. Q_c is low for small ρ and rises steeply to a maximum at about $\rho = 4$ and then, at higher values of ρ, it oscillates around the large particle value of 2 because of interference phenomena related to the wave nature of light. Simplistically, these oscillations may be regarded as arising because of constructive and destructive interference between the waves passing through the particle (retarded in their travel because of the lower speed of light within the particle compared with that in the water) with those waves diffracting round the particle. (More detail is given by van de Hulst (1957).) Both dispersion of particle sizes ('polydispersion') and light absorption by particles tend to smear out these oscillations (except the first, major one, Fig. 2.9A), so that they are usually not of practical importance for natural waters which typically contain suspensions of particles that are both absorbing and polydisperse.

Values of ρ for natural water particulates can range widely owing to variation in physical size and composition (and hence refractive index relative to that of water). However both 'optically small' ($\rho << 4$) particles, such as macromolecules, colloids and bacterial cells, and 'optically large' ($\rho > 4$) particles, such as mineral grains or organic floccules, are typically present, together with intermediate-sized (and maximally light-attenuating) algal cells and clay mineral particles.

The attenuation per unit mass concentration, the so-called *attenuation cross-section* (units: $m^2\ g^{-1}$), for an individual particle is simply the attenuation efficiency multiplied by the average projected surface area, s, and divided by the particle mass, m:

$$C = \frac{Q_c s}{m} \qquad (2.10)$$

For a sphere, projected area is simply the cross-sectional area, so $s = \pi d^2/4$; and mass m = density times volume = $\rho_s \pi d^3/6$, where ρ_s is sphere density. Substituting in equation (2.10) we obtain

$$C = \frac{1.5 Q_c}{d \rho_s} \qquad (2.11)$$

Evidently C is inversely proportional to both the diameter and the material density of the sphere, as well as being proportional to the attenuation efficiency, Q_c. For optically large spheres of given composition (and therefore, density), C varies as d^{-1} because the attenuation efficiency is nearly constant (at 2) whilst the area-to-mass ratio varies as the inverse of the diameter (as a consequence of the mass of a sphere being proportional to the diameter cubed whereas the cross-sectional area, and therefore the light attenuation, is proportional to the diameter squared).

Fig. 2.9B shows the attenuation cross-section of spheres of quartz and of 'typical' organic matter plotted against diameter. The attenuation cross-section was calculated from the curve for Q_c for non-absorbing particles all of the same size (monodisperse) in Fig. 2.9A using equation (2.11). (Q_c, as given by equation (2.9), was first re-expressed as a function of diameter, d, for a given particle composition (i.e. density, ρ_s, and refractive index, n).) Attenuation cross-section is not very sensitive to particle shape, and attenuation as a function of size follows the patterns for spheres shown in Fig. 2.9B, if the characteristic dimension of the non-spherical particle under consideration is taken to be the diameter of a sphere of equal volume. Actually, the curves shown in Fig. 2.9B are approximately correct even for particles as highly aspherical as clay mineral platelets.

Fig. 2.9B shows that the attenuation cross-section increases with decrease in particle size for optically large particles as predicted from equation (2.11). C reaches a maximum at characteristic sizes in the order of a few μm, the precise position depending on particle composition. At smaller diameters, attenuation cross-section decreases abruptly owing to a rapid fall in attenuation efficiency which is not compensated by the continuing increase in area-to-mass ratio of the particles.

For quartz particles, the maximum of the attenuation cross-section occurs at around 1.2 μm diameter (Fig. 2.9B). This explains why clay mineral particles, with a density and refractive index similar to those of quartz and a characteristic size around 1 μm, are very efficiently light attenuating. For organic particles the maximum occurs at appreciably larger sizes (about 5 μm in Fig. 2.9B—for a refractive index relative to water of 1.04) because both their refractive index relative to water and their density are much lower than those of quartz and other mineral materials, mainly because of their high water content. Very small organic particles attenuate light more weakly than mineral particles of the same characteristic dimension, but organic particles larger than a few μm in diameter attenuate light more than twice as efficiently as mineral particles, simply because the density of organic particles is less than half that of mineral particles of the same size.

Particles that contribute to light attenuation in waters must stay suspended for a reasonable length of time and, therefore, are fairly slow-settling. The settling velocity of fine particles, v_s, is given by Stokes' law (e.g. Lerman 1979):

$$v_s = \frac{\alpha g d^2 (\rho_s - \rho)}{18\eta} \qquad (2.12)$$

where α is a dimensionless 'shape factor' (= 1 for a sphere), g is the gravitational acceleration, d is a characteristic linear dimension (diameter for a sphere), ρ_s is the density of the particle, ρ is the density of the water, and η is absolute viscosity (1 g m^{-1} s^{-1} at 20°C) of the water. This relationship shows that settling velocity depends mainly on particle size and density relative to water, although shape (deviation from sphericity as measured by α) also has an effect.

Stokes' law explains why solid matter suspended in waters is usually either fine grained (small d) or has a density comparable with that of water (low $(\rho_s - \rho)$) such that its settling velocity is very low, say less than a few mm s^{-1}. Particles with higher settling velocities will settle out of the water column in standing water bodies to form bed deposits, or, in running waters, will be present as a 'graded suspension' in which concentration significantly varies with depth in the water column (e.g. ASCE 1975, pp. 74–81). Part of the total suspended load of rivers is typically composed of sand-sized material ($d > 50$ μm, $v_s > 2$ mm s^{-1} for quartz density materials) moving in graded suspension, as well as silt and clay-sized particles which are more nearly uniformly distributed with depth (ASCE 1975, p. 317). However, unless they are present at very high concentrations, coarse particles contribute little to total light attenuation since their optical cross-sections are comparatively small by comparison with the uniformly distributed small ('fine') particles.

Mineral particles typically have densities similar to or greater than quartz (density relative to that of water = 2.65) and generally only remain in uniform suspension in natural waters if particle diameters are smaller than the sand–silt boundary at 50 μm (with corresponding settling velocities < 2 mm s^{-1} in water at 20°C). Some of the silt-sized material

(2–50 μm diameter, $3 \ \mu m \ s^{-1} < v_s < 2 \ mm \ s^{-1}$ for quartz-density particles) eventually settles out in standing waters such as lakes. There is a strong selection towards removal from the water column of the coarser silt size ranges because v_s depends on the square of d in Stokes' law (equation (2.12)). A particle at the clay–silt boundary, settling at around $3 \ \mu m \ s^{-1}$, takes about 4 days to settle 1 m. On that time scale there is usually ample opportunity for lake water turbulence induced by wind action or convective cooling to mix the water so that the finer silt sizes, say $d < 10 \ \mu m$ ($v_s < 70 \ \mu m \ s^{-1}$), may never settle.

Spherical particles of quartz density in the size range of efficient light attenuation (say, 0.2 to 5 μm diameter) have settling velocities in the range 0.03–20 μm s^{-1}. Particles at the high end of this size range settle sufficiently fast for limited removal to occur in very quiescent waters, but it is evident that simple gravitational settling will seldom produce much reduction in the overall light attenuation of suspensions containing these particle sizes.

Clay minerals, being layer silicates (e.g. van Olphen 1977), usually occur as highly aspherical plates ($\alpha \ll 1$), and therefore settle more slowly than would be predicted from their characteristic linear dimensions, such as can be measured in electron micrographs. Lerman (1979) has summarized Stokes' law velocities for different geometric shapes for which drag forces can be theoretically calculated. Disc-shaped particles (highly oblate spheroids), which are a reasonable geometric model of clay mineral grains, settle at less than half the speed of spheres of the same volume and roughly 1/10 the speed of spheres of the same diameter, depending on disc thickness (ratio of long to short dimension).

Clay particles tend not to settle even in the most quiescent of lake waters. For example, a particle of smectite (a common, relatively fine-grained, clay mineral) settling at 3×10^{-10} m s^{-1} (= 3×10^{-4} μm s^{-1}) would theoretically take 100 years to settle 1 m. Upwards-directed currents of much greater speeds occur in all surface waters owing to turbulence related to winds and thermal phenomena, and these counteract gravitational settling. The fact that clays are (sometimes) removed from lake waters reflects agglomeration into larger, and much faster-settling, flocs rather than settlement as discrete particles (Weilenmann *et al.* 1989). Co-flocculation with algae and organic detritus together with associated bacterial polyelectrolytes (e.g. Avnimelech *et al.* 1982; Leppard 1984a,b) is thought to be important in natural clarification of lake waters. In saline waters, notably estuaries, the high ionic strength of the water screens surface electrical charges on particles so promoting agglomeration into floccules.

Organic particles and flocs of organic and mineral particles containing trapped water typically have low densities (e.g. Chase (1979) reported a density relative to water of 1.035 ± 0.011 for freshwater particles). From Stokes' law we would expect settling velocities to average only 2% (= 100 (1.035 – 1.0)/(2.65 –1.0)) of that of quartz density particles of the same size, although the ratio will vary strongly with density and particle shape. However, Chase (1979) found experimentally that organic-rich natural particles settle much more rapidly (up to 10 times faster) than predicted from Stokes' law, although still more slowly than similarly sized mineral particles. The apparent reduction in drag forces (reduced skin friction) varies in a non-linear manner with particle size and is attributable to chemically adsorbed organic matter.

Even allowing for such refinements to a simple Stokes' law calculation, a 10 μm diameter organic particle at the upper end of the efficiently light-attenuating size range, is likely to have a settling velocity $< 20 \ \mu m \ s^{-1}$. Therefore, efficiently light-attenuating organic particles settle at speeds similar to or slower than efficiently light-attenuating mineral particles (0.2 to 5 μm). Unaided gravitational settling seldom produces much reduction in overall light attenuation because the 'optically efficient' particulates in waters, of both inorganic and organic composition, fall so slowly. The turbulence present in most natural surface waters tends to maintain turbid suspensions for long periods unless particle agglomeration occurs.

2.4.3 Phytoplankton

Phytoplankton are a particularly important special case of aquatic suspensoid. The low clarity of eutrophic (= nutrient-enriched) lakes with their high concentrations of phytoplankton usually results from the light attenuation by the organic material derived from algal production together with algal cells themselves (Plate 3). A very useful index of the concentration of living phytoplankton in a water is the concentration of the universal photosynthetic pigment, chlorophyll a (Table 2.1). Chlorophyll a in dead algal cells degrades rapidly to phaeopigment, so the sum chlorophyll a + phaeopigment (known as 'total pigment') is an index of the algal biomass plus algal-derived detritus.

Although all algae contain chlorophyll a, different algal groups have very different assemblages of accessory photosynthetic pigments (e.g. Dring 1990) with the result that they have different colours—hence green algae (the chlorophytes) versus the blue–greens (cyanophytes). Another important group, the diatoms, are often yellow in colour owing to absorption of blue–green light by carotenoid pigments. Fig. 2.10 shows the absorption cross-section (absorption coefficient per unit total pigment concentration) measured on cultures of three freshwater phytoplantonic algae representative of these three groups. Absorption peaks at 440 nm in the blue and at 676 nm in the red part of the

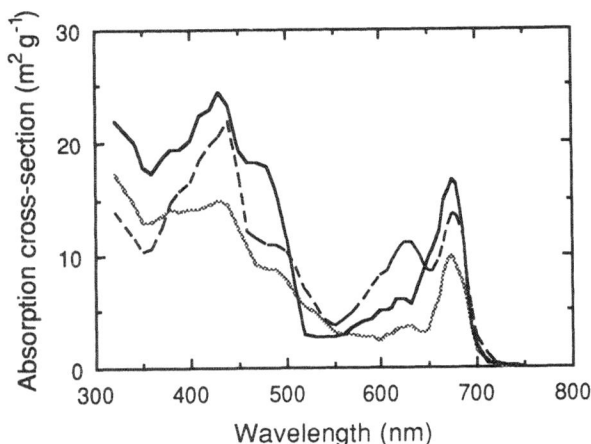

Fig. 2.10. Absorption spectra of cultures of some freshwater phytoplanktonic algae measured by spectrophotometer. (Redrawn from the data of Davies-Colley *et al.* (1986).) The spectra are given as absorption cross-sections (absorption coefficient per unit chlorophyll a concentration). ——, *Scenedesmus bijunga* (Trup.), a green alga; – – –, *Anabaena oscillaroides*, a blue–green alga; ——, *Navicula* sp., a diatom.

spectrum, both being attributable to chlorophyll *a*, are common to all the spectra in Fig. 2.10. Otherwise, the absorption spectra differ appreciably in shape.

As well as the variation in *spectral pattern* of absorption by phytoplankton, the *magnitude* of the absorption (per unit pigment concentration) varies. Light absorption by pigmented particles, such as phytoplankton cells, is always lower than if the absorbing pigment were distributed homogeneously in the water—to an extent dependent on the degree of 'packaging' of pigment in particles (Duysens 1956). (The term 'packaging' in this context was first used by Kirk (1975a,b).) The packaging effect can be conceptualized as the result of self-shading of chromophores: removal of light from a beam by absorption at the leading edge of a particle results in less light reaching the chromophores located deeper in the package. The greater the degree of packaging of pigment in phytoplankton, that is, the larger the cells or cell aggregates and the greater the intracellular pigment concentration, the more the absorption cross-section is decreased. Up to five-fold variation in absorption cross-section at the red peak (676 nm) can be expected with typical ranges in phytoplankton cell or cell aggregate size and intracellular pigment concentration. As well as reducing the magnitude of absorption at any given wavelength, the packaging effect also tends to reduce spectral variation in absorption (i.e. to flatten out the spectrum) because absorption at the peaks is reduced more than in troughs.

Fig. 2.11 shows 'average' absorption spectra for marine phytoplankton expressed per unit total pigment concentration, and the dispersion of absorption is indicated by the standard deviation as a function of wavelength. Unfortunately no equivalent average spectra appear to have been reported for freshwater phytoplankton for which there is a paucity of data (Davies-Colley *et al.* 1986), although the marine spectrum is probably broadly applicable. The spectral absorption attributable to the phytoplankton in a water

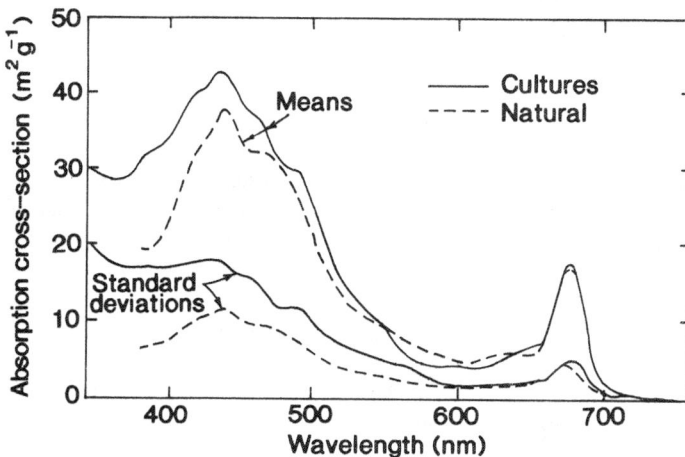

Fig. 2.11. Indicative spectral absorption cross-sections of phytoplankton. Dashed lines give the mean and standard deviation for phytoplankton of the Peru upwelling measured *in situ* and the solid lines give the mean and standard deviation for 13 different marine phytoplankton species in culture. (From Bricaud & Stramski (1990), with permission.)

can be estimated (roughly, owing to both the operation of the packaging effect and variation in pigment composition) by multiplying the spectrum in Fig. 2.11 by the total pigment (chlorophyll *a* + phaeopigment) concentration of that water.

Although phytoplankton are strongly pigmented particles, their scattering cross-sections (scattering coefficient divided by pigment concentration) are typically an order of magnitude higher than their absorption cross-sections. Thus light beam attenuation by phytoplankton, and, therefore, their influence on visual water clarity, depends more strongly on scattering than absorption. The scattering cross-sections of phytoplankton are even more variable than their absorption cross-sections. This variability is again dependent on variation in cell size and intracellular pigment concentration, but also expresses variation in refractive index of cells. Morel (1987) has summarized laboratory and field data in relation to a theoretical framework to demonstrate an order of magnitude variation in the scattering cross-section of phytoplankton (from 60 to 600 m^2 (g pigment)$^{-1}$). Small algae (in the 3 to 8 μm range of cell diameter) are generally efficiently light scattering (and light attenuating), whereas larger cells (and cell colonies) decrease in attenuation cross-section in proportion to increase in diameter—in accord with the trend for organic particles shown in Fig. 2.9B. The scattering of blue–green bacteria (cyanophytes) may be greatly enhanced by the presence within their cells of gas vacuoles which act as scattering subparticles. For example, Ganf *et al.* (1989) found that the scattering cross-section of *Microcystis* cells dropped abruptly from 190 to 21 m^2 g^{-1} when gas vesicles collapsed under pressure.

2.4.4 Organic detritus and yellow substance

The attenuation cross-section of solid particles of organic detritus can be expected to vary mainly with size, broadly according to the curve for organic matter in Fig. 2.9. Some of the attenuation of light by detritus in natural waters is actually attributable to natural populations of bacteria which are weakly absorbing, but scatter light appreciably in spite of small size relative to the wavelength of light and rather low scattering efficiency. Stramski & Kiefer (1990) reported measurements and calculations of light attenuation by marine bacterioplankton which suggest that bacteria may be an important, or even dominant, light-attenuating constituent in oligotrophic waters.

Spectral absorption of light by organic detritus may depend on particle size (owing to operation of the 'packaging' effect again) but, in addition, can be expected to depend on the original pigment composition of the once-living material and the extent of degradation, ultimately to humus, by bacterial and fungal action. Aquatic humus, like the soil humus to which it is closely related chemically, is not a well-defined chemical compound but is a random polymer of many organic substances and is characterized by acidic properties, colloidal-sized macromolecular structure and strong light absorption (Thurman 1985).

The filter-passing fraction of aquatic humus (operationally the 'dissolved' fraction) is known as 'yellow substance' (German: *Gelbstoff* (Kalle 1966)), a reference to the yellow hue imparted by this material to waters. Kirk (1976, 1983) proposed the alternative term 'gilvin', a name which is apparently in common use in Australia. The more widely familiar 'yellow substance' will be used here. Kirk (1976) has suggested that the absorption coefficient (units m^{-1}) of membrane-filtered water at 440 nm (g_{440}, g standing for gilvin) can be taken as a convenient index of the concentration of yellow substance.

Yellow substance is the only filter-passing constituent of waters, other than water itself, that is significantly light attenuating. This organic material is optically and chemically similar to the water-soluble, yellow-coloured, fraction of soil humic matter known as fulvic acid. Indeed a large fraction of the yellow substance, at least in freshwaters, may be derived from soil leaching (Kirk 1983; Thurman 1985). The remainder is presumably derived from organic matter produced *in situ* by aquatic plants (e.g. Davies-Colley & Vant 1987).

Yellow substance has a characteristic exponential-shaped absorption spectrum (Fig. 2.12) which appears to be of similar shape in very different natural waters ranging over more than 3 orders of magnitude in dissolved organic carbon (DOC) concentration (Bricaud *et al.* 1981; Davies-Colley & Vant 1987). In the ultraviolet region of the spectrum at about 270 nm, yellow substance typically exhibits a small shoulder (indicated on Fig. 2.12), but the UV–visible spectrum of absorption is otherwise featureless except for the approximately exponential decline with increasing wavelength. In waters of comparatively high yellow substance concentration, say $g_{440} > 1$ m^{-1}, yellow substance is usually responsible for most of the light absorption, and this absorption, being selective for blue light, is what causes the observed yellow hue (Plate 4) and dark appearance (low reflectance) in such waters. Scattering by yellow substance is

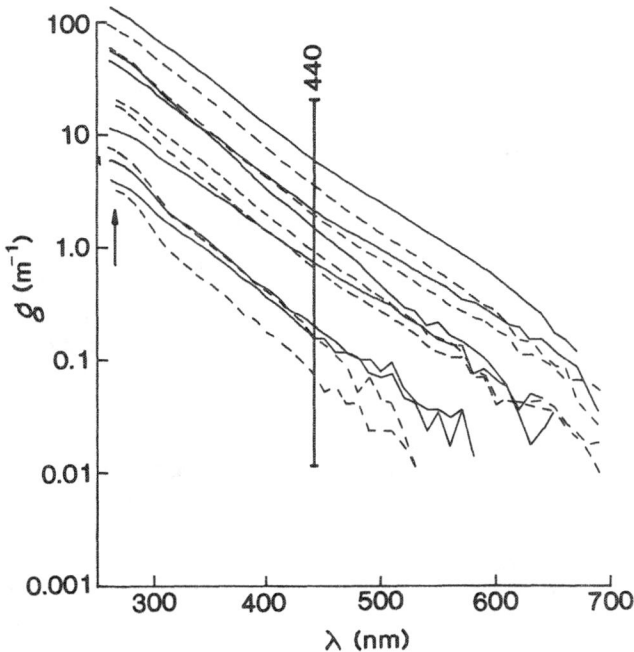

Fig. 2.12. Absorption spectra of yellow substance in 12 New Zealand lake waters. (From Davies-Colley & Vant (1987).) Every second line (at 440 nm) is dotted to facilitate following the slope by eye. The arrow indicates a small spectral feature at 270 nm. The bar at 440 nm shows the inferred range of g_{440} compiled from the literature by Kirk (1983).

very low, compared with that by particulate constituents, and can be neglected in natural waters. Thus waters dominated by yellow substance can be visually clear although appreciably coloured.

Particulate organic detritus in waters may include yellow substance chemically adsorbed on mineral surfaces (e.g. Tipping 1981; Kirk 1983) or present in co-flocculated aggregates (e.g. Leppard 1984a,b) as well as degraded cellular material. Light-absorbing particulate organic matter in waters has been called 'tripton' by Kirk (1979, 1980). Since 'detritus' can refer to organic matter in a wide range of chemical states (degree of humification), the term 'tripton' would be useful if it were assigned a more specific meaning: 'particulate aquatic humus'. Certainly, the highly coloured and turbid Australian lakes studied by Kirk (1979, 1980) would have contained much aquatic humus in the particulate fraction, although some of his suspensoid absorption spectra show peaks attributable to algal pigments. Similarly, Lake Waahi (Plate 5) is strongly coloured by tripton as well as yellow substance, and is highly turbid ('muddy' coloured) owing to a high content of suspended clays.

The absorption spectrum of tripton is rather poorly known but appears to be similar to that of yellow substance in being approximately exponential in shape but with a generally lower exponential slope (e.g. Kishino et al. 1985; Bricaud & Stramski 1990). The absorption spectra of particles of organic detritus which are only partially degraded usually have small peaks of phaeopigment (a degradation product of chlorophyll—Section 2.4.3) superimposed on the exponential spectrum of aquatic humus (e.g. Iturriaga & Siegel 1989).

2.5 CLARITY OF WATERS

Colour and clarity are the two main directly observable features of natural waters. Water clarity is probably simpler to understand than water colour because clarity, to a first approximation, depends only on broad band optical properties. Consideration of colour, as we shall see in the next section, requires knowledge of the spectral dependence of optical properties. If we are concerned with human visual range in water, the appropriate 'light' to consider is luminous flux measured with a sensor of spectral sensitivity identical to that of the human eye, which has a peak sensitivity at 555 nm in the green part of the spectrum (Fig. 2.19 below). If the concern is with the vision of animals in water, the appropriate spectral response will generally be different from that of the human eye. If we are concerned with clarity in terms of light penetration for aquatic plant growth we are concerned with so-called 'photosynthetically available radiation' (PAR), the photon flux in the 400–700 nm wavelength band.

We have already hinted at the two main aspects of water clarity with which we are concerned. Firstly we have visual range in water, for humans and for aquatic animals. Secondly we may be concerned with the penetration of diffuse light (more properly referred to as irradiance) into water for photosynthesis or vision. These facets of clarity are related only distantly, and, as we shall see, it is very important to be clear about which we are dealing with! There are other aspects of water clarity which may be of concern in some contexts. For example, in remote sensing applications the depth to which the remote sensor 'sees' into the water is of concern if there is any vertical stratification of the surface water layers. These other aspects will not be considered further.

2.5.1 Visual clarity

Submerged objects must be sufficiently well lit (illuminated) before they can be seen by animals or man. So the first consideration with vision underwater is with the penetration of illuminance (Appendix 1) into the water column (Fig. 2.13). It is the illuminating light reflected from objects that conveys a recognizable image to an observer's eye.

To form an image, reflected light rays must travel from the object to the observer by straight lines, or at least lines 'bent' in an orderly way by refraction in an optical system. From this argument we would intuitively expect that the beam attenuation coefficient, c, which quantifies attenuation of an idealized, perfectly collimated, monochromatic light beam, due to both absorption and scattering, would correlate closely with the underwater visual range. This turns out to be correct. The visual range of a given object, for example a Secchi disc, is approximately inversely proportional to the beam attenuation coefficient, weighted by the spectral sensitivity of the human eye. This is the beam attenuation of *luminance,* the psychophysical equivalent of radiance (refer to Appendix 1 for definitions of quantities used for measuring radiation). Fig. 2.5A shows that the Secchi depth correlates closely (inversely) with the beam attenuation coefficient in New Zealand lakes.

Further consideration of the problem of vision in water suggests that the characteristics of the viewed object will also influence visual range. We would generally expect a dark object to disappear at a smaller range than a bright, reflective object such as the Secchi disc. We might also expect the visual range to depend on size, a larger object being more 'visible' (seen at a greater distance) than a small object of the same size and surface character. Finally we might expect visual range to depend on lighting level. The human eye is adaptable to a very wide range of illumination (more than eight orders of magnitude), but even the most dark-adapted eye can not see in the absence of light. The

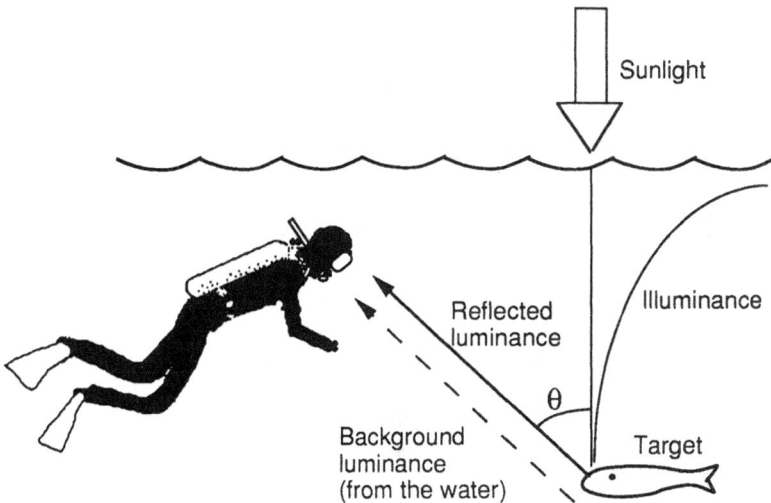

Fig. 2.13. Vision underwater. (Refer to Appendix 1 for definitions.)

general problem of human vision underwater (or into water), and indeed the vision of aquatic animals, will involve consideration of background lighting (and therefore viewing direction), and size and surface character of the object, as well as beam attenuation which expresses the 'inherent clarity' of the water medium.

We see an object, henceforth referred to as a 'target', because it contrasts with its surroundings. Inherent contrast is defined as

$$C_0 = \frac{B - B_b}{B_b} \tag{2.13}$$

where B_b is background luminance and B is luminance of the target. Inherent contrast can vary from -1 for a black object to $+\infty$ for a reflective object seen against a black background. In an optically clear medium, such as air, the visual range of the target is most commonly limited by diminishing apparent size. However, we are all familiar with smoke and fog, and know how these atmospheric contaminants greatly reduce visual range. Water is broadly analogous to foggy air as an optical medium, and the underwater 'fog' is familiar to divers. Visual range in water, as in fog, is determined by attenuation of apparent contrast to a threshold below which the human eye cannot resolve the image from the background.

Duntley (1963) has shown that attenuation of contrast in water is approximately exponential under rather robust and general assumptions. The apparent contrast at range r is given by

$$C_r = C_0 \exp\left[-(c + K\cos\theta)r\right] \tag{2.14}$$

where c is the beam attenuation coefficient for luminance (i.e. the beam attenuation weighted by the spectral sensitivity of the human eye), K is the attenuation coefficient for illuminance (rather than luminance), and θ is the angle of the path of sight to the vertical (Fig. 2.13). Without labouring through its formal derivation (e.g. Duntley 1963; Jerlov 1976, pp. 157–159), equation (2.14) can be understood as follows. As we have seen, the transmission of an image depends on beam attenuation, which is why c appears in the expression. If viewing is horizontal ($\theta = 90°$, $\cos\theta = 0$) the second term in the argument of the exponential disappears and apparent contrast, and thus visibility, as could be expected, depends only on c. However, at viewing angles other than horizontal, apparent contrast will be affected by the gradient of background luminance in the water. When an object is viewed deeper in the water than the observer ($0° < \theta < 90°$, $\cos\theta > 0$), apparent contrast at a given range, r, is reduced by the increase in background luminance from the water between the depth of the viewed object and that of the observer (Fig. 2.13). This background luminance at angle θ varies with depth at the same rate as the illuminance (quantified by K), so its variation along the inclined path of sight is as $K\cos\theta$. Thus the term $K\cos\theta$ in equation (2.14) accounts for the decrease in apparent contrast owing to the exponential increase in background luminance with distance along the path of sight from object to observer. When an object is viewed higher in the water than the observer ($90° < \theta < 180°$, $\cos\theta < 0$, $K\cos\theta < 0$), apparent contrast at a given range, r, is increased because of the diminution of background luminance along the inclined path of sight from object to observer.

As we have seen, the rate of attenuation of apparent contrast with range varies with viewing angle. However, the inherent contrast C_0 also varies with viewing angle. A bright, reflective, object such as a Secchi disc has a high value of C_0 viewed vertically down (roughly 40 (Tyler 1968)), but viewed vertically up the disc face is in shadow and will appear darker than the water luminance (C_0 negative). At some intermediate inclined viewing angle the disc will have zero contrast and, at least theoretically, will be invisible at any range. Thus equation (2.14) is deceptively simple and belies the true complexity of visibility at arbitrary angles.

The threshold apparent contrast at which a target just disappears depends on background (adaptive) lighting and angular size. For the human eye, threshold contrasts of circular targets are well known as a result of exhaustive studies by Blackwell (1946). The threshold contrast approaches a minimum ($C_T \sim \pm0.0066$) asymptotically for targets of diameter greater than about $1°$ (= 0.017 rad) viewed under relatively bright adaptive lighting conditions (say brighter than about 0.1% 'bright sunlight') (Taylor 1964).

Given an appropriate value for threshold contrast we can predict the visual range of relatively large targets of known inherent contrast viewed under daylight in near-surface water using (2.14). Since both object luminance, B, and background luminance, B_b, vary (independently) with angle of view, θ, as well as depth in water, the inherent contrast given by (2.13) also varies with angle of view. Thus it is only possible to generalize regarding visual range for particular angles of view.

One important application of the contrast attenuation theory was Tyler's (1968) theory of Secchi disc visibility. When the Secchi disc is viewed vertically, as is usual, $\cos \theta = 1$ in (2.14), and setting range $r = z_{SD}$, the Secchi depth, when the apparent contrast, C_r, is reduced to the threshold value, C_T, we obtain

$$z_{SD} = \frac{\ln\left(C_0/C_T\right)}{c + K} = \frac{\Gamma}{c + K} \qquad (2.15)$$

The inherent contrast, C_0, of the Secchi disc (white surfaces) viewed vertically is approximately

$$C_0 = \frac{R_{disc} - R}{R} \qquad (2.16)$$

in which R is the reflectance coefficient (for illuminance) of the water and R_{disc} is the inherent reflectance of the white-painted surface of the Secchi disc. Tyler estimated $C_0 \sim$ 40 based on an assumed average ocean water reflectance of 2%, and 82% reflectance of the white paint on the Secchi disc, but C_0 varies with reflectance of the background water which can range from <1% to >30% (Davies-Colley & Vant 1988). Putting $C_T =$ 0.0066, Tyler obtained $\Gamma = 8.69$. Højerslev (1986) verified this threshold contrast empirically for the Secchi disc. Being a logarithmic function, Γ is rather insensitive to assumed values of C_0 and C_T (Preisendorfer 1986) and it is often satisfactory to assume $\Gamma \sim 9$.

Equation (2.15) shows that the Secchi depth varies inversely with the sum $c + K$ rather than with c alone. Because K is an apparent optical property, so too is z_{SD}. Kirk (1982, 1985) has referred to $c + K$ as the contrast attenuation coefficient and $1/(c + K)$ (units: m) as the contrast attenuation depth. The correlation of z_{SD} and c indicated in Fig. 2.5A results from the covariation of c and K. However, the product cz_{SD} is only approximately constant (at about 6) and can vary by 50% or more (Gordon & Wouters 1978).

When the Secchi disc is viewed horizontally the angle $\theta = 90°$ ($\cos \theta = 0$) and the relevant contrast attenuation coefficient is simply c. The reduction in contrast attenuation from $c + K$ to c tends to increase visual range compared with vertical viewing. However, this is partly compensated by the appreciably lower inherent contrast for horizontal viewing because the water background is brighter than when viewed vertically. The result is that the horizontal Secchi disc range, y_{SD}, is typically only slightly greater than the vertical range, z_{SD}. Unfortunately, the inherent contrast of the Secchi disc, and therefore the sighting range, depends on horizontal direction with respect to the sun (at least at shallow depths in the water column), so y_{SD} is not as useful as z_{SD} as an index of visual water clarity.

Unlike reflective targets such as the Secchi disc, black targets reflect no light, and are seen as silhouettes against the background light in water. The inherent contrast of such targets is constant at -1, irrespective of angle of view, as can be seen from consideration of the definition of inherent contrast in (2.13). The visual range of black targets in water can therefore be generalized in a way that is not possible for bright targets such as the Secchi disc. From consideration of (2.14) we obtain the following expressions for the visual range of a black target (e.g. a black disc, denoted BD) viewed horizontally,

$$y_{BD} = \frac{\ln(C_0/C_T)}{c} = \frac{\ln(-1/C_T)}{c} = \frac{\Psi}{c} \tag{2.17a}$$

and for vertical viewing,

$$z_{BD} = \frac{\ln(-1/C_T)}{c + K} = \frac{\Psi}{c + K} \tag{2.17b}$$

Blackwell (1946) showed that the threshold contrast for dark objects of negative inherent contrast was identical to that of targets of numerically equivalent positive contrast. Therefore we can assume $C_T \approx -0.0066$ giving $\Psi \approx 5$. Duntley (1963) has reported that the horizontal visual range of a great variety of dark submerged objects is about $(4-5)/c$ (i.e. 4–5 attenuation lengths, where $1/c$ is the attenuation length). For example, fish, being large dark (countershaded) targets, are usually seen by divers at about 4–5 attenuation lengths, although the flash of light reflected from their scales may be visible at greater distances. Davies-Colley (1988a) found experimentally that the value of Ψ is about 4.8 for submerged, black-painted discs, corresponding to a threshold contrast averaging 0.008. This provides a very useful means for estimating c as we shall see in Section 3.4.2.

2.5.2 Light climate for aquatic plants

Concepts involved in description of the underwater light climate for aquatic plants have been reviewed by several authors, most notably by Kirk (1983). Here we merely give a brief overview of the important concepts. The reader is referred to Kirk (1983) for more detail.

Almost all natural ecosystems depend for their energy on capture of light from the sun by plants. We now know sufficient about the photosynthesis of plants to specify accurately the *light climate* for plant growth. It turns out that plants respond identically to light photons captured by their photosynthetic pigments, irrespective of energy (wavelength) in the 400–700 nm range of wavelength. That is, a captured red photon ($\lambda \sim 700$ nm) is just as useful to a plant as a captured blue photon ($\lambda \sim 400$ nm) in spite of the fact that the blue photon contains 50% more energy. However, the probability of capture of a given photon by absorption does depend on wavelength. For example, Figs 2.10 and 2.11 show that green light ($\lambda \sim 550$ nm) is generally less strongly absorbed by photoplankton than blue light.

Since plants respond to photons rather than energy in the 400–700 nm (visible) waveband the appropriate measure of 'light' to specify the light climate for plants is quantum irradiance in the photosynthetically available band or *photosynthetically available radiation* (PAR) (units: moles of photons m^{-2} s^{-1}, where 1 mole of photons = Avogadro's number of photons = 6.0225×10^{23} photons).

The appropriate measure of water clarity in the context of aquatic plant production is the irradiance attenuation coefficient for scalar PAR, K_o(PAR), which, in practice, is very similar in magnitude to the irradiance attenuation coefficient for downwelling vector PAR, K_d(PAR) (Fig. 2.4). K_o is, of course, an apparent optical property and therefore is dependent to some extent on the characteristics of the ambient light field as well as the optical properties of water. However, it is often convenient to think of a particular water as having a particular K_o (or K_d) value, and to compare different waters as to clarity, as if K_o was dependent only on water composition.

The 'quasi-inherent' character of the irradiance attenuation coefficient has been discussed by Baker & Smith (1979) who showed that K_d varies only weakly through the day with change in solar elevation.

A useful, if approximate, index of the depth above which PAR is sufficient for aquatic plants to grow is the *euphotic depth*, z_{eu}. This is conventionally taken as the depth at which PAR, ideally measured as scalar irradiance, E_o, has declined to 1% of the surface irradiance (Kirk 1983, p. 113). That is, the euphotic depth occurs where $E_o(z_{eu}) = 0.01E_o(0-)$, in which $(0-)$ denotes a null depth, just below the water surface. The euphotic depth, so calculated, is indicated on Fig. 2.4. The water column above the euphotic depth is known as the euphotic zone. Since PAR typically attenuates nearly exponentially, a good approximation for euphotic depth is simply $z_{eu} \sim \ln(100)/(\bar{K}_o) = 4.6/\bar{K}_o$, the overbar indicating an average obtained by linear regression of $\ln E_o$ on z in the euphotic zone (Kirk 1981a). This expression may give a slightly biased result owing to the weak depth dependence of PAR attenuation.

A more physiologically meaningful concept than the euphotic depth is the *compensation depth*, the depth below which long-term consumption of oxygen by aquatic plant respiration exceeds photosynthetic production of oxygen. The compensation depth corresponds to a time-averaged PAR value which is just sufficient to maintain biomass with no growth. Kirk (1983, Table 10.1) has compiled a useful list of compensation irradiances.

The photosynthetic fixation of carbon by plants, including aquatic plants, is typically a non-linear function of quantum irradiance (Fig. 2.14). At low PAR levels, rate of photosynthesis, P, is proportional to PAR. At higher irradiances, photosynthesis becomes progressively less dependent on PAR and approaches an upper maximum light-saturated value, P_{max}. At all irradiances the plant consumes fixed carbon in respiration, and at zero irradiance this can be measured. The production curve intersects the PAR axis at the *compensation irradiance*. Typically this is of the order of 10 μmol m^{-2} s^{-1}, but it varies appreciably with plant type and physiological state (e.g. with season). Obviously, the depth in a water body at which irradiance falls to the compensation value (i.e. the

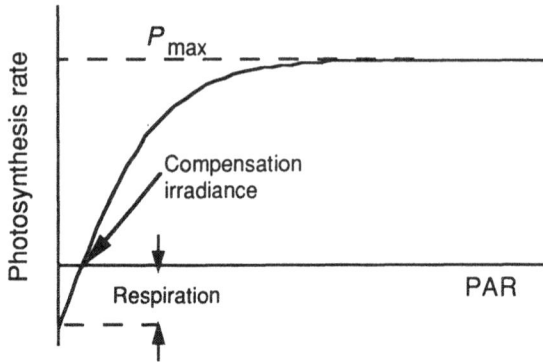

Fig. 2.14. Rate of photosynthesis by plants as a function of quantum irradiance (PAR). At zero irradiance, respiration is measured. The intercept of the production curve with the abscissa gives the compensation irradiance value.

compensation depth) will depend on ambient light and water clarity (the irradiance attenuation coefficient) as well as the photosynthetic characteristics of the plants.

The maximum depth of macrophyte colonization of lake waters (symbol z_c) is often determined by light penetration (Spence 1976). Thus z_c is the depth at which the irradiance is reduced to the compensation value, unless some factor other than light is limiting.

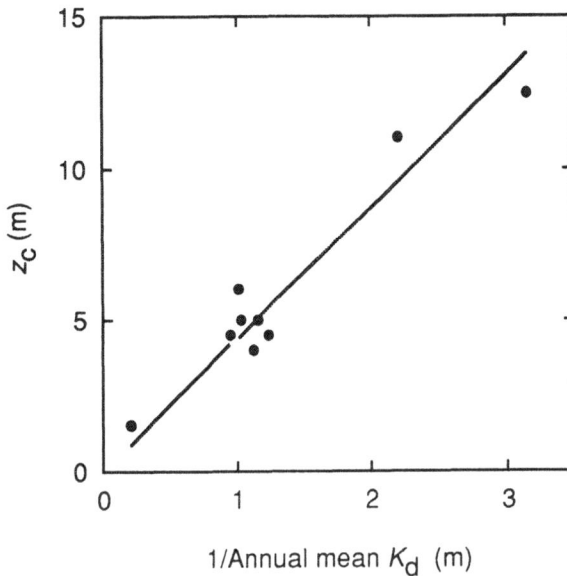

Fig. 2.15. Dependence of maximum depth of macrophyte colonization, z_c, on annual average light penetration in nine New Zealand lakes. (From Vant et al. (1986).) Light penetration is scaled as the reciprocal of the irradiance attenuation coefficient, $1/K_d$, sometimes known as the attenuation depth. The equation of the regression line is $z_c = 4.3/K_d$.

Vant *et al.* (1986) demonstrated that the observed z_c in nine New Zealand lakes correlated inversely with K_d (Fig. 2.15), and that z_c corresponded on average to the depth of 1.7% of surface PAR, in fair agreement with the rule of thumb that growth occurs down to the 1% PAR level. It should be noted, however, that macrophytes vary somewhat in their 'shade tolerance' and the actual z_c depends to some extent on light-harvesting characteristics of the particular colonizing plant. Also, factors other than light can sometimes determine z_c, including hydrostatic pressure (Coffey & Kar-Wah 1988), substrate stability and grazing by aquatic animals.

Phytoplankton, unlike macrophytes and the less well-studied benthic algae ('periphyton'), are not fixed to a substrate but move with turbulence and water currents in the

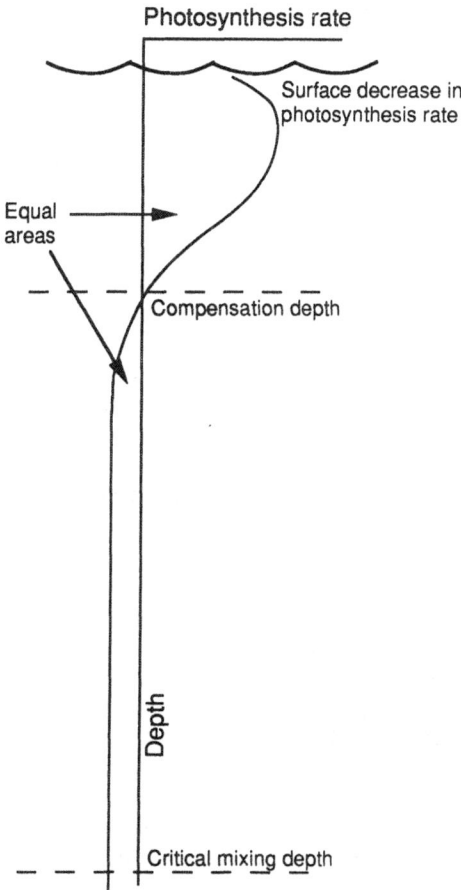

Fig. 2.16. Idealized photosynthesis rate versus depth in a water body. The compensation depth occurs where the net production is zero, and the critical mixing depth occurs where gains from production just balance losses from respiration in the overlying water column. Apparent photoinhibition at shallow depths is indicated.

mixed layer of a water body. If this mixed layer extends deeper than the compensation depth the phytoplankton cells spend some time 'in the dark' with respiration losses of organic carbon exceeding photosynthetic gains. At a particular critical ratio of mixed depth to compensation depth (about 5, e.g. Talling 1971), no net growth of phytoplankton occurs owing to light limitation. Kirk's (1983, pp. 254–260) review of mixing and phyotosynthesis suggests that light limitation may be more common than is often recognized, occurring particularly in deep, stratifying, lakes during winter turnover under low ambient light levels, but also in shallow lakes with highly light-attenuating waters.

If most of the light attenuation in water is by the algal cells themselves or by their recent organic products ('detritus'), the phytoplankton are referred to as 'self-shading'. Self shading is common in so-called hypertrophic ('over-enriched'—by nutrients) lakes and in waste stabilization ponds. In a self-shaded water, any increase in phytoplankton biomass causes a decrease in water clarity (increase in K_d). Water column-averaged respiration then exceeds photosynthesis until the biomass declines back to the steady state level where light is just sufficient for production to balance respiration and net growth is zero again. Actually, the real situation is more complicated, particularly because the feedback between water clarity and biomass is not instantaneous, and considerable variation can occur in the phytoplankton biomass of self-shaded systems.

Fig. 2.14 suggests that photosynthesis rate of phytoplankton should be constant (maximal) in surface waters where irradiance is saturating, and decline exponentially in parallel with irradiance at greater depths where irradiance is limiting. Such photosynthesis–depth curves are actually measured with bottles incubated at a range of depths in lakes and the ocean (Fig. 2.16). A simple model of this photosynthetic pattern was given by Talling (1957). The compensation depth and critical mixing depth for light limitation are both indicated in this figure. A fall in production rate near the surface is typically observed owing to photoinhibition of algae at high irradiance (if not related to gravitational settling of algae in quiescent water). However, photoinhibition as observed in surface bottles may be something of an artefact since unconstrained phytoplankton constantly circulate to the mixing depth with water turbulence (see Kirk (1983, p. 233) for a discussion).

2.6 WATER COLOUR

2.6.1 Intrinsic water colour

When a water body is viewed from above, what is seen is a complex mixture of light reflected from the water surface and light 'reflected' from within the bulk of the water by the processes of backscattering and multiple forward scattering. In shallow water, as in some rivers, light reflected from the sediment bed may also contribute to the total signal.

The surface reflectance is perhaps the single most characteristic feature of water as an element of scenery, but it does not relate to the quality or optical nature of the water itself. Indeed, a waste stabilization pond would appear identical to a clear natural lake if it were not for the markedly different character of the component scattered back from within the water in the two systems. The surface reflectance, for all its contribution to the scenic appeal of water, is merely an image, variously distorted by ripples and waves, of the sky, clouds and surroundings, and of the sun itself (glitter). No interaction with the water occurs to modify the spectral distribution of this component, which is therefore not considered to contribute to water colour as such.

The component reflected off the bottom of shallow water bodies has interacted with the water itself (by selective absorption) and may give an impression of colour of the water as well as that of the sediment bed, notably where the bed sediment is composed of neutrally reflective white sand. The bottom-reflected component typically varies spatially in a complex way in shallow rivers, and near-shore zones of lakes and the coast, depending on water depth and reflectance characteristics of the bottom material (sand, gravel, macrophytes or periphyton).

The component arising from scattering of light in the bulk of the water is the only part of the total signal that can be unequivocally regarded as expressing 'colour' of water, viewed from above. Thus we may define 'intrinsic' colour of a water of given optical character as follows:

Water colour is the perception of light scattered back from within the water mass as observed when viewing downwards at a near-vertical angle.

A familiar phenomenon is the change in the relative importance of the surface and bulk reflection components with change in viewing angle. Viewed at a low angle to the horizontal (high angle of observation) water appears largely as a broken image of the sky or surroundings. When viewed at an acute angle approaching vertical (zenith viewing) more impression of intrinsic water colour is obtained.

Kirk (1982) has analysed this dependence of the appearance of water on viewing angle. The main reason for the change in appearance with angle is the dependence of surface reflection on angle of incidence (angle of reflection). Fig. 2.17, plotted from the data of Austin (1974), shows that reflectance is low (2–3%) up to about 45° and thereafter it increases rapidly to reach 100% at grazing incidence (90°). Fig. 2.17 also shows that the surface reflectance depends on sea state. When there are waves, the scale of which depends on wind speed, surface reflectance at small to intermediate angles (up to around 67°) is higher than in still conditions, but at large angles surface reflectance is lower than in still conditions. This explains why an observer viewing at oblique angles sometimes

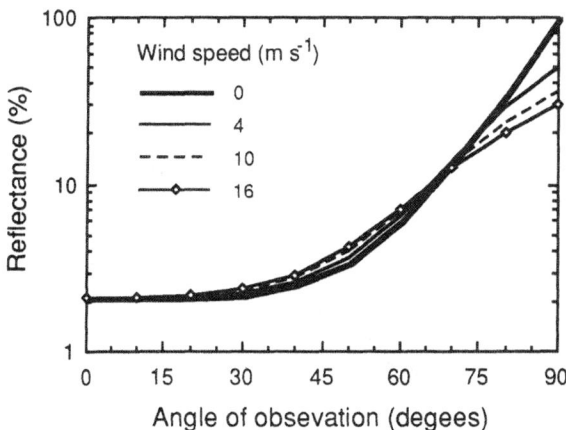

Fig. 2.17. Reflectance of the water surface as a function of the angle of observation (= angle of incidence, measured from the zenith) for different wind speeds. (Plotted from the data of Austin (1974).)

gains a more definite impression of intrinsic water colour of a lake or the sea with choppy waves than in still conditions (e.g. Plate 5). However, more generally an acute angle
of view is preferable to observe intrinsic water colour as we now demonstrate.

The total signal at a particular viewing angle can be quantified as the total radiance, L,
the sum of the radiance emerging from within the water, L_w, and the surface-reflected
radiance, L_s:

$$L = L_w + L_s \qquad (2.18)$$

Kirk (1982) used the still water curve from Fig. 2.17, together with consideration of the
angular dependence of upwelling radiance emerging from within the water, to prepare
Fig. 2.18 showing the ratio L_w/L as a function of observation angle for a still water surface.

As the angle of observation increases, the water-derived proportion of total radiance
barely changes up to about 40° but thereafter drops rapidly to zero at near-horizontal
viewing. Thus the appearance of the water changes little up to about 40°, and the colour
of the water is easily seen in this range. However, the water signal tends to be masked by
the surface-reflected signal at larger angles. The ratio L_w/L depends on characteristics of
the incident lighting as quantified by the parameter D, the ratio of diffuse skylight to
total irradiance. Under overcast skies when D = 1.0, L_w/L is lower than under clear sun

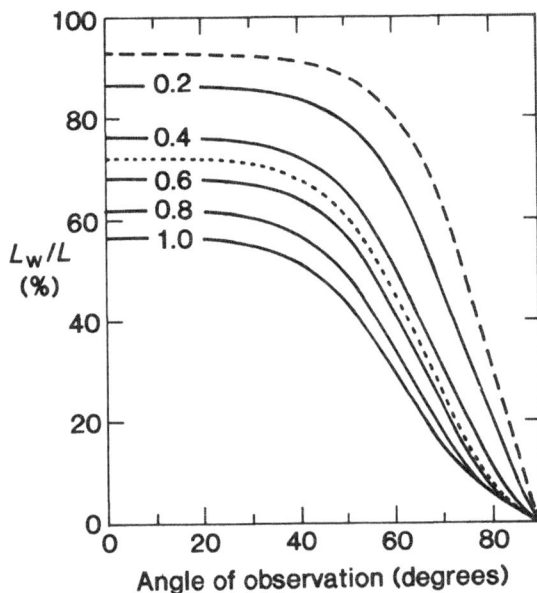

Fig. 2.18. Ratio of water-derived radiance to total radiance emanating from water as a function of angle
of viewing (measured from zenith). (From Kirk (1982), with permission.) The ratio of skylight to total
incident irradiance, D, ranges from 0.2 (clear sun) to 1.0 (overcast sky) and irradiance reflectance, $R(0) =$
5% (a typi-cal value). Curves are also given for $R(0) = 2\%$ (·····) and $R(0) = 10\%$ (– – –), both for D = 0.2.

(D ~ 0.2). As expected, L_w/L also depends on the irradiance reflectance coefficient (henceforth simply the 'reflectance') of the water, $R = E_u/E_d$ (Fig. 2.18).

2.6.2 Irradiance reflectance

It is the irradiance reflectance coefficient (Section 2.3) which relates the water colour to the composition of the water. Several workers have shown from modelling of light penetration into water that the reflectance immediately below the water surface is approximately proportional to a simple function of the absorption coefficient, a, and the backscattering coefficient, b_b. Morel & Prieur (1977) gave the following simple expression for the reflectance just below the water surface, $R(0-)$ (the negative sign after the zero indicating a null depth in the water), as a function of b_b and a:

$$R(0-) = 0.33 b_b/a \qquad (2.19)$$

Equation (2.19) is consistent with physical intuition. Once light is absorbed by the water it is no longer available for backscattering and cannot contribute to bulk reflectance, and hence the inverse dependence of $R(0-)$ on a. The greater the backscattering, the greater the upwelling light field in the water, and hence $R(0-)$ is proportional to b_b.

$R(0-)$, like both b_b and a, is a function of wavelength. The remarkable blue–violet colour of optically pure natural waters, such as Crater Lake, Oregon (Plate 1), arises because the spectral absorption, $a(\lambda)$, is essentially that of pure water, whilst pure water also contributes significantly to the spectral backscattering, $b_b(\lambda)$. As we saw in Section 2.4.1, backscattering by pure water, b_{bw}, is half of the total scattering by water, which is proportional to the –4.3 power of wavelength (Fig. 2.6). This explains why backscattering by optically pure waters is enhanced at the short wavelength, blue–violet end of the spectrum.

Kirk (1991) showed that the irradiance reflectance at zero depth is given quite accurately by equation (2.19) for a wide range of the angular dependence of scattering as quantified by the volume scattering function, and therefore a wide range of the ratio of backscattering to total scattering, b_b/b. Since backscattering is usually about 2% of total scattering (although the ratio depends on the form of the volume scattering function, $\beta(\theta)$), we can see that $R(0-)$ is approximately proportional to the ratio of total scattering to absorption.

2.6.3 The concept of colour applied to water

Colour is rather more conceptually complex than water clarity. This is mainly because water colour depends on the wavelength dependence of the absorption and scattering coefficients of a water, whereas clarity, as we have seen in Section 2.5, to a good approximation depends only on broad-band average values of absorption and scattering. Colour has three main attributes, all of which must be specified to describe satisfactorily particular colours, including water colours.

The most characteristic attribute of colour is the *hue*, the perception described as blue, green or red, for example. Hue, more than any of the concepts we have introduced so far, depends on the wavelength distribution of the light, since blue light, for example, has a different range of wavelengths from red light. (Fig. 2.19 shows the wavelength dependence of different colours in the visible spectrum in relation to the spectral sensitivity of

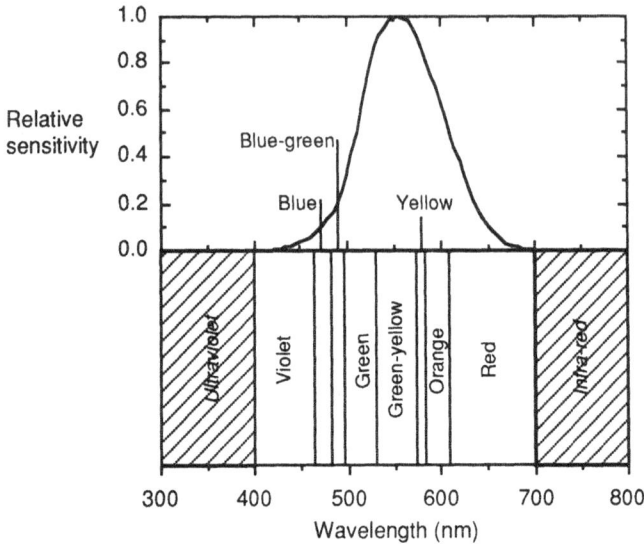

Fig. 2.19. Wavelength range of light producing different hues in the visible spectrum. The relative sensitivity curve for the 'standard' light-adapted human eye is also shown.

the human eye.) The hue of water depends on the wavelength distribution of the light scattered back from within the water. This in turn, depends on the wavelength selectivity of the processes of absorption and backscattering as they affect the spectral reflectance. The hue perception induced by a given distribution of energy can be identified with the hue of a particular monochromatic source. Thus the hue is specified by the dominant wavelength, λ_d, equal to the wavelength of the matching monochromatic source (Anon 1966).

Hue is undoubtedly the dominant colour attribute and is often considered synonymous with 'colour'. However, hue does not define colour completely. Two colours of the same hue can appear quite different because of differing colour saturation (*purity*) or different *brightness*. The *saturation* is the perception of colour purity which is related to the distribution or spread of energy around the dominant wavelength. Saturation can vary from the maximum purity of a spectral line to the neutral grey of a completely polychromatic energy distribution (equal energy at all visible wavelengths). Saturation is specified as 'per cent monochromatic', on a scale from 0% for neutral greys, to 100% for a spectral line.

Brightness is the perception associated with the 'amount' of light reflected from a coloured object as weighted by the spectral sensitivity of the human eye (Fig. 2.19). This is the luminous flux (unit, lm; Appendix 1). When considering colour of an object rather than colour of a luminous source of light, brightness is associated with a quantity known as luminous reflectance (Anon 1966, p. 232). In water the appropriate reflectance to consider is the illuminance reflectance defined as

$$R_I = I_u/I_d \qquad (2.20)$$

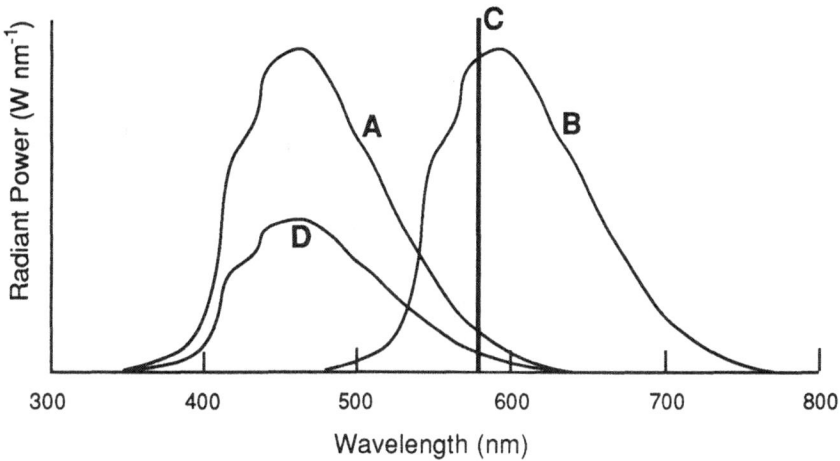

Fig. 2.20. Schematic of reflectance spectra illustrating the concepts of hue, saturation and brightness of colour. Hue is determined by the spectral distribution of the light. Thus spectral distributions A and B, while of identical spectral shape, have different hues because of the difference in central tendency, A being a blue and B an orange. The hue of B could be specified as the dominant wavelength, λ_d, the wavelength of a spectral line (C) which best matches the hue sensation. Because A and B have identical shape they would be expected to have similar colour purity or saturation. (The spectral line C has a colour purity of unity.) Spectra A and D, having identical shape, also have identical saturation but D is attenuated by the same factor at all wavelengths, resulting in a lower brightness. Note that, in general, the dominant wavelength is not the same as the mode of the radiant power distribution.

where I_u is the upwelling illuminance and I_d the downwelling illuminance (units: lm m^{-2} = lux). These three aspects of colour (hue, saturation and brightness) can be understood after study of Fig. 2.20.

2.6.4 Chromaticity

Colour of water, as of any object, is very conveniently specified with the standard colorimetric system of the Commission Internationale de l'Eclairage (CIE 1957). Specification of a colour in this system amounts to mathematically evaluating the proportions of standard, red, blue and green, primary colours which need to be mixed to match the colour under consideration. (Note that the physical primary colours are red, blue and green, in contrast to the subtractive 'artists' primaries: magenta, cyan and yellow.) The primary colours could, in principle, be any three different stimuli, but in the CIE system they are chosen so that the mixing curve or *chromaticity curve* for one of the primaries (green) is identical with the sensitivity curve for the light-adapted human eye (Fig. 2.19). The standard CIE chromaticity curves are the so-called '*tristimulus functions*', given in Fig. 2.21. For any given wavelength, these mixing curves (x, y, z) give the relative amounts of red, green and blue primary which need to be mixed to match the colour of a spectral line (monochromatic light) at that wavelength. The luminosities (brightnesses, measured in lumens) are additive in the mixture so, since the green primary has the same luminosity as the matched colour, the blue and red primaries have zero luminosity; that is, they are invisible! This underlines the abstract character of the standard stimuli in the CIE system.

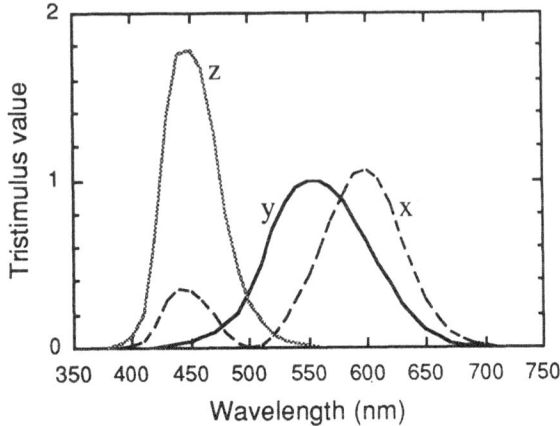

Fig. 2.21. Tristimulus functions (mixing curves) for the hypothetical standard red (x), green (y), and blue (z) primaries in the CIE colorimetric system.

These primary 'colours' (stimuli) are purely imaginary constructs, chosen for computational convenience (Anon 1966). In practice it is not necessary to understand these subtleties fully to use the CIE system for specifying colour. The interested reader should consult Anon (1966), 'The science of color' or other texts if more information is required.

The amount of red (X), green (Y) and blue (Z) primary required to be mixed to match a colour under consideration (spectral radiant flux, $\Phi(\lambda)$, in W nm^{-1}) are computed by the integrations

$$X = K_{max} \int_\Lambda \Phi(\lambda)x(\lambda)\,d\lambda \tag{2.21a}$$

$$Y = K_{max} \int_\Lambda \Phi(\lambda)y(\lambda)\,d\lambda = \text{luminous flux} \tag{2.21b}$$

$$Z = K_{max} \int_\Lambda \Phi(\lambda)z(\lambda)\,d\lambda \tag{2.21c}$$

where K_{max} is the maximum luminous efficiency of 680 lm W^{-1} (for 555 nm light, Appendix 1) and the integrations are over all wavelengths (in practice, from 380 to about 760 nm—Fig. 2.21). Equation (2.21b) gives, by definition, the luminous flux (in lumens, Appendix 1) and therefore defines the brightness of the colour.

In the case of an object which is not self-luminous, colour is associated with *reflected* spectral radiant flux. Chromaticity of water colours can be calculated using the spectral irradiance, $E(\lambda)$, in place of radiant flux, $\Phi(\lambda)$ (e.g. Smith *et al.* 1973). However, if the intrinsic colour is required, independent of the colour of the light source (i.e. independent of incoming sunlight—which can vary somewhat depending on solar altitude and atmospheric quality), chromaticity is calculated using the spectral reflectance $R(\lambda)$ (e.g. Morel & Prieur 1977).

Only brightness depends on the quantity of light; the other two attributes of colour, saturation and hue, depend only on the *distribution* of the light energy in the visible spectrum. Hue and saturation are specified by computing the proportions of red, green and blue primaries as:

$$x = \frac{X}{X+Y+Z} \qquad (2.22a)$$

$$y = \frac{Y}{X+Y+Z} \qquad (2.22b)$$

$$z = \frac{Z}{X+Y+Z} \qquad (2.22c)$$

Since $x + y + z = 1$ it is usual to report only (x, y) to specify fully the hue and saturation. In practice this is done by plotting (x, y) in the standard CIE chromaticity diagram (Fig. 2.22, Plate 6) on which spectral colours plot as a horseshoe-shaped curve. Dominant wavelength is obtained by projecting (x, y) from the achromatic point (C in Fig. 2.22) onto the spectrum locus. The saturation of the colour is expressed as the distance from C to (x, y)

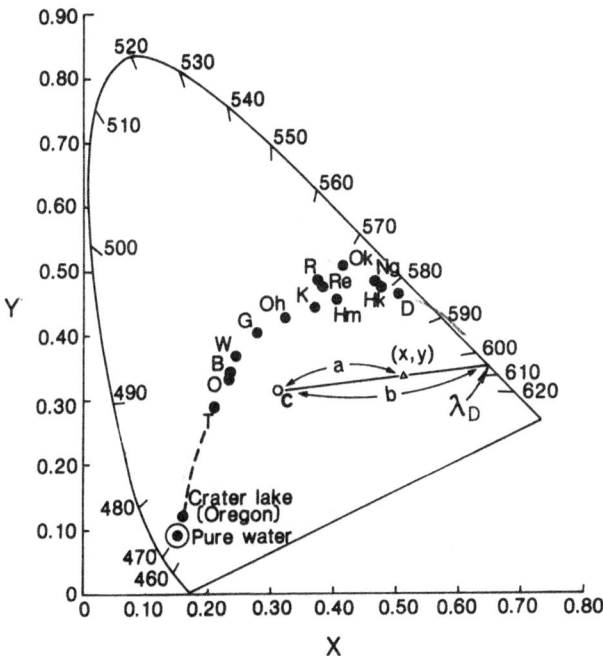

Fig. 2.22. The standard CIE chromaticity diagram (Anon 1966). Chromaticities of 14 New Zealand lake water colours are plotted together with the chromaticity of Crater Lake and that of pure water. (From Davies-Colley *et al.* (1988), with permission.) The achromatic point is 'standard illuminant C' (Newhall *et al.* 1943).

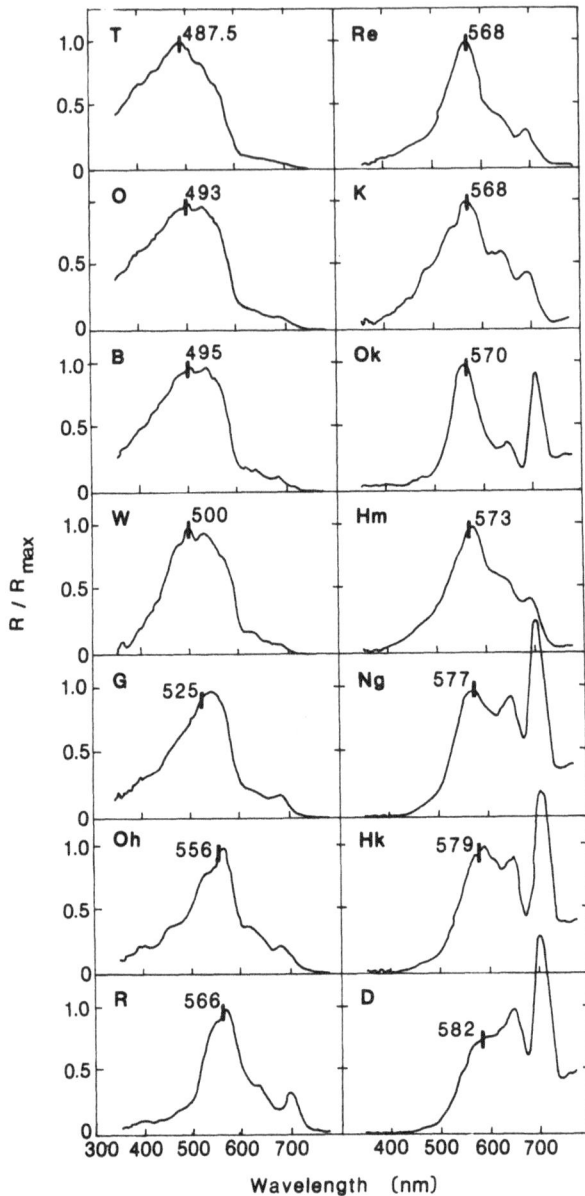

Fig. 2.23. Reflectance spectra of the 14 lakes for which chromaticity coordinates are plotted in Fig. 2.22. (From Davies-Colley *et al.* (1988), with permission.) The letter codes refer to the names of the lakes, the best known being Lakes Taupo (T), Waikaremoana (W) and Rotorua (R). To facilitate comparison of reflectance spectral shape among colours that vary appreciably in brightness the spectra are plotted as relative reflectance, R/R_{max}, where R_{max} refers to the maximum spectral reflectance in the 400–600 nm range of wavelength. The dominant wavelength (i.e. the wavelength of monochromatic light that speci-fies the hue) is indicated on each spectrum.

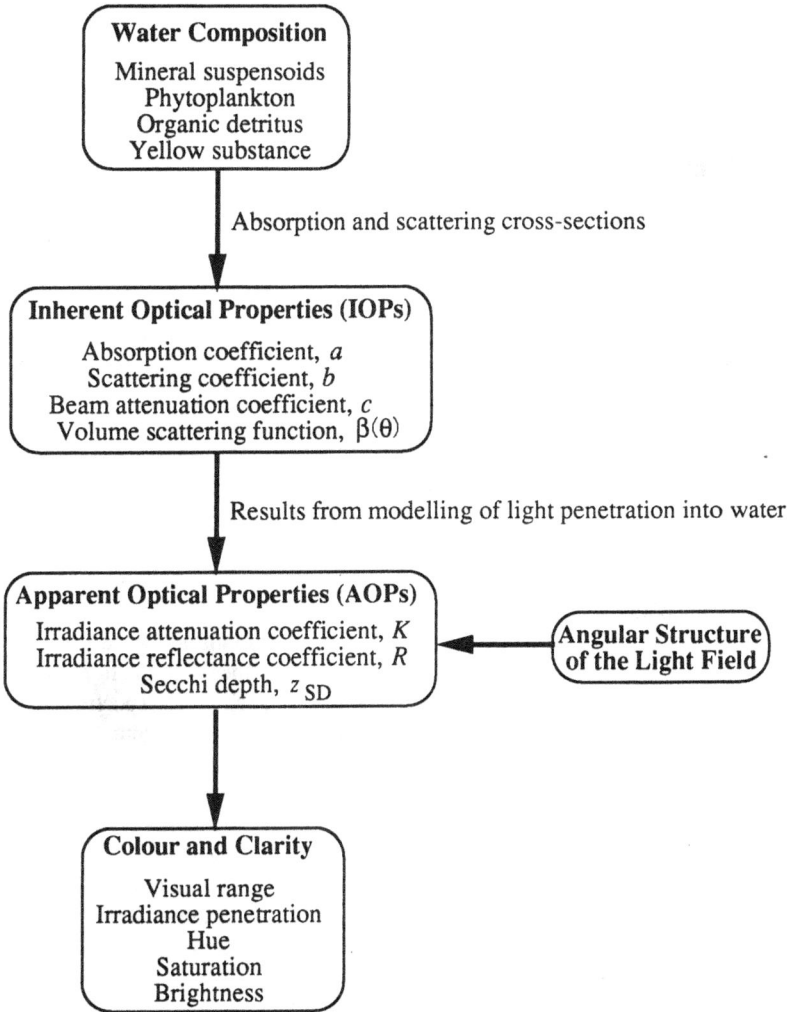

Fig. 2.24. Conceptual relationships between water colour and clarity, and water composition. (Modified after Vant & Davies-Colley (1984).)

as a proportion of total distance from C to the spectrum locus (ratio a/b in Fig. 2.22).

Fig. 2.23 shows the spectral reflectances of 14 New Zealand lakes as measured by underwater spectroradiometer. The lakes ranged from very clear and blue–green coloured Lake Taupo to yellow, humic-stained Lake D. Chromaticity analysis applied to these reflectance spectra yielded the chromaticity coordinates plotted in Fig. 2.22 (Davies-Colley *et al.* 1988). The chromaticity coordinates of the different waters appear to fall in a fairly narrow band arcing from the high purity blues of the purest natural waters (e.g. Crater Lake) through the low saturation greens of waters containing moderate concentrations of organic matter, to the high saturation yellows and even oranges of eutrophic or humic-stained waters. Similar trends have been observed in chromaticity analysis of seawaters (Jerlov 1976, p. 170). However, orange colours, such as those commonly encountered in humic-stained freshwaters, are probably rare in marine environments.

2.7 SYNTHESIS

In this chapter the nature of water as an optical medium has been surveyed, the optical properties that quantify aspects of light transmission have been introduced, the constituents of natural waters which modify its optical character have been discussed, and the concepts of colour and clarity have been introduced. It is appropriate at this point to consider the linkages, direct and indirect, between the optical water quality (i.e. the colour and clarity) and the composition of the water, as shown in Fig. 2.24. Morel & Bricaud (1988) have referred to this system of relationships as a 'cause–effect chain' since the light field in the water and therefore the colour and clarity are ultimately controlled by the light-attenuating constituents. They point out that this 'chain' of relationships between quantities can be 'explored in both directions'.

The light-attenuating constituents, particularly the suspensoids but also the yellow substance and water itself, determine the inherent optical properties of waters. These inherent properties (IOPs) largely determine the apparent optical properties (AOPs) such as the irradiance attenuation coefficient, and the reflectance coefficient. However, the angular and spectral properties of the light from the sun also influence the apparent properties for given values of the IOPs. Finally the AOPs, together with the light field, determine the optical quality of the water—its colour and clarity. The spectral reflectance coefficient, together with the spectral irradiance, determines the colour, and the irradiance attenuation coefficient quantifies the light penetration. Visual clarity relates very closely to the beam attenuation coefficient (an IOP), but, with the exception of the special case of a black body viewed horizontally, visual clarity also depends on the illuminance attenuation in the water and the angular structure of the light field.

In general, the system of relationships implied in Fig. 2.24 cannot be 'short circuited' (i.e. no more than one step should be moved at a time). For example, it is not appropriate to attempt to establish direct relationships between the apparent optical properties, such as the irradiance attenuation coefficient, and water constituents, such as the phytoplankton (i.e. to bypass the inherent optical properties), although this is often attempted. Such relationships will be flawed, if only because of the influence of the ambient light field, but also, usually, because the underlying relationships with the IOPs that have been bypassed are non-linear.

Although short-circuiting in the chain of relationships shown in Fig. 2.24 is to be discouraged, fairly robust relationships are available between the various components at each step as a result of progress in aquatic optics over the past several decades. For example, the Secchi depth can be expressed as a simple function of the attenuation coefficients K and c as discussed above (Section 2.5.1). The optical properties of the main natural constituents of water which are important influences on colour and clarity are fairly well known. Thus it is possible to proceed in two steps from the water composition to predict the resulting colour and clarity. As we shall see in Chapter 5, this system of relationships provides a powerful tool for management of the colour and clarity of natural waters.

3

Measurement

3.1 INTRODUCTION

In this chapter we discuss the measurement of the quantities introduced and defined in Chapter 2 and Appendix 1. Appendix 2 gives 'recipe-style' instructions for the use of particular instruments and procedures for the measurement of those quantities most likely to be of concern to aquatic scientists and water managers.

Firstly we consider characterization of the light field in water and the calculation of various apparent optical properties (AOPs). The AOPs are generally more easily measured than the intrinsic optical properties (IOPs), but are less valuable for analysing water colour and clarity and solving problems.

The IOPs, by contrast, are comparatively difficult to measure, with the exception of the beam attenuation coefficient, c. Special optical instrumentation with its own light source is usually required. However, the IOPs are more desirable quantities for characterization of water than the AOPs because they are rigorously additive and depend only on the composition of the water (Section 2.2). Fortunately it is possible to estimate the IOPs with fair precision from the AOPs using the results of modelling of light penetration into water (Fig. 2.24).

The chapter goes on to discuss measurement of water clarity—both visual clarity as it relates to the vision of sighted aquatic organisms and man, and light penetration into waters for aquatic plant photosynthesis and the illumination of submerged features. Some of the measurements, while simple field or laboratory procedures, are difficult to relate to actual field appearance of waters. Others provide a fairly complete picture of water appearance in the field but require special instrumentation. Measurement of water colour is difficult and requires special instrumentation (and considerable computation) but valuable observations (as opposed to measurements) of colour can be made in the field.

We then consider measurement of those water constituents which are primary determinants of water colour and clarity: the dissolved yellow substance and the various organic and inorganic particulates. We also consider the optical cross-sections of these constituents, which can be used to predict optical character of waters from their composition.

Finally, we briefly consider how optical sensors can be used to monitor water quality—
continuously or remotely.

3.2 APPARENT OPTICAL PROPERTIES

The apparent optical properties that we are concerned with in practical problems of water
management are those properties pertaining to ambient daylight in water. These AOPs
are fairly readily measured using various light (radiance, irradiance, spectral irradiance,
quantum irradiance, illuminance) sensors deployed in water during daylight hours.

Ideally, the spectral response of a sensor should be appropriate to the objective of the
measurement (Tyler 1973). Thus in the context of lighting for human vision, light flux is
measured (in lumens) by a sensor with a spectral response approximating that of the
human eye (Fig. 3.1). For some purposes a sensor with a uniform response to all wavelengths may be appropriate so as to measure total irradiance in W m^{-2}.

In the context of plant photosynthesis a sensor that is equally sensitive to all photons
in the PAR band (400–700 nm) should be used since plants respond equally to light
quanta in the visible range once absorbed, irrespective of wavelength (note though that
light absorption by plants is wavelength dependent). The energy content of photons is
inversely proportional to wavelength (equation (A1.1), Appendix 1), such that photons at
the red end of the visible spectrum have less energy than blue photons. To achieve equal
sensitivity to photons of all wavelengths, the spectral sensitivity of PAR quantum sensors
is made proportional to wavelength—as shown in Fig. 3.1.

Where the spectral response of concern is unknown or is very different from that of
available sensors with a broad-band spectral response, the problem of light measurement

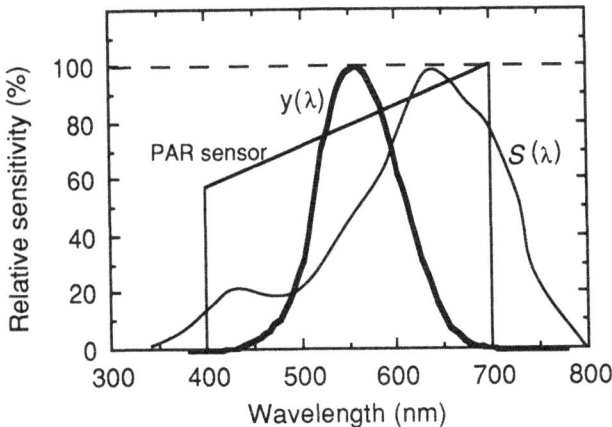

Fig. 3.1. Ideal spectral response curves for various sensors of light. The heavy curve is the spectral
response of an ideal luminous flux sensor (lux sensor) which mimics the spectral sensitivity of the day-
light-adapted human eye (photopic curve, y(λ)). The thin curve, $S(\lambda)$, is a hypothetical spectral sensitivity curve for a biological organism. Also shown is the ideal response curve of a PAR-band quantum
flux sensor, the sensitivity of which is made proportional to wavelength because the energy content of
photons is inversely proportional to wavelength (Appendix 1). An ideal energy sensor would have a
uniform response spectrum, i.e. the sensor would be equally sensitive to all wavelengths.

is more difficult. This might be the case, for example, if the concern was with the vision of a particular fish with a visual sensitivity function very different from the photopic function of the light-adapted human eye (Lythgoe 1979). In this situation only spectral irradiance, $E(\lambda)$, measurements are meaningful. The response of the organism can then be calculated by integrating over the spectrum (denoted by Λ):

$$\text{response} = \int_{\Lambda} E(\lambda)S(\lambda)\,d\lambda \qquad (3.1)$$

where $E(\lambda)$ is the measured spectral irradiance (radiance would ideally be measured instead if a particular narrow direction was of concern), and $S(\lambda)$ is the spectral response function, that is, the function quantifying the relative sensitivity to light of different wavelengths (Fig. 3.1).

Fig. 3.2 shows examples of irradiance sensors (in this case, sensors of quantum irradiance in the photosynthetic band, i.e. PAR sensors). For most purposes flat plate type, vector irradiance sensors are more useful (as well as being typically cheaper and more robust) than spherical sensors of scalar irradiance. The latter can be used, in conjunction with vector irradiance sensors, to provide an index of the degree of diffusion of the light field in waters (average cosine; refer to Appendix 1). Scalar irradiance sensors are particularly useful when the total irradiance must be measured, particularly the PAR available to aquatic plants during measurements of rates of photosynthesis in natural waters or the laboratory.

Fig. 3.2. Quantum irradiance (PAR) measuring equipment. The spherically shaped sensor on the left is a submersible scalar irradiance sensor. Vector irradiance sensors are shown mounted on a support structure so as to record both up- and downwelling irradiance. These sensors are connected via underwater cables to the readout instrument (in this case a datalogger with digital display). Also shown is a non-submersible sensor ('terrestrial sensor', for use in air only), mounted on a levelling base. Photograph courtesy of Li-Cor Inc., P.O. Box 4425, Lincoln, Nebraska 68504, USA.

3.2.1 Vector irradiance

Under clear skies or uniformly overcast skies, ambient irradiance is constant, and measurement of underwater irradiance is comparatively straightforward, the only variation in irradiance being with depth into the water body. The sensor support with mounted sensors is suspended in the water on the sunny side of the boat, preferably from a derrick or crane, and a number of readings, perhaps ten, are taken at different depths down to about the euphotic depth (1% light level). Such a procedure, under clear sky conditions, was used to obtain the irradiance data plotted in Fig. 2.4. Care needs to be taken with depth measurements, particularly in very light-attenuating waters (where irradiance changes rapidly with depth) and in choppy wave conditions (when the average depth is difficult to estimate). In highly light-attenuating waters, and in rivers, mounting the sensor on a rigid frame held by the field worker is preferable to use of a suspended support structure such as that shown in Fig. 3.2. A design for a light sensor frame is given by Westlake (1986).

Near-surface downwelling irradiance measurements are typically 'noisy' owing to wave action. Wavelets act as lenses and their passage induces 'flicker' in the irradiance readout. Switchable damping or switchable integration times for averaging the flicker are desirable features of the readout instrument.

Where water currents are running, the support structure must be weighted to ensure (a) that the sensors are held in the horizontal plane and (b) that the depth measurement is not in error owing to lateral displacement from the vertical of the cable and suspending line. In rivers, measurement of irradiance profiles is often difficult because of limited depths as well as water currents. In larger rivers which are optically deep, measurements from a boat drifting with the current are feasible, although hazardous if there is a possibility of fouling a snag.

Upwelling irradiance (E_u) measurements should be made as well as downwelling (E_d) after simply reversing the sensor orientation. (In very light-attenuating water a correction will need to be made for the small displacement of the sensor element in relation to the upward-looking level.) Unfortunately many users of irradiance sensors, particularly those whose main concern is light climate for aquatic plants, fail to measure upwelling as well as downwelling profiles. From the upwelling and downwelling measurements taken together, valuable extra information on the optical nature of the water can be obtained (Section 3.3.4).

The reflectance coefficient quantifying the 'brightness' of the water is obtained immediately:

$$R = E_u / E_d \tag{3.2}$$

The irradiance attenuation coefficients (at average depth $z_{av} = (z_2 + z_1)/2$) are calculated by the following discrete approximation to the defining relationships given in Chapter 2:

$$K(z_{av}) = \frac{\ln E(z_1) - \ln E(z_2)}{z_2 - z_1} = \frac{\ln[E(z_1)/E(z_1)]}{z_2 - z_1} \tag{3.3}$$

where z_1 and z_2 are two successive depths at which irradiance is measured, and E may be upwelling irradiance, E_u, downwelling irradiance, E_d, net vector irradiance, $E_d - E_u$, or scalar irradiance, E_0. If, as is typical, the profile of irradiance appears linear with the exception of some surface points, linear regression (of ln $E(z)$ on z) can be used to estimate an average value of the attenuation coefficient over some meaningful depth interval, such as the euphotic depth.

In practice we have found it very convenient to plot observations of irradiance in the field, directly they are made, onto four-cycle semilog paper. (The semilog grid is printed onto plastic sheet 'wet note' material so that wetting with rainwater or spray is not a concern.) Field plotting permits the data quality to be checked immediately, before moving off site. Fig. 3.3 shows a field plot of irradiance data for the Waikato River, North Island, New Zealand, when patches of high cirrus cloud partially occluded the sun which until then had been unobstructed. The profile of downwelling irradiance connected by broken lines is immediately recognized as flawed because of its non-linearity. When the profile was repeated sometime later, again under clear sun, a much more stable (linear) profile was obtained (Fig. 3.3). The direct field plotting of irradiance data also permits checking that the profile of E_u is parallel to that of E_d at appreciable depth (as is theoretically expected (Preisendorfer 1958)), and this increases confidence that calculated values of the reflectance coefficient ($R = E_u/E_d$) are accurate. Furthermore, the direct plotting of irradiance data in the field permits the recognition of change in slope of the profile due to real variation in the value of the irradiance attenuation coefficient with depth that is related to stratification of the water column. If data are plotted on site, there is an opportunity to confirm such changes by further measurements.

Under changeable ambient lighting conditions, particularly with patchy cloudy skies, measurement of irradiance profiles suitable for estimation of K_d, K_u, and R can be frustratingly difficult. Fig. 3.3 illustrates this problem. Ambient irradiance values can change in a matter of seconds by a factor of 5 between dense cloud and clear sun—the worst (most variable) conditions being under patchy cumulus cloud (e.g. Davies-Colley et al. 1984).

Solution to these difficulties lies in the fact that the AOPs of interest (K_d, K_u, and R) are all independent of irradiance. That is, they do not include the units of irradiance (W m^{-2} or μmol m^{-2} s^{-1}) because the irradiances used in their calculation (equations (3.2) and (3.3)) appear as *ratios*. This suggests that relative irradiance measurements can be used just as well as absolute irradiance measurements for the calculation of K_d, K_u, and R.

It has long been recognized that correction for the change in ambient lighting can be achieved by measuring in-water irradiance relative to that sensed by a reference sensor on the boat deck (e.g. Westlake 1986). Ideally the two sensors are compared immediately (i.e. electronically) rather than by interpolating in a logged record of the deck cell output for example. Unfortunately, two-channel, direct ratio readout instruments are not commonly available. However, multiple-channel dataloggers, such as that shown in Fig. 3.2, can be used fairly conveniently to measure relative irradiance calculated as $E(z)/E_r$, where $E(z)$ is the irradiance (E_u, E_d or E_0) at depth z, and E_r is the downwelling irradiance recorded by the reference (deck) sensor.

Davies-Colley et al. (1984) demonstrated that a further improvement in stability of relative irradiance measurements is achieved if the reference sensor is placed in water rather than in air. This is because refraction across the air–water interface changes the angular structure of light and thus the vector irradiance. Thus the ratio of two in-water

Date _10/3/83_ Time(NZST)_1500_ Site _Waikato R. at Tuakau_

Cloud _2/8_ Sun angle ___—___ Sea ___—___
Colour ___—___ Clarity _Secchi = 1.0 m_

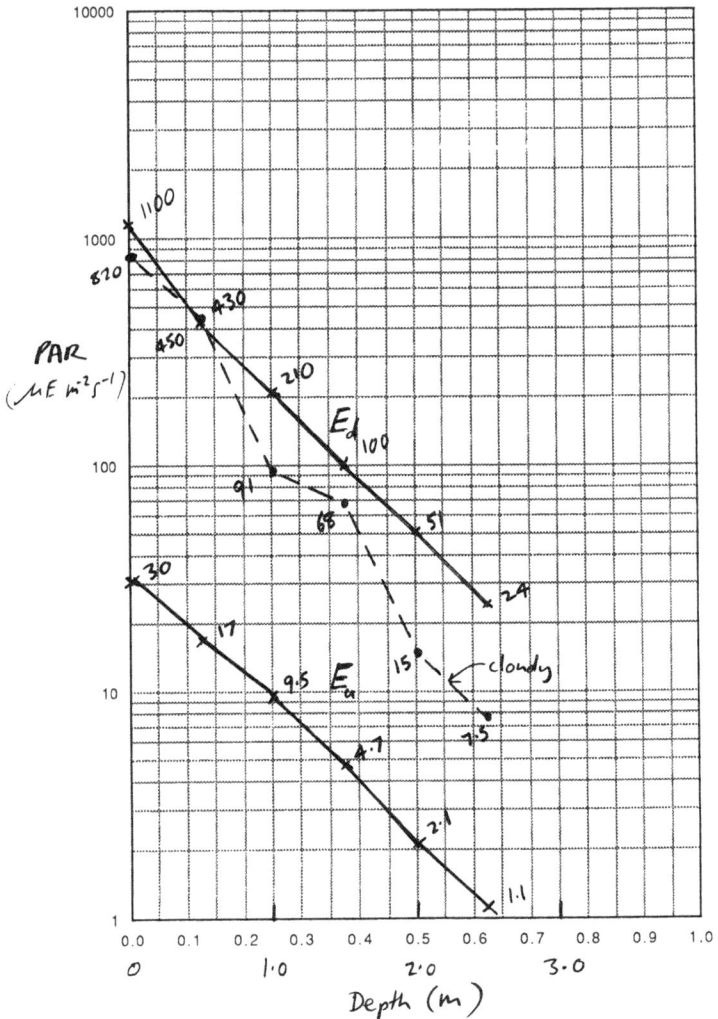

Fig. 3.3. Direct plotting of irradiance data in the field on semilogarithmic paper. The effect of changing ambient light conditions was immediately recognized in the field. The initial downwelling irradiance profile (– – –) is non-linear and has a spurious slope owing to shadowing by advancing clouds. Repetition of the profile later, under clear sun, gave a much more satisfactory profile, parallelling the upwelling irradiance profile. (Authors' unpublished data for the Waikato River near Tuakau, New Zealand.)

signals is more stable in the presence of changeable ambient light than the ratio of an in-water signal to an in-air signal.

A two-channel instrument (e.g. Davies-Colley *et al.* 1984), or a multichannel logger, can be used for direct measurement of K using a support structure designed to hold two identical sensors, positioned with identical orientation, but separated by a distance Δz (Fig. 3.4). Ideally the displacement of the sensors should be variable so as to optimize measurement over a range of water clarities. The ratio of the irradiances detected by the two sensors is the irradiance transmittance,

$$T_{\mathrm{E}} = \frac{E(z + \Delta z)}{E(z)} \tag{3.4}$$

from which the attenuation coefficient (a weak function of depth) is calculated:

$$K_{\mathrm{d}} = \frac{\ln(1/T_{\mathrm{E}})}{\Delta z} \tag{3.5}$$

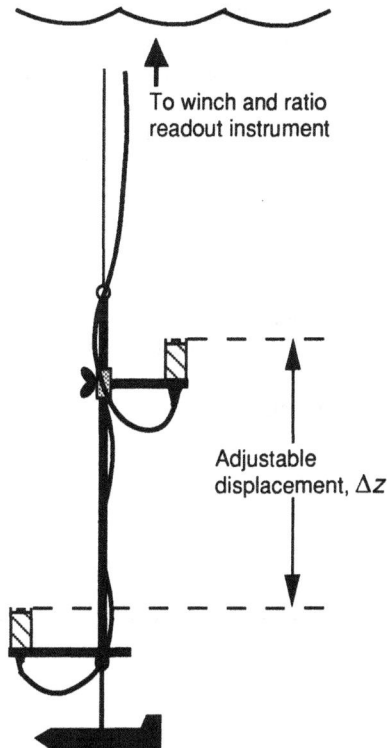

Fig. 3.4. Measurement of irradiance transmittance with two identical (and identically orient-ed) irradiance sensors mounted, displaced by depth difference, Δz, on a rigid frame.

Used in this mode the two-channel 'ratio meter' is essentially an *irradiance transmis-someter*, analogous to the *radiance transmissometer* (beam transmissometer) discussed in Section 3.3.1. The mathematical operation given in equation (3.5) can, of course, be performed electronically, whereupon the instrument becomes a direct reading *irradiance attenuance* meter (e.g. Westlake 1986).

A two-channel system can be used with vector irradiance sensors oriented oppositely so as to read out directly the irradiance reflectance according to equation (3.1). For example, the datalogger in Fig. 3.2 can be configured to read out reflectance directly as calculated from signals received from the vector irradiance sensors (also shown) measuring up- and downwelling irradiance at the same depth.

Jordon (1988) has reported the design of a string of 13 custom-made scalar irradiance sensors for continuous measurement of the irradiance attenuation coefficient, irrespective of ambient light conditions, in oceanographic productivity work.

Appendix 2.1 gives a checklist and brief 'recipe' for the measurement of underwater irradiance suitable for calculation of the AOPs.

3.2.2 Scalar irradiance and spectral irradiance

Less often measured than vector irradiance is the scalar irradiance (E_o). Vector irradiance, as we have seen, is useful for optically characterizing a water body. Scalar irradiance is more appropriate where absolute values of irradiance are required. It is of no consequence to a plant cell whether the radiant flux reaches it from above, below or sideways, and therefore a spherical sensor of scalar irradiance 'sees' the light field more in the manner of a plant cell than does a vector irradiance sensor. A scalar irradiance sensor is, therefore, the sensor of choice for productivity studies, particularly those in shallow rivers or in weed beds (or in laboratory incubators) where the spatial structure of the light field is not one dimensional (i.e. light varying only with depth).

More information regarding the light field in a water can be obtained with a scalar irradiance sensor together with a vector irradiance sensor than with the vector sensor alone. The ratio of net vector irradiance, $\vec{E}\,(= E_d - E_u)$, to scalar irradiance, E_o, is the average cosine, μ (Appendix 1), which is a useful measure of the degree of diffusion of the light field in the water. From conservation of energy considerations the following expression for the average cosine can be derived:

$$\mu = \frac{E_d - E_u}{E_o} = \frac{a}{K_E} \qquad (3.6)$$

where a is the absorption coefficient (an IOP) and K_E is the attenuation coefficient for net vector irradiance.

So far we have assumed that PAR sensors will be used for measurement of AOPs, but, for some purposes, sensors of different spectral sensitivity are required. The PAR sensor is the sensor of choice for measurements in the context of plant production in water. It is not appropriate to use a lux sensor as is sometimes still done to measure photosynthetic light. Lux sensors are appropriate for measurements connected with human vision (Tyler 1973). For other (special) purposes still other sensors may be required. Ultraviolet sensors may be required in studies of the biocidal action of natural UV radiation.

Where there is special interest in a particular region of the visible or near-visible spectrum of light in water, it will usually be necessary to take *spectral* irradiance measurements. Most suitable for this purpose is a *spectroradiometer*, a device which scans the visible spectrum (and perhaps a portion of the adjacent UV and IR) as required using a motor-driven monochromator (e.g. LICOR 1800UW, Fig. 3.5) or an array of sensors of different peak spectral sensitivity (e.g. Biospherical Instruments MER 1000). These are expensive instruments, generally only justified for research purposes, although of great value for investigations of particular problems.

Spectral irradiance measurements at a few selected wavelengths can be made fairly readily with ordinary flat-plate (vector) irradiance sensors fitted with narrow bandpass filters. Fig. 3.6 shows spectral irradiance measurements made in Lake Waikaremoana using a PAR sensor equipped with filters of this kind compared with spectral irradiance measurements by spectroradiometer at the peak transmission wavelengths of the filters. The slight mismatch of the blue filter profile can probably be attributed to the appreciably wide bandwidth of the blue filter (compared with the ~4 nm band width of the LI-1800 UW spectroradiometer) in a spectral region where the attenuation changes fairly rapidly with wavelength. Such irradiance profiles at a few selected wavebands can give a rough 'feel' for spectral variation of water penetration by sunlight, but are otherwise of dubious value unless a particular narrow spectral region is physiologically or otherwise dominant.

Fig. 3.5. Submersible spectroradiometer (LICOR L1-1800 U/W). The white diffuser on top of the black cylinder provides radiation to the monochromator and detector. The cylinder also contains a computer for controlling the optics and for data collection. Communication to the instrument is by cable via the portable terminal shown or, alternatively, a personal computer. Photograph courtesy of Li-Cor Inc., P.O. Box 4425, Lincoln, Nebraska 68504, USA.

3.3 INHERENT OPTICAL PROPERTIES

3.3.1 Beam attenuation

Most easily measured of the IOPs, and perhaps that IOP likely to feature most in practical management of optical water quality, is the beam attenuation coefficient, c. Duntley (1963) discusses several different ways of measuring or estimating c but the most practical of these is by use of beam transmissometers. These devices project a narrow collimated beam of light across a known path length, r, in water and measure the transmitted flux, Φ_r. The beam transmittance (symbol T_c, where the subscript indicates collimated light), is, by definition

$$T_c = \Phi_r / \Phi_0 \qquad (3.7)$$

in which Φ_0 is the flux of the incident beam. The beam attenuation coefficient is

$$c = \frac{\ln(1/T_c)}{r} \qquad (3.8)$$

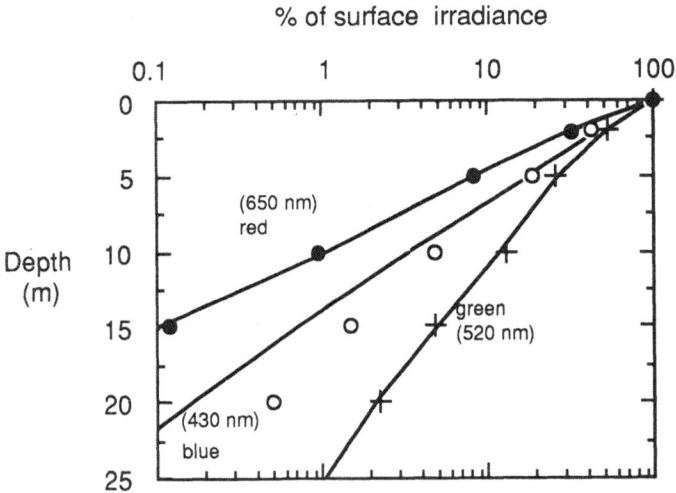

Fig. 3.6. Comparison of irradiance profiles made using PAR sensors equipped with optical filters ($\bullet$, $\circ$, +) with spectral irradiance profiles calculated from spectroradiometric scans (——). Measurements were made in Wairau Basin, Lake Waikaremoana, at about 1000 hr on 7 August 1984. The optical filters had the following peak transmission wavelengths (and half-power bandwidths): blue filter (Schott BG-12), 430 nm (400–468 nm); green filter (VE-9), 520 nm (482–564 nm); red filter (RG-10), 650 nm (605–700 nm). (Authors' unpublished data.)

A beam transmissometer of suitable optical design for the measurement of c (Petzold & Austin 1968) is illustrated in Fig. 3.7. Some commercial transmissometers suffer to a significant extent from detection of light scattered at low angles to the incident beam, and thus overestimate T_c and underestimate c—to an extent dependent on the half-angle of acceptance of scattered light and on the volume scattering function (e.g. Jerlov, 1976, p. 48). Practical transmissometers should detect light scattered no more than about 1° from the incident beam.

Note that a laboratory spectrophotometer does not yield an estimate of c since the transmittance includes light scattered up to about 5° from the incident beam. Scattering between 0° and 5° includes typically about 50% of total scattering at all angles (Petzold 1972). Spectrophotometric methods for measurement of c exist, but they either are indirect (as discussed below) or involve special modification of standard laboratory spectrophotometers in order to reduce greatly the angle of acceptance at the detector (e.g. Bricaud *et al.* 1983).

There is a need for an *in situ* recording transmissometer suitable for routine measurements in water resources work (e.g. McCluney 1975).

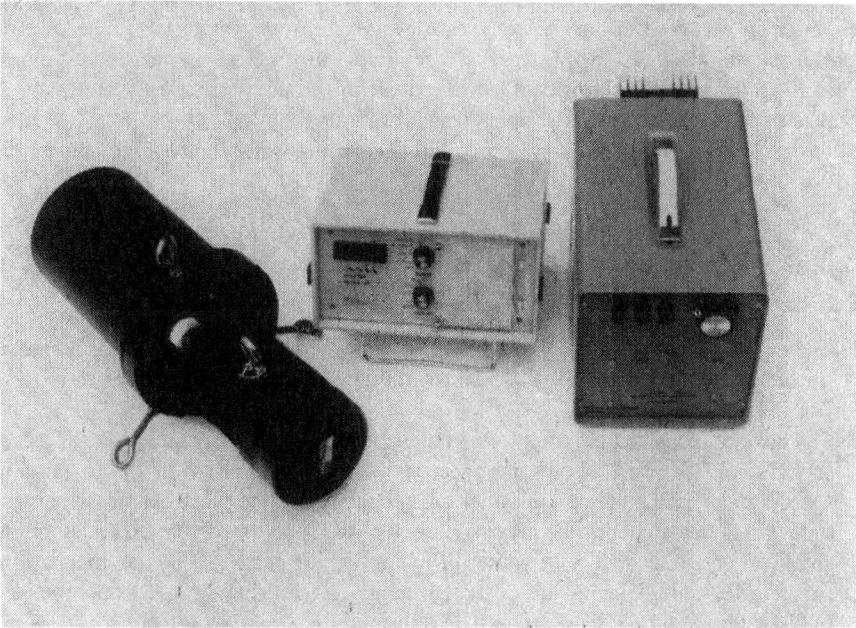

Fig. 3.7. Transmissometer (Martek XMS). An *in situ* sensor with a 250 mm light path is shown on the left. This sensor is suitable for relatively turbid waters (c: 0.4–18 m^{-1}); a 1 m light path sensor suitable for clearer waters (c: 0.1–4.6 m^{-1}) is also available. The cylindrical pressure housing on the sensor unit holds the optical elements and light source and detector. The light path through the water is folded (2×125 mm) using a Porro prism located in the 'nose'. The digital display unit (centre) is connected to the sensor via hydrographic cable (not shown) and reads out directly in beam transmittance units. The unit on the right powers the light source and electronics at 12 V.

Appendix 2 (Section A2.2) gives a method for the measurement of c by beam transmissometer. Errors arising from detection of light scattered at low angles (leading to underestimation of c) have been discussed by Petzold & Austin (1968) and Bartz *et al.* (1978).

3.3.2 Absorption measurements by spectrophotometer

Less easily measured than the beam attenuation coefficient is the absorption coefficient, *a*. *In situ* absorption meters have been developed (e.g. see the review in Kirk (1983, pp. 45–49)), but these are specialized instruments more suitable for oceanographic work in connection with remote sensing than for water management purposes. It is commonly thought that an irradiance sensor measures absorption with depth into a water body, but the quantity actually measured, as we have seen in Section 3.2, is an irradiance attenuation coefficient which is weakly dependent on scattering as well as strongly dependent on absorption. Nevertheless, estimation of the absorption coefficient from measurements of K and some other AOPs is feasible as will be described in Section 3.3.4.

Laboratory spectrophotometers are designed for measuring absorption of light, and are common instruments, available in most water analytical laboratories. However, laboratory spectrophotometers are designed for non-scattering liquid samples such as are generated in many colorimetric analyses of water and other materials. Filtration of water using glass fibre or membrane filters removes most scattering, but the absorption of the filtrates is often significantly lower than in the unfractionated water sample because of removal of the particulates which are generally light absorbing as well as scattering. Measurement of absorption by light-scattering water samples requires modification of normal spectrophotometric procedures (Fig. 3.8).

Since most scattering in natural waters is at small angles, it is possible to collect most scattered light simply by increasing the angular 'field of view' of the detector (photomultiplier) in the spectrophotometer with respect to the water sample contained in the cuvette. There are three main methods (Fig. 3.8).

(1) The cuvette is simply shifted much closer to the photomultiplier tube using a special 'turbid sample position' (Fig. 3.8B) available on some spectrophotometers (e.g. Bricaud *et al.* 1983).
(2) The cuvette is located adjacent to the entrance slot of an 'integrating sphere' accessory (Fig. 3.8C) available on some spectrophotometers (e.g. Kirk 1980). Integrating spheres incorporate a photomultiplier tube connected to the spectrophotometer electronics. The remainder of the interior surface of the sphere (with the exception of the entrance slot) is coated with white diffuse reflecting material such as magnesium oxide. Integrating spheres are designed mainly for reflectance spectrophotometry (with opaque solid samples) rather than transmission spectrophotometry.
(3) A diffuser plate ('opal glass') is placed between the cuvette and the detector close to the face of the cuvette (Fig. 3.8D). The intense diffusion of light by the opal glass in nearly all directions is independent of the angular structure of the incident light (and so is the same whether it has come through the non-scattering distilled water blank or through the highly scattering natural water sample) (Shibata 1959).

All three procedures markedly increase the efficiency of collection of scattered light (typically detecting better than 90% of total scattered light). Procedures (1) and (2) above

A. Normal sample position

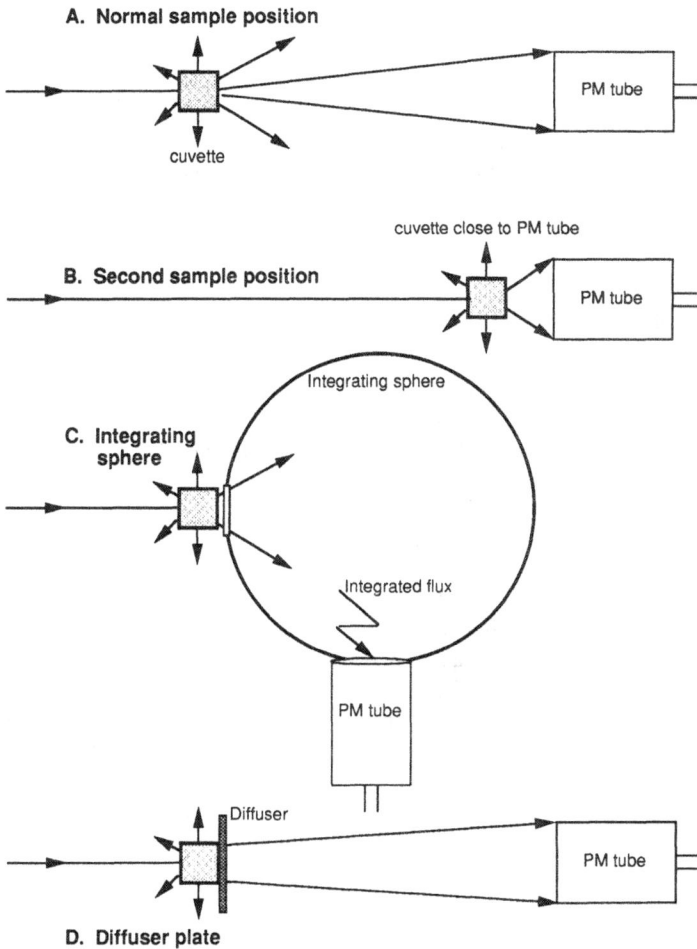

cuvette

PM tube

cuvette close to PM tube

B. Second sample position

PM tube

Integrating sphere

**C. Integrating
   sphere**

Integrated flux

PM tube

Diffuser

PM tube

D. Diffuser plate

Fig. 3.8. Methods for increasing the scattered light collection efficiency in a laboratory spectrophotometer for measurement of absorption in light-scattering natural water samples. (A) Cuvette in the normal position in the spectrophotometer sample compartment. (B) The cuvette is moved much closer to the photomultiplier (preferably an end window type) and positioned in the so-called 'second sample position' or 'turbid sample holder'. (C) An integrating sphere is used to collect diffuse transmitted light. (D) A light diffuser ('opal glass') is placed on the detector side of the cuvette to produce intense diffusion of light, irrespective of the light field in the cuvette.

are preferable to (3) which involves loss of most light to the detector and therefore produces a 'noisy' readout requiring long integration times or slow scanning speeds.

The apparent absorption coefficient measured by one of the above procedures can be calculated:

$$\chi = \frac{\ln(1/T)}{r} \qquad (3.9)$$

where T is the transmittance and r is cuvette path length. Equation (3.9) has identical form to (3.8), but the transmittance appearing in the two equations (T versus T_c) refers to different optical geometries (partly diffuse light in equation (3.9) as opposed to collimated light in equation (3.8)). If the spectrophotometer reads out in absorbance units (as is typical) χ is calculated:

$$\chi = \frac{2.303D}{r} \qquad (3.10)$$

where D is absorbance ($= -\log_{10} T$) and the factor 2.303 ($\log_e 10$) converts from $\log_{10}$ units to $\log_e$ units.

Some light is lost to the detector by scattering at large forward angles or backscattering, and a correction must be made to obtain the true absorption coefficient. When the scattering length, $1/b$, is large in relation to the cuvette path length, r (i.e. $rb \ll 1$) we can write (Bricaud *et al.* 1981)

$$\chi = a + (1 - \varepsilon)b \qquad (3.11)$$

where b is the scattering coefficient and ε is the detected fraction of scattered light. The term $(1 - \varepsilon)b$ can be measured in a region of the spectrum where true absorption by water constituents (other than water itself) is known to be practically negligible, for example in the near IR region around 750 nm. By making reasonable assumptions about the spectral dependence of scattering, the whole visible spectrum can be corrected to give an estimate of the true spectral absorption coefficient from equation (3.11). In practice, equation (3.11) is a reasonable approximation if $rb < 0.3$ (e.g. van de Hulst 1957), so that if a 10 mm cuvette is used ($r = 0.01$ m) b should be < 30 m^{-1}, a high scattering coefficient corresponding to fairly 'turbid' or 'dirty' water.

Filtration with glass fibre or membrane filters removes most scattering from natural water samples, since almost all the material passing such filters has a particle size smaller than the wavelength of light (400–700 nm, i.e. 0.4–0.7 µm). Thus measurement of the absorption coefficient in filtered water samples can be performed using almost any spectrophotometer by neglecting scattering as a first approximation. Residual scattering due to filter-passing (colloidal) particles can be corrected for, using measurements of absorbance at near IR wavelengths (e.g. 740 nm). The filtrate absorption coefficient (denoted by subscript f) is then

$$a_f(\lambda) = g_\lambda = \chi_f(\lambda) - \chi_f(740)\frac{740}{\lambda} \qquad (3.12)$$

where the second term on the right-hand side corrects for the scattering by any filter-passing colloids, assuming (1) that all absorbance at 740 nm is due to scattering, and (2) that this scattering is inversely proportional to wavelength (Bricaud *et al.* 1981). The symbol g, denoting 'gilvin' or '*Gelbstoff*' following Kirk (1976), will be used henceforth for the absorption coefficient of yellow substance in water filtrates.

If the absorption by yellow substance is low (say $g_{440} < 1 \text{ m}^{-1}$), the measurement may need to be made in long path length cuvettes, or extrapolated to the visible from ultraviolet measurements (e.g. at 340 nm (Davies-Colley & Vant 1987)) assuming an exponential slope factor in the equation (Bricaud *et al.* 1981):

$$g_\lambda = g_\Lambda \exp\left[S(\Lambda - \lambda)\right] \tag{3.13}$$

where λ and Λ are two different wavelengths. The exponential slope, S, averaged 0.0187 nm^{-1} in 11 diverse New Zealand lakes (Davies-Colley & Vant 1987), but other workers have reported somewhat lower values. For example, Bricaud *et al.* (1981) found $S = 0.014 \text{ nm}^{-1}$ in a wide range of seawaters, and the average value in 77 river waters sampled in the New Zealand National Water Quality Network (Smith & McBride 1990) was 0.0158 nm^{-1} (authors' unpublished data).

When the water is relatively highly light attenuating, absorption coefficients of the total constituents (dissolved and particulate) can be measured directly on unfractionated water samples using one of the three systems described above which improve scattered light collection efficiency to 90% or better. A correction of such measurements for scattering is essential. The absorption coefficient of all water constituents (denoted by subscript c) is calculated:

$$a_c(\lambda) = \chi_c(\lambda) - \chi_c(740)\frac{740}{\lambda} \tag{3.14}$$

where, again, the second term corrects for scattering, assuming that all absorbance at 740 nm is due to scattering and that scattering is inversely proportional to wavelength. In fact, more generally, the scattering is not inversely proportional to wavelength and a better estimate of the spectral absorption coefficient, $a_c(\lambda)$, can be obtained if the scattering spectrum is measured, if only on a relative scale, as described in Section 3.3.3.

In clearer or less light-attenuating waters, absorption by water constituents is rather low through most of the visible spectrum, and is difficult to measure directly by spectrophotometer. In such waters it is better to measure absorption on a volumetric concentrate of the particulates, and to add this (divided by the concentration factor) to the absorption of the yellow substance to give the total constituent absorption.

The available concentration procedures are time consuming and not entirely satisfactory because the physical, and thus optical, nature of the particles may be changed and recovery may be significantly lower than 100%. One method (e.g. Kirk 1980; Davies-Colley 1983) is simply to resuspend the residue captured on a 0.2 μm membrane filter in distilled water. A recovery of about 90% can be achieved routinely—as indicated by measurements of chlorophyll *a* (a convenient tracer of total suspensoids, at least in lake waters) on the filter after resuspension. Possibly a better method is to concentrate the particulates by filtration in a stirred cell (e.g. an ultrafiltration unit, but using standard membrane filters) which avoids the resuspension step. Low temperature/low pressure concentration in a rotary evaporator might also be feasible, although time consuming. (We have not attempted this.)

Another method for obtaining the particulate absorption spectrum, first proposed by Yentsch and coworkers (e.g. Truper & Yentsch 1962), is to carry out spectrophotometric absorption scans on the suspensoids collected by filtration onto a glass fibre filter, rather than in aqueous suspension. Absorbances (D) are measured relative to an unused, wet glass fibre filter in the reference beam. The absorption coefficient of the particulates (denoted by subscript p) is calculated as follows (e.g. Bricaud & Stramski 1990):

$$a_p(\lambda) = \frac{2.303 D_p(\lambda) S_f}{V_f \beta} \qquad (3.15)$$

where S_f is the filtration area of the filter, V_f is filtered volume of the water sample, and β is the so-called 'path amplification factor' that corrects for the diffusion of light in the filter. (β is essentially the reciprocal of the average cosine of light paths in the filter residue, and is typically around 2–4 in magnitude.) This method is simple and more suitable for routine use than concentration methods, but the path amplification factor must be estimated, and this is somewhat arbitrary. Bricaud & Stramski (1990) suggested that β is a weak power function of absorbance, and therefore of wavelength: $\beta(\lambda) = 1.63 D_p(\lambda)^{-0.22}$.

To obtain the total spectral absorption coefficient of a water from spectrophotometric measurements it is necessary to add the absorption of pure water to that of the water constituents. This is because the 'blank' or reference measurement in transmission spectrophotometry is high quality distilled water such that the instrument measures absorbance by the water constituents only, not the water itself. Thus the total absorption coefficient can be written

$$a = g + a_p + a_w = a_c + a_w \qquad (3.16)$$

where subscript p denotes 'particulates', and w, 'water'. By way of example, Fig. 3.9 shows the separate absorption spectra of the filterable and particulate constituents of water from a hypertrophic lake.

Appendix 2 (Sections A2.3–A2.6) summarizes the procedures for measurement of the spectral absorption coefficients appearing in equation (3.16).

3.3.3 Scattering measurements

Direct measurement of the scattering coefficient is difficult because it requires special instrumentation which can collect all scattered light, usually by scanning with angle. Jerlov (1976) and Kirk (1983) review some methods which have been used by researchers, but, apparently, there are no instruments suitable for routine water management applications. In the absence of absorption, the scattering coefficient is equal to the beam attenuation coefficient and can therefore be measured directly by transmissometer. Thus a transmissometer designed to work at a near-IR wavelength (e.g. 740 nm where absorption by water constituents other than water itself is practically negligible) could be used to calculate the scattering coefficient:

$$b(740) = c(740) - a(740) \approx c(740) - a_w(740) \qquad (3.17)$$

where c is the attenuation coefficient calculated from the measured beam transmittance. For most practical purposes it is more convenient to estimate the scattering coefficient as the difference $c - a$ (where both c and a are measured quantities) rather than to attempt direct measurement of b.

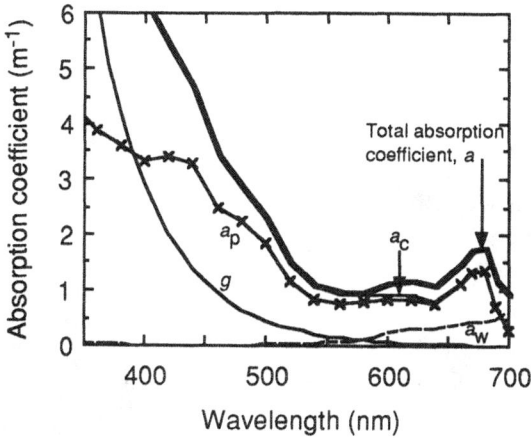

Fig. 3.9. Absorption spectra for water constituents in hypertrophic Lake Hakanoa. The total absorption coefficient (———) is the sum of the absorption coefficient of pure water (a_w, — — —) and that of the particulate constituents (a_p) and yellow substance (g). The absorption coefficient of all constituents, other than water itself, is a_c ($= a_p + g$). (Redrawn after Davies-Colley (1983), with permission.)

Although direct measurement of the total scattering coefficient in absolute units is not readily achieved, relative scattering measurements at a fixed angle are easily and commonly performed with an instrument termed a nephelometer. Fig. 3.10 shows the laboratory nephelometer perhaps most used in water quality work: the Hach 2100A. This instrument measures scattering in a cone centred on 90° and reads out in arbitrary nephelometric turbidity units (NTU). Usually nephelometers of this type are standardized using formazin, a polymer which is almost negligibly absorbing in the visible region (appearing milky white), but is intensely scattering, and can be readily made up in the laboratory to standard specifications (e.g. APHA 1989).

Turbidity measurements using nephelometers are commonly made in process control (e.g. of suspended solids removal in sedimentation tanks) and in water quality monitoring. Continuous-recording nephelometers are particularly valuable for water and wastewater monitoring. Unfortunately turbidity measurement by nephelometry has some severe limitations for work on the optical character of waters (e.g. McCluney 1975).

(1) Turbidity is highly instrument specific and very different values may be obtained on different nephelometers even though calibrated identically to formazin.
(2) Turbidity measurement suffers interferences from absorption in highly coloured water samples and from sources of scattering other than the particulates, such as dirty cuvettes or bubbles.
(3) Turbidity is an arbitrary, relative scattering measurement, which does not relate simply to scattering in absolute units or to optical concerns in water management such as visual clarity. Also, turbidity does not relate simply to suspended solids concentration, although reasonable correlations will be obtained in waters in which the particulate composition and size range are not too variable.

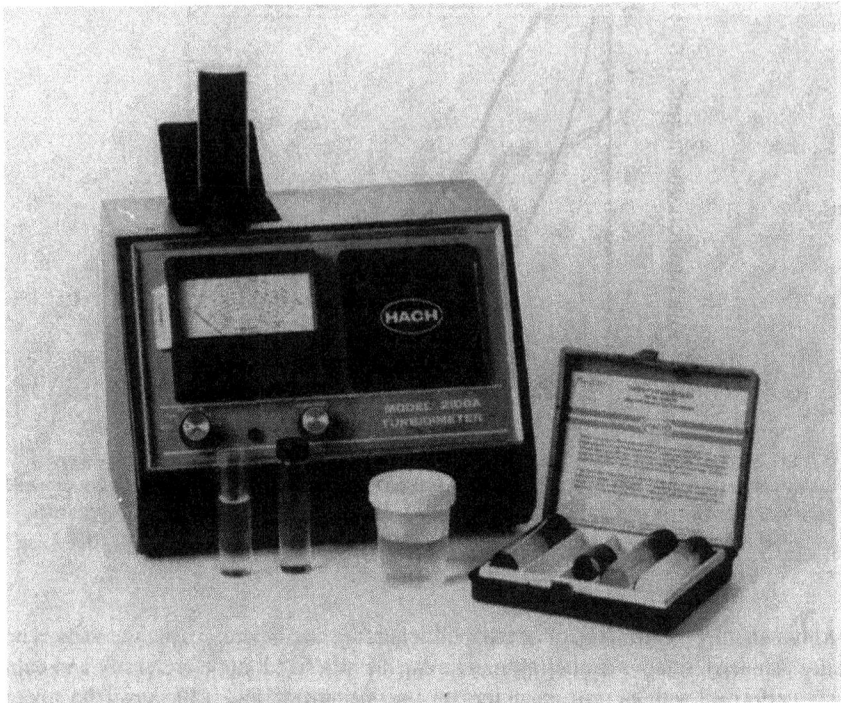

Fig. 3.10. Hach 2100A Nephelometer ('Turbidimeter'). Rubber latex standards (calibrated to formazin) are shown, one for each of the four ranges of the instrument: 0–1000 NTU, 0–100 NTU, 0–10 NTU and 0–1 NTU. The samples, contained in cylindrical glass cuvettes, are placed in position under the cylindrical black cover. Light from an incandescent lamp shining from below is scattered by particles and the light scattered through a range of angles centred on 90° is recorded by a photomultiplier.

The instrument specificity is perhaps the most severely limiting problem. This reflects different optical geometry, orientation and different spectral characteristics of the light source and spectral sensitivity of the detector in different instruments. McGirr (1974) analysed turbidity results obtained by different laboratories on four samples in a laboratory intercomparison exercise and showed that correlation between the results obtained with different turbidimeters was very poor, with discrepancies of up to an order of magnitude. However, agreement was fair between laboratories using the same instrument. The coefficient of variation was about 20% for samples of about 3 NTU as measured by the laboratories using the Hach 2100 and 2100A nephelometers.

Evidently nephelometric turbidity measurements cannot be compared if made on different instruments. To a large extent turbidity measurement is now redundant, being superseded by other relatively 'low-tech' measurements available to the water manager. Turbidity should be regarded merely as an optional extra, measured only where a check on consistency of other optical data (e.g. visual clarity) is desirable (Appendix 2, Section A2.7).

Having said that, there is one important caveat. Turbidity measurement by nephelometry can be readily automated so that continuous monitoring is feasible with commercially

available instruments (Section 3.7.1). In the authors' opinion, however, continuous monitoring of beam transmittance is preferable to nephelometry.

Austin (1973) and McCluney (1975) have pointed out that nephelometric measurements would be much more valuable if manufacturers were to develop a fixed angle
(preferably a forward angle, e.g. 15°) scattering meter which could be calibrated in absolute units of the volume scattering function, $\beta(\theta)$. Apparently their recommendations have
been ignored and at present there is no commonly available commercial instrument suitable for measuring absolute scattering in routine water quality work. Until and unless the
water industry demands such instruments from manufacturers, little progress is likely to
be made towards more satisfactory instrumentation. McCluney (1975) recommended that
the term 'turbidity' be dropped except as a qualitative descriptor of water clarity as in
'highly turbid water'. Instrumental measurements related to what was formerly called
'turbidity' should refer explicitly to the basis of measurement, for example 'beam attenuation', 'diffuse attenuation', '90° scattering'. This recommendation is followed in this
book, except that 'turbidity' is also used to refer to nephelometric turbidity measurements.

A laboratory spectrophotometer, equipped with one of the accessories described in
Section 3.3.2 for improving efficiency of scattered light collection, can be used to obtain
rough, but useful, estimates of the spectral dependence of scattering. The difference
between the absorbance, D, measured in the normal mode (Fig. 3.8A) and the (lower)
absorbance, D^*, measured in one of the 'wide scattering collection modes' (Figs
3.8B–3.8D) is entirely due to the scattering in an intermediate range of angles (Davies-
Colley 1983). By assuming a particular angular dependence of scattering, and therefore,
ε values in equation (3.11), we can estimate the proportion of total scattered light represented by the difference $D - D^*$. The scattering coefficient is then estimated from the
expression

$$b = \frac{\chi - \chi^*}{\varepsilon^* - \varepsilon} = \frac{2.303(D - D^*)}{r(\varepsilon^* - \varepsilon)} \qquad (3.18)$$

which follows from equations (3.10) and (3.11). For example, Davies-Colley *et al.*
(1986) estimated that $\varepsilon - \varepsilon^*$ was about 50% for a Varian 635 spectrophotometer
equipped with an integrating sphere, so that, in this case, $\chi - \chi^* \sim 0.5b$.

Estimates of the *spectral trend* of scattering as indicated by the difference spectrum
$\Delta\chi(\lambda)$ ($\Delta\chi = \chi - \chi^*$ (Latimer & Eubanks 1962)) are possibly of more value than the absolute scattering. Measurement of $\Delta\chi$, termed the 'relative scattering coefficient' by
Davies-Colley (1983), permits correction of the spectral absorption coefficient of water
samples for scattering. Thus equation (3.14) is replaced by

$$a_c(\lambda) = \chi_c(\lambda) - \chi_c(740)\frac{\Delta\chi_c(\lambda)}{\Delta\chi_c(740)} \qquad (3.19)$$

where it is assumed that scattering out of the angular range of the detector is proportional
to scattering in the intermediate range of angles between that of the 'wide scattering
collection mode' and that of the normal mode. Again, absorbance at 740 nm is assumed
to be entirely due to scattering. Equation (3.19) reduces the spectral distortion which can
result from use of equaion (3.14).

3.3.4 Estimation of the IOPs from measurement of the AOPs

The IOPs are more useful than the AOPs for analysing colour and clarity in waters, but, as we have seen, their measurement generally requires special instruments or procedures. Fortunately there are means available for the approximate calculation of IOPs from the AOPs.

Some particularly useful relationships between IOPs and AOPs have been reported by Kirk (1981a,b, 1984a, 1989) from Monte Carlo simulations of light penetration into water. Kirk (1981b) gave nomograms from which the absorption and scattering coefficients can be estimated from measurements of the irradiances, E_u and E_d. Kirk (1984a) later reported more detailed Monte Carlo studies of the influence of solar altitude on the relationships. Fig. 3.11 shows the dependence of reflectance, $R(z_m)$ at the midpoint of the euphotic zone on the ratio b/a. Using this nomogram with the measured reflectance, b/a can be estimated as can the average cosine ($\mu = \vec{E}/E_0$). A restatement of equation (3.6) gives

$$a = K_E \mu = K_E \frac{\vec{E}}{E_0} = K_E \frac{E_d - E_u}{E_0} \qquad (3.20)$$

The absorption coefficient, a, can be calculated from equation (3.20) and then the scattering coefficient, b, from the ratio b/a with known a. The estimate of b obtained in this way is very useful for the optical characterization of waters and usually agrees tolerably well with independent estimates. However, the estimate of a, being a PAR-band average, is rather less useful since a typically varies appreciably with wavelength.

A worked example of the calculation of b and a from measured values of irradiance using Kirk's (1981b) procedure is given in Table 3.1. The calculation involves estimating b/a and μ from the reflectance calculated for the midpoint of the euphotic zone, $R(z_m)$, as illustrated in Fig. 3.11.

An alternative method for estimating the scattering coefficient from measurements of both K_d and R has been given by Kirk (1989) based on analysis of the upwelling light stream in Monte Carlo simulations of light transmission through natural waters. The method involves calculation of a quantity termed by Kirk (1989) the diffuse backscattering coefficient at the midpoint of the euphotic zone: $b_{bd}(z_m) = 3.5R(z_m)K_d(z_m)$. The backscattering coefficient, b_b, is then obtained from a nomogram in Kirk (1989) (giving $b_{bd}(z_m)/b_b$ as a function of $R(z_m)$). Finally, the total scattering coefficient is estimated as $53b_b$. This method can be used to check the estimate of b obtained from Kirk's (1981b) method. The value of b calculated using this method on the data in Table 3.1 agreed tolerably well with the value obtained by the Kirk (1981b) method.

Appreciably different values might be obtained using the two different methods for estimating the scattering coefficient where a water has a significantly different angular dependence of scattering from that assumed by Kirk (1981a,b) and Kirk (1989) in his Monte Carlo simulations (i.e. $\beta(\theta)/b$ as measured in San Diego Harbor by Petzold (1972)—Fig. 2.3), particularly at high angles. Some insights into the influence of the volume scattering function on the relationships between the IOPs and AOPs can be obtained from the work of Gordon et al. (1975) and by Kirk (1991). Unfortunately there are as yet no simple means for refining estimates of IOPs that can be obtained from measured AOPs.

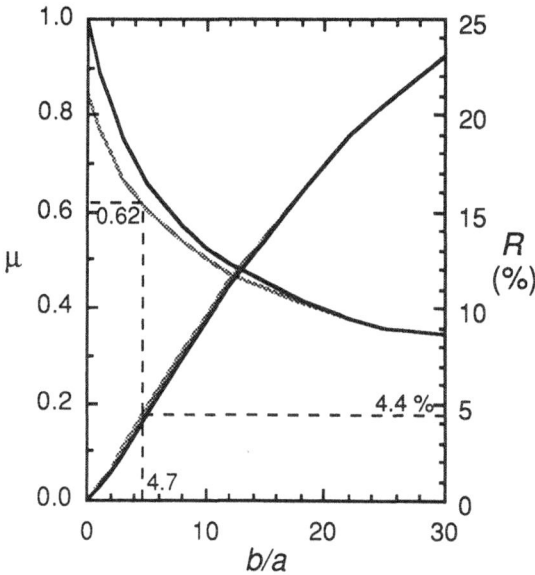

Fig. 3.11. Reflectance, R, and average cosine, μ, at the midpoint of the euphotic zone (depth z_m) of a water as a function of the scattering to absorption ratio, b/a. The curves, redrawn after Kirk (1981a, with permission), are based on the results of his Monte Carlo simulations of photon penetration into water. ——, vertically incident light; ——, light incident at 45°. Interpolation of b/a and μ values for a water with $R = 4.4\%$ is indicated (refer to the worked example in Table 3.1 for the Waikato River data plotted in Fig. 3.3).

A very useful means for estimating the spectral absorption coefficient from near-surface spectral irradiance measurements was derived by Pelevin (1965). The following equation, exact for any depth, z, and any given wavelength, can be derived from the equation for conservation of radiant energy in the water column (equation (3.6)):

$$a = K_d \left(1 - R + \frac{1}{K_d} \frac{dR}{dz} \right) \frac{E_d}{E_o} \qquad (3.21)$$

The last term in the parentheses is very small and can be safely neglected. Reasonable approximations for the ratio of irradiances (E_d/E_o) in equation (3.21) can be made for the near-surface water layer to yield the following equation for calculating the spectral absorption coefficient from spectral irradiance (E_u and E_d) measurements with a spectroradiometer:

$$a = \frac{K_d(1-R)\cos j}{0.6 + (0.47 + 2.5R)\cos j} \qquad (3.22)$$

where j is the angle to the vertical of the direct solar beam in water.

Table 3.1. Calculation of the PAR-band average optical properties from quantum irradiance measurements

Worked example for the Waikato River at Tuakau, 10 March 1983, 1500 hr.
Data given below are plotted in Fig. 3.3 (upwelling and downwelling quantum irradiance under clear skies, solar altitude 40° above the horizon).

z (m)	E_d	E_u	$E = E_d - E_u$	$R = E_u/E_d$	K_d	K_u	K_E
0	1100	30	1070	0.027			
					1.87	1.14	1.90
0.5	430	17	413	0.040			
					1.43	1.16	1.45
1.0	210	9.5	200.5	0.045			
					1.48	1.41	1.49
1.5	100	4.7	95.3	0.047			
					1.35	1.61	1.33
2.0	51	2.1	48.9	0.041			
					1.51	1.29	1.52
2.5	24	1.1	22.9	0.045			

The euphotic depth, z_{eu} = 3.04 m, and the midpoint of the euphotic zone is z_m = 1.52 m.
(Note that $R(z_m)$ is the ratio of interpolated irradiances at z_m, i.e. $R(z_m) = \hat{E}_u(z_m)/\hat{E}_d(z_m)$, where the caret over-marks denote interpolated—predicted—values.)

$R(z_m) = 0.044$ (i.e. 4.4%)
$K_E(z_m) = 1.44$ m^{-1}
$K_d(z_m) = 1.43$ m^{-1}

Calculation of the scattering coefficient (an IOP) from R and K_E at z_m.
(1) *Method A* (Kirk 1981a,b)
Kirk's (1981a) nomogram (Fig. 3.14) giving R and μ as functions of the ratio b/a applies at the midpoint of the euphotic zone (see also, Kirk 1981b).
From the nomogram with $R(z_m)$ = 4.4% (45° line) we have b/a = 4.7 and μ = 0.62 (Fig. 3.11).
The absorption coefficient, $a = \mu K_E = 0.62 \times 1.44$ m^{-1} = 0.89 m^{-1}.
The scattering coefficient, $b = (b/a) \times a = 4.7 \times 0.89$ m^{-1} = 4.2 m^{-1}.
(2) *Method B* (Kirk 1989)
Kirk (1989) showed that the diffuse backscattering coefficient, $b_{bd}(z_m) = 3.5R(z_m)K_d(z_m)$. Substituting the above values for $R(z_m)$ and $K_d(z_m)$ gives:
$b_{bd}(z_m) = 3.5 \times 0.044 \times 1.43$ m^{-1} = 0.22 m^{-1}.
From Kirk's nomogram giving $b_{bd}(z_m)/b_b$ as a function of $R(z_m)$, with $R(z_m)$ = 0.044, we have
$b_{bd}(z_m)/b_b = 2.6$.
Therefore $b_b = b_{bd}(z_m)/b_{bd}(z_m)/b_b = 0.22/2.6$ m^{-1} = 0.0846 m^{-1}.
Finally, since $b \sim 53b_b$, $b \sim 53 \times 0.0846$ m^{-1} = 4.5 m^{-1}.

Thus method B yields a value for b in fair agreement with that from method A.

3.4 WATER CLARITY

We have seen in Chapter 2 that there are two main aspects of water clarity: sighting distance through water (visual clarity) and penetration of diffuse irradiance from the sun into water. Both aspects are dependent on the optical properties of water but very different techniques are required for the measurement of quantities associated with these aspects. Visual clarity is best measured directly by noting the extinction of the image of a standard target. Penetration of irradiance into water is best measured with submersible irradiance sensors (or illuminance sensors in the context of human vision). Thus clarity of water in the sense of irradiance penetration is effectively quantified by the irradiance attenuation coefficients, the measurement of which was discussed in Section 3.2. Further discussion is given in this section on the interpretation of illuminance and irradiance attenuation coefficients. First, however, we consider the measurement of visual range in water.

3.4.1 The Secchi disc
Visual clarity has long been measured with a *Secchi disc*, named after Dr A. Secchi, the nineteenth century Italian scientist who carried out the first scientific investigation with the device (Tyler 1968; Preisendorfer 1986). In marine waters and clearer lake waters a 300 mm diameter white disc has been used traditionally, but in most lake studies a 200 mm diameter disc painted in black and white quadrants is favoured since it is considered to give a sharper extinction point than an all-white disc. The disc is weighted to hold it vertically in the water column, suspended from a graduated line (e.g. Lind 1977).

The optical basis of the Secchi observation was briefly reviewed in Section 2.5.1. Use of the Secchi disc is very simple. The disc is lowered into the water from a boat or suitable structure until the image just disappears at depth z_1. The disc is then slowly raised and the depth, z_2, at which it reappears is noted. The *Secchi depth* is simply the average: $(z_1 + z_2)/2$.

There are probably more Secchi depth data available on natural waters than any other optical characteristic, including nephelometric turbidity. The full range in natural waters probably exceeds three orders of magnitude, from >50 m in the clearest seawaters to <0.05 m in extremely turbid waters.

Some workers have referred to the Secchi depth observation as 'rough' or 'crude' (e.g. Brezonik 1978; Anon 1984); possibly because the measurement is simple and apparently subjective, and is considered dependent on ambient lighting conditions. Also the Secchi depth does not correlate perfectly with other variables of interest such as chlorophyll *a*. However, the Secchi disc is no more 'rough' in terms of precision level than many other measurements in water quality and is more precise than many—including nephelometric measurement which is often considered preferable! Nephelometric turbidity can seldom be measured to much better than 20% relative error (e.g. McGirr 1974), whereas different observers will generally agree to better than 10% in Secchi depth observations (Højerslev 1986).

The Secchi depth, contrary to popular opinion, is only weakly dependent on the ambient lighting conditions, so long as there is sufficient light for the threshold contrast of human vision to be independent of background luminance, and so long as the observer takes the time to allow eye adaptation to the luminance of the upwelling light field.

Secchi depth observations should not be made too close to sunrise or sunset when the light (illuminance) is low (resulting in an increase in threshold contrast) and much light is incident at large zenith angles. The slight dependence of Secchi depth on directional structure of ambient lighting follows from consideration of Tyler's (1968) equation for the Secchi depth (refer also to Preisendorfer (1986)):

$$z_{SD} = \frac{\ln(C_0/C_T)}{c+K} = \frac{\Gamma}{c+K} \qquad (3.23)$$

first introduced in Chapter 2 (equation (2.13)). Threshold apparent contrast, C_T, is independent of the ambient light field, as is the IOP, c. However, both the inherent contrast of the disc, C_0, and the illuminance attenuation coefficient, K, are dependent on directional structure of ambient lighting, and so, therefore, is the Secchi depth. Equation (3.23) suggests that z_{SD} can be regarded as an apparent optical property (AOP) with even greater 'quasi-inherent' character than K_d.

One of the reasons for the rather low esteem in which the Secchi disc measurement is held may be that many of the z_{SD} data collected over the years have not been taken with sufficient care and attention to important operational details as is desirable, and this has given rise to appreciable 'noise'. In particular the disc should be viewed only on the sunny side of the boat, not the shaded side as is often recommended to reduce the interfering effect of reflection from the water surface (e.g. Lind 1977). The Tyler equation (3.23) no longer applies to the disc viewed in, or through, the boat's shadow (Højerslev 1986). The relationship between the Secchi depth and the optical properties appearing in equation (3.23) is degraded by viewing in the shade. However, it is particularly important that surface reflections are reduced to a minimum. Tyler (1968) has suggested that observations are ideally made by a snorkel diver but this might be inconvenient for routine monitoring work. The next best approach, and one that is recommended as a standard part of the Secchi procedure (Appendix 2, Section A2.8), is to observe the disc through a viewing box or tube. A seal around the observer's face to screen ambient light is a very desirable feature.

Precise Secchi depth measurement in turbid water is often difficult during wave action. One approach under such conditions is to have a snorkel diver make the observations, although this is an unattractive prospect for the individual assigned the task. Effler (1985) has suggested that, when conditions are adverse for Secchi observations, z_{SD} can be estimated indirectly from irradiance measurements using equation (3.23). The irradiance measurements give K (ideally this would be converted to the attenuation coefficient for illuminance as described in Section 3.5.2), and c ($= a + b$) is estimated following Kirk's (1981b) method.

A brief recommended method for taking Secchi disc observations is given in Appendix 2.5.

3.4.2 The black disc

The visual range of most objects of interest underwater is much shorter than the Secchi depth because their inherent contrast, C_0, is much lower than that of the white disc (for which $C_0 \sim 40$ (Tyler 1968)). Duntley (1963) has reported that the visual range of many submerged objects is about 4–5 attenuation lengths (i.e. 4 to 5 times $1/c$) suggesting that inherent contrast is often close to unity. For this reason the visual range of a black target seen as a silhouette ($C_0 = (B - B_b)/B_b = (0 - B_b)/B_b = -1$, where B is the luminance of the target and B_b that of the water background; Section 2.5) usefully characterizes practical sighting ranges through water.

The ideal black target would be a light trap (e.g. Pilgrim et al. 1989), but a more convenient target is a disc, constructed identically to a Secchi disc, but painted completely black (Davies-Colley 1988a). As well as relating more closely to sighting ranges of practical importance in waters, the black disc (referred to as 'the black spot') has important theoretical and practical advantages over the Secchi disc (Davies-Colley 1988a).

In Section 2.5.1 we showed that the horizontal visual range of a perfectly black target is inversely proportional to the beam attenuation coefficient, weighted by the photopic sensitivity function of the human eye (i.e. the photopic beam attenuation coefficient). Thus the photopic beam attenuation coefficient can be calculated from measurement of the horizontal sighting range of a black disc,

$$c = \frac{\Psi}{y_{BD}} \qquad (3.24)$$

Similarly, the vertical sighting range of the black disc, z_{BD}, gives an estimate of the sum of the photopic beam attenuation coefficient and the illuminance attenuation coefficient:

$$c + K = \frac{\Psi}{z_{BD}} \qquad (3.25)$$

Davies-Colley (1988a) found that the dimensionless constant of proportionality, Ψ, was about 4.8 but varied weakly with the apparent size of the disc.

If the sighting range of the black disc is measured in both the vertical and horizontal directions, an estimate of an important AOP (K) as well as a fundamental IOP (c) can be obtained from equations (3.24) and (3.25). The Secchi depth, by contrast, can only yield a rough estimate of c, and very rough estimate of K, since Γ (= $\ln(C_0/C_T)$ in equation (3.23))is not constant, but varies depending on the value of inherent contrast, C_0 which, in turn, depends on the reflectance of the water (Preisendorfer 1986).

The main practical advantage of the black disc over the Secchi disc is that measurements can be made with the former target in shallow, clear rivers and ponds. In these 'optically shallow' systems, black disc measurements cannot be made vertically, but horizontal observations are possible and these yield a useful estimate of c from equation (3.24). In optically deep waters, including some rivers, vertical black disc observations can (and should) be made as well as horizontal observations so that light penetration can be estimated (albeit with rather low precision (Davies-Colley 1988a)) from equations (3.24) and (3.25).

Black disc observations are readily made by snorkel divers, working in pairs for the horizontal observations, with one holding the disc and one making the observation. Deployment of snorkel divers is not always convenient, so a special underwater viewer constructed of optically clear polycarbonate is used (Fig. 3.12). A mirror at 45° permits the observer to view horizontally through a side window just under the water surface while remaining above water. The viewer is easily used in small boats or at wadeable depths in rivers (Plate 7). With the mirror removed the observer can sight vertically into the water through the end window for vertical black disc measurements or standard

Secchi disc observations. (The viewer is also useful for observing underwater objects such as fish or the bed material of rivers.)

For horizontal measurements in clearer waters the black disc is conveniently mounted on a black-painted pole held by the observer's assistant (Fig. 3.12A). In more turbid water, with a visual range less than 0.5 m, the disc is mounted on a black-painted support structure held, together with the viewer, by the observer (Figs 3.12B and 12C).

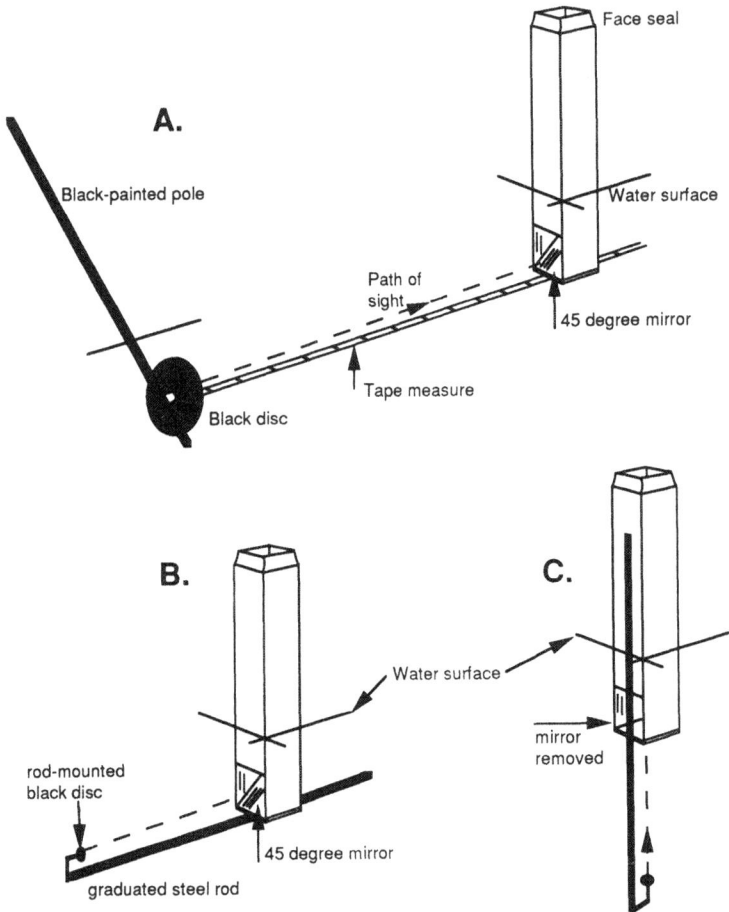

Fig. 3.12. Measuring visual water clarity with black disc equipment. An underwater viewer with both side and end windows, and a removeable 45° mirror, is used for the observations. (A) Horizontal observations in natural waters with discs of larger sizes fitted to a pole. Two people are normally needed for this measurement: one to hold the disc and one to make the observation. Vertical observations with the larger black discs are made by one person in the same way as Secchi disc observations. (B) Horizontal observations in turbid water with a small disc fixed permanently to a blackened steel rod. (C) Vertical observations in turbid water with a small rod-mounted black disc. Only one operator is needed in (B) and (C).

For satisfactory characterization of the visual clarity of a wide range of waters (and for estimation of the attenuation coefficients, c and K) it is important that the factor Ψ in equations (3.24) and (3.25) remain constant. However, at small apparent sizes the disc may no longer be an 'optically large' target, and the threshold contrast, C_T, increases. At large apparent sizes the shadowing of the disc can locally distort the light field, increasing the apparent visual range. Davies-Colley (1988a) has suggested that, to ensure the near-constancy of Ψ, the angular size of the black disc should be kept nearly constant, say in the range 2°–10° of arc; that is, larger discs should be used in clearer water. (This 'constant angular size' rule should also be applied to Secchi observations, but seldom is —which further degrades the value of Secchi data.)

Fig. 3.13 shows disc diameter within the angular range 2°–10°, as a function of water clarity. For example, visual ranges between 1.5 and 5 m (at the outside about 10 m) should be observed with a 200 mm disc. A 600 mm disc can be used up to about 30 m maximum—a very clear natural water, rarely encountered except in the deep sea. A 6 mm diameter disc (e.g. a 6 mm cylinder viewed end-on) has been used by the authors down to about 40 mm visual range, but the underwater viewer itself seems to interfere significantly with the light field in such turbid water, thereby degrading direct clarity measurements.

In order to extend the range of application of the black disc method to very turbid waters and effluents Davies-Colley & Smith (1992) developed a method for measuring visual clarity off-site on a diluted water sample contained in a trough. A correction for the imperfect visual clarity of the dilution water (of order 10 m for tap water—as estimated from beam transmissometry) is desirable. The *in situ* clarity of the turbid water sample is calculated from the mass balance on light attenuation cross-section (i.e. beam attenuation coefficient multiplied by volume):

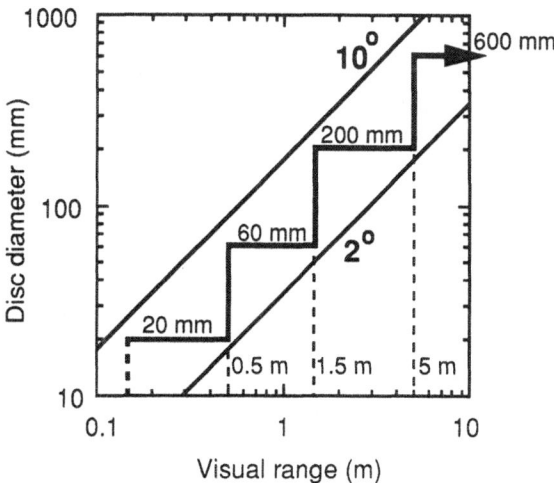

Fig. 3.13. Black disc (or Secchi disc) size as a function of sighting range. The apparent size of the disc should be kept in the range 2°–10°. The recommended range of application of 20 mm, 60 mm and 200 mm diameter discs is shown, and also part of the range for a 600 mm diameter disc.

$$c_{mix}V_{mix} = c_s V_s + c_{dil}V_{dil} \qquad (3.26)$$

where V is volume and the subscripts denote the mixture (mix) of dilution water (dil) and sample (s). Solving for the unknown sample attenuation coefficient and sample visual clarity, with dilution factor, $F = V_{mix}/V_s$, we obtain

$$c_s = c_{mix}F - c_{dil}(F-1) \qquad (3.27a)$$

and

$$\frac{1}{y_{BDs}} = \frac{F}{y_{BDmix}} - \frac{F-1}{y_{BDdil}} \qquad (3.27b)$$

Clarity of very turbid waters ($y_{BD} < 100$ mm) is probably best calculated from measurements on diluted samples using equation (3.27b) rather than measured directly.

Appendix 2 (Sections A2.9 and A2.10) gives a recommended 'standard method' for black disc observations.

3.4.3 Light penetration into water bodies

The various irradiance attenuation coefficients, and the illuminance attenuation coefficient, quantify the penetration into water of light of different spatial character and spectral quality. The AOPs discussed in Section 3.2 can, therefore, be regarded as quantifying light penetration into water bodies.

The attenuation of illuminance with depth into water effectively quantifies the amount of light available for human vision. This is probably of minor concern in water management, but is of considerable practical concern to divers. At some depth in the water, depending on clarity, there is insufficient illumination for normal *photopic* colour vision (retinal cone cells being the light receptors) and the human visual system is forced to 'switch over' to *scotopic* vision using retinal rod cells which see only shades of grey. As illuminance falls still further, visual range is reduced owing to increase in the threshold contrast. The illuminance attenuation coefficient is fairly readily calculated from depth profiles of illuminance (I), measured by lux meter. Alternatively, the illuminance attenuation coefficient can be estimated from the attenuation coefficient for quantum (PAR) irradiance if the hue of the water is known (Section 3.5.2).

The attenuation with depth of illumination may be significant to the ecology of aquatic animals. The illuminance penetration may not adequately characterize the illumination for particular animals, which, in general, have eyes with a spectral sensitivity different from that of the standard human eye (Fig. 3.1). Generally, therefore, the spectral attenuation coefficient in a water body must be measured in order to interpret ecological ramifications for animals with eyes of given spectral sensitivity.

With regard to the light climate for growth of aquatic plants, it is the attenuation with depth, not of visible light (illuminance), but of quantum irradiance in the PAR band that is of primary interest. This is easily calculated from irradiance measurements with a PAR sensor, ideally a scalar irradiance sensor. As we saw in Section 2.5.2, a useful rule-of-thumb is that plants will grow in water down to the euphotic depth, the depth at which irradiance is reduced to 1% of the value at null depth (i.e. the depth at which irradiance is $0.01E(0-)$).

Estimation of the value of K_d(PAR), and thus of z_{eu}, from Secchi depth observations is often attempted, following the pioneering work of Poole & Atkins (1929) (e.g. Idso & Gilbert 1974; Idso 1982; Walker 1980,1982). The relationship between these quantities is assumed to be a simple reciprocal one:

$$K_d(PAR) = \kappa/z_{SD} \qquad (3.28)$$

where κ is assumed to be a constant (~1.7 according to Poole & Atkins (1929)). A rough overall correlation between K_d(PAR) and z_{SD} exists of course; highly light-attenuating waters usually have poor visual clarity (and vice versa). However, the precision with which K_d(PAR) can be estimated from measured values of z_{SD} using equation (3.28) is poor, and errors of 50% or more are common. This is because κ is not a constant but varies, mainly with the reflectance of the water (Davies-Colley & Vant 1988). Fig. 3.14, in which K_d(PAR) is plotted against z_{SD} measured in 28 different lakes, illustrates the problem. The range of clarity (as measured either by K_d(PAR) or z_{SD}) in this dataset covers more than 2 orders of magnitude, sufficient to give a good correlation between the variables ($r^2 = 0.92$) that belies the poor precison of estimation (coefficient of variation = 32%). Davies-Colley & Vant (1988) showed that most of the variation in the product, $\kappa = K_d z_{SD}$, derives from variation in the scattering-to-absorption ratio, b/a, or, equivalently, in the reflectance coefficient, R. They warned against uncritical assumption of a particular value of κ for the purposes of estimating light penetration from visual clarity measurements.

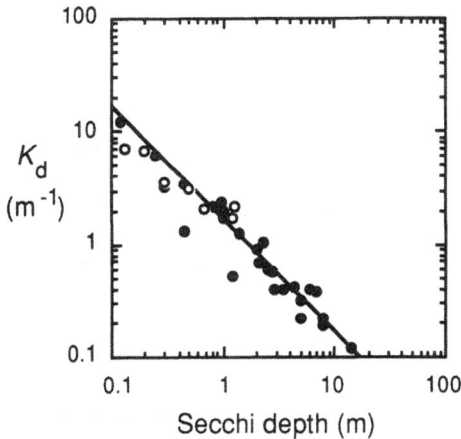

Fig. 3.14. Irradiance attenuation coefficient (for downwelling quantum irradiance) plotted against Secchi depth for 28 New Zealand lakes. ●, data for 27 lakes reported by Vant & Davies-Colley (1984); ○, data for Lake Whangape (Davies-Colley & Vant 1988). A general correlation is observed, but with considerable scatter around the line of Poole & Atkins (1929): $\kappa = 1.7$.

3.5 WATER COLOUR

In Section 2.6 we introduced the concept of 'intrinsic' water colour, which for practical purposes can be thought of as that colour observed when looking vertically down in optically deep water. This water colour can be described, as can any colour, by specifying the hue, brightness and saturation (colour purity) associated with the spectral reflectance. Thus to measure water colour we are concerned with measuring the spectral reflectance, $R(\lambda) = E_u(\lambda)/E_d(\lambda)$. This section will be concerned mainly with measuring or otherwise specifying the water colour associated with $R(0, \lambda)$, the spectral reflectance at zero depth in the water, but first some traditional measures, still recommended in handbooks, will be reviewed.

3.5.1 Traditional measures of water 'colour'
The traditional platinum–cobalt colour (Hazen) measurement first introduced by Hazen (1892) is still given in the 17th edition of 'Standard methods' (APHA 1989). In this method the yellowness of a water sample contained in a glass Nessler tube is matched to that of a standard solution of chloroplatinate ion ($PtCl_6^{2-}$) tinted with cobalt. Glass filters are more convenient to use as secondary standards than the chloroplatinate primary standard. In standard Nessler tubes, Pt–Co colour of water can be matched with a precision of about 2.5 Pt–Co colour units (2.5 CU) (using standards at 5 Pt–Co colour unit intervals) in the range up to 70 CU above which dilution of water samples is required. Typically Hazen colour is measured on filtered water samples ('true colour') as well as on unfiltered samples ('apparent colour').

The Hazen colour measurement is still commonly used in water management. In the paper industry Hazen colour is used for convenient process control of Kraft process pulp 'black liquor' and bleaching plant effluents whose yellow colour is similar to the naturally occurring coloured organic material in waters known as yellow substance (although much more concentrated—of the order of 1000 CU).

Notwithstanding the common use of the Pt–Co colour technique, there are some severe problems with this measurement. Firstly, Pt–Co colour bears little relation to water appearance in the field. Secondly, the references to 'true colour' and 'apparent colour' are misnomers. As discussed in Chapter 2, the colour of water depends on spectrally selective absorption and to a lesser extent, scattering, by both filterable and particulate constituents of water. Thirdly, the precision of Pt–Co colour matching is rather low, especially for waters of low colour say, <10 CU, which probably includes most relatively clear freshwaters. Finally, at low Pt–Co colour levels (say <5 CU), the hue of waters is typically green or blue–green rather than yellow, and it is difficult to obtain a Pt–Co colour match, even in long Nessler tubes.

Increasingly the Pt–Co 'true' colour measurement is being replaced by measurement of optical density using a spectrophotometer (i.e. the filtrate absorption measurements discussed in Section 3.3). Typically the result is reported only as an absorbance for a given wavelength (usually in the blue or near-UV), and cuvette path length, but preferably the spectrophotometric data are reported as an absorption coefficient (suitably corrected for any residual scattering) which is unambiguous as regards path length (Appendix 2, Sections A2.3–A2.6). In most waters it is possible to obtain a good correlation between 'true' Pt–Co colour and the filtrate absorption coefficient, although this

varies somewhat between waters. For example, Bowling *et al.* (1986) found a good correlation between g_{440} and Pt–Co colour for 320 samples from Tasmanian waters ranging
in colour from <5 CU to extremely dark waters of ~ 600 CU (regression equation: g_{440} =
0.081(Pt–Co colour) + 0.40, r^2 = 0.98).

A traditional method for specifying hue of lake waters which is apparently still in use,
at least in Europe, is the Forel–Uhle system (Hutchinson 1957, pp. 415–416). This
involves matching the hue of water, viewed through an underwater viewer, to one of 21
standard coloured inorganic salt solutions ranging from blue to orange ('brown'). No
consideration is given to brightness or colour purity, or, for that matter, to spectroradiometric characterization of the hues. In spite of the lack of radiometric standardization,
this would appear to be a better means of specifying water colour as it appears in the
field than the Pt–Co colour measurement.

3.5.2 Brightness
The brightness of a colour depends on the luminous flux reaching the observer's eye and
is therefore determined by the luminous reflectance. The brightness of water colour
observed from above can be defined in terms of the illuminance reflectance just below
the water surface (denoted 0–):

$$R_I(0-) = \frac{I_u(0-)}{I_d(0-)}$$

$$(3.29)$$

If a submersible 'lux meter' (illuminance meter) is available, the luminous reflectance
can be obtained very simply from equation (3.29) using measurements of upwelling and
downwelling illuminance. Alternatively quantum irradiance sensors can be used to provide a satisfactory estimate of illuminance reflectance when fitted with a suitable green
optical filter to approximate the photopic sensitivity function. Illuminance reflectance R_I
can also be roughly estimated from the quantum irradiance reflectance, R_{PAR}, at zero
depth using Fig. 3.15, if the hue of the water colour is known. Generally $R_{PAR} < R_I$ and
the greatest deviation (about 20%) occurs, as expected, for green–yellow coloured
waters. The lines of best fit in this figure could be used to estimate R_I from measured
R_{PAR} (and vice versa) if water colour is known.

The illuminance reflectance can, in turn, be used to estimate the Munsell 'value' (V),
a useful index of brightness. Tables are available for the interpolation of Munsell value
from measured luminous reflectance (ASTM 1980). As a rough guide we may characterize the brightness of waters as in Table 3.2.

3.5.3 Chromaticity analysis
The hue and purity of a colour depend on the distribution of radiant power in the spectrum of reflected visible light, unlike the brightness which depends only on the amount
of light (luminous flux). Thus the hue and purity of water colour depend on the shape of
the irradiance reflectance spectrum, $R(\lambda)$, whereas the brightness of water depends on
one number: the magnitude of the illuminance reflectance, R_I. In order to define the hue
and colour purity of water completely we need to measure the reflectance spectrum and
to apply chromaticity analysis (Section 2.6.4).

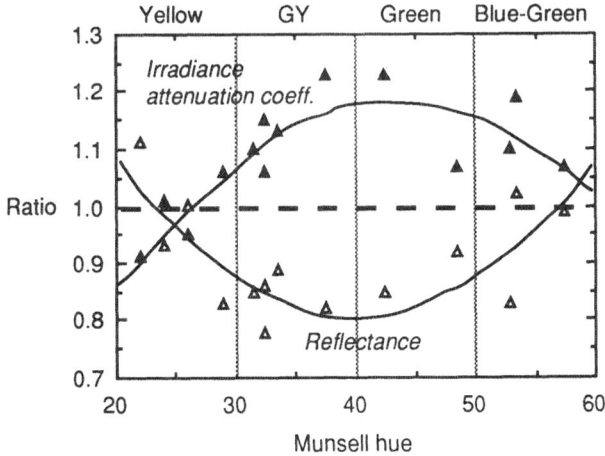

Fig. 3.15. Ratio of attenuation and reflectance coefficients for PAR to those for illuminance plotted against Munsell colour (hue). Data are for 14 New Zealand lake waters from Davies-Colley *et al.* (1988, and unpublished.). ▲, ratio of PAR attenuation coefficient to illuminance attenuation coefficient; △, ratio of PAR reflectance to illuminance reflectance. Some of the scatter around the trend lines (non-linear least squares) can be attributed to measurement error, but some derives from differences in colour purity. The coefficients for PAR and illuminance are evidently most divergent for green-coloured waters. If the hue of a water is known the curves can be used to estimate irradiance attenuation coefficients from measured illuminance attenuation coefficients (or vice versa) or to estimate irradiance reflectance from measured illuminance reflectance (or vice versa).

The only way in which the reflectance spectrum of a water body can be measured directly is by use of a spectroradiometer which can 'look' downward to scan the upwelling irradiance spectrum, $E_u(\lambda)$, as well as the downwelling irradiance spectrum, $E_d(\lambda)$ 'seen' by the sensor pointing upwards. Spectroradiometric measurement of underwater irradiance was discussed briefly in Section 3.1. Spectroradiometers are really research instruments that are unlikely to become widely used in water management, except perhaps where the colour of a particularly valuable water body, such as a sensitive lake, is at risk.

There is a less direct means for obtaining the reflectance spectrum suitable for chromaticity analysis of water colour. Gordon *et al.* (1975), Morel & Prieur (1977) and Kirk (1981a, 1984a) have shown that the irradiance reflectance just below the water surface is approximately proportional to the backscattering coefficient divided by the absorption coefficient (equation (2.19), Section 2.6.2), and these coefficients, or at least their spectral trend, can be estimated from spectrophotometric measurements. The spectral absorption coefficient, $a(\lambda)$, as we have seen, can be measured on water samples with a spectrophotometer (Section 3.3.2). A spectrophotometer can also be used to measure the scattering coefficient, albeit only roughly. Alternatively, a modified spectrophotometer can be used to measure the spectral beam attenuation coefficient, $c(\lambda)$, and the spectral scattering coefficient can then be calculated by subtracting the spectral absorption coefficient: $b(\lambda) = c(\lambda) - a(\lambda)$. If we assume that the spectral trend of backscattering is similar to that of the total scattering we can calculate a *relative reflectance* spectrum:

Table 3.2. Suggested description of water brightness in terms of illuminance reflectance

Brightness descriptor	Illuminance reflectance R_I	Approximate Munsell value (V)
'Very bright'	>10%	>3.7%
'Bright'	5–10%	2.6–3.7%
'Average'	2.5–5%	1.7–2.6%
'Dark'	1.25–2.5%	1.0–1.7%
'Very dark'	<1.25%	<1.0%

$$R_r(\lambda) = b(\lambda)/a(\lambda) \tag{3.30}$$

This spectrum reproduces the shape of the reflectance spectrum but with some distortion deriving from spectral differences between total scattering and backscattering.

Once the reflectance spectrum has been obtained, either by spectroradiometry or from spectrophotometry, chromaticity analysis can be carried out on this spectrum to extract the hue and purity of the colour as outlined in Section 2.6.4. The relative amounts of standard red (x) and green (y) primaries are plotted in the standard horseshoe-shaped CIE chromaticity diagram (Fig. 3.16) given, for example, by APHA (1989). The hue is specified as a dominant wavelength, and the saturation as a measure of colour purity, as shown in the figure. If required, the colour can be expressed as a code in the Munsell system of colour by transferring the plot (x, y) to a CIE diagram on which Munsell correlations have been superimposed (Newhall et al. 1943; ASTM 1980) (Fig. 3.16). The Munsell colour system is briefly described in Section 3.5.4.

APHA (1989) gives a method for measurement of water colour based on manual chromaticity analysis of spectrophotometric transmittance measurements using the 'selected ordinate' method. In this method the spectrophotometric measurements are made at wavelengths chosen so as to sample the spectrum more intensively in regions where the tristimulus (mixing curve) values are high (Fig. 2.21). The transmittances are then simply summed to give values of x, y, and z which are proportional to the amounts of red, green and blue primary stimuli. The colour so calculated is a *transmission* colour (as seen, for example, by a diver looking up through water), rather than the *intrinsic* colour (reflectance colour) which relates more closely to water management concerns. The selected ordinate method could, of course, be used for chromaticity analysis of reflectance (intrinsic) colour. This would involve calculation of $R_r(\lambda)$ at 30 wavelengths (10 for each of the three primary stimuli). For spectrophotometric analysis, this would require 60 discrete measurements on each water sample, 30 in 'wide scattering collection mode' and 30 in normal mode, followed by considerable computation by computer. This procedure is probably too complex and time consuming for routine use, but may be warranted in the case of special investigations of colour impact on a water body.

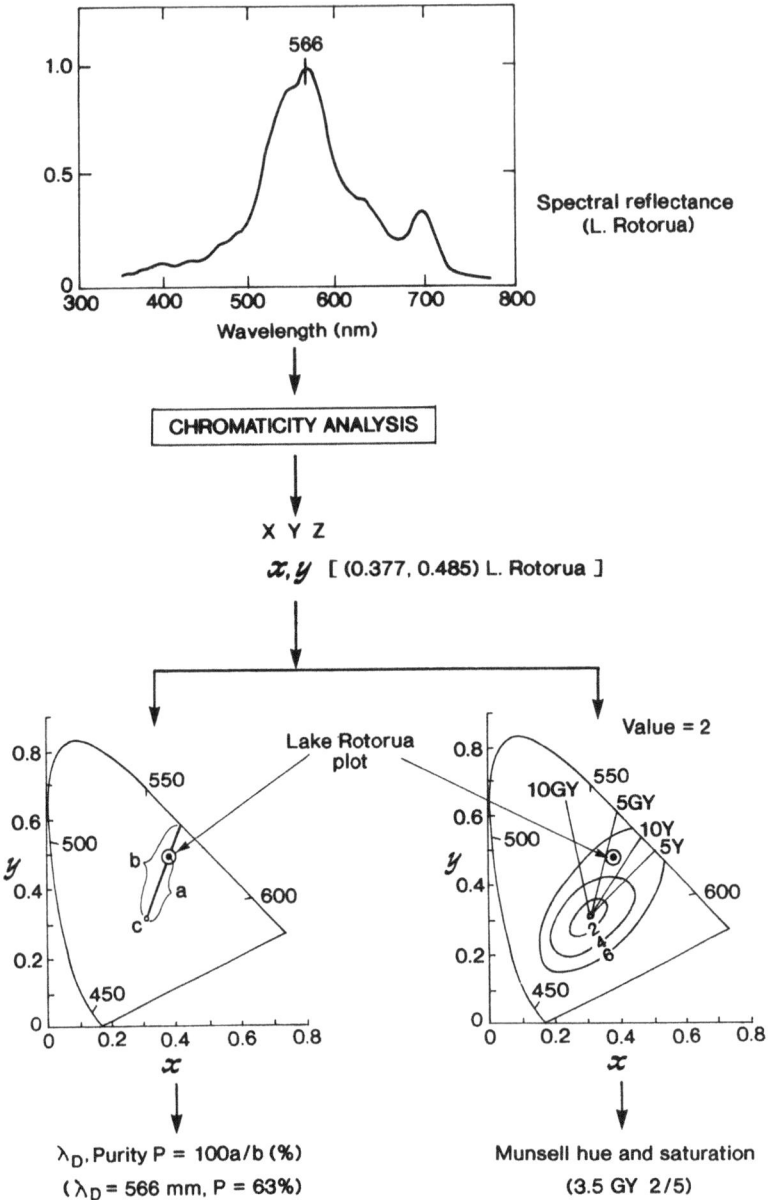

Fig. 3.16. Chromaticity analysis of the spectral reflectance curve of a water. The hue and colour purity can be specified as dominant wavelength, λ_D, and per cent saturation, P, respectively. Alternatively, if the brightness of the water colour is also known, the Munsell colour specification can be given in the form H V/C where H is hue, V is Munsell value (brightness) and C is Munsell chroma (colour purity).

3.5.4 Matching water colour

Measurement and prediction of water colour, as we have seen, are difficult and require special instrumentation and much computation. Fortunately, however, direct matching of colours to colour standards is feasible. Human observers respond very subjectively to colours, but they can match colours to colour standards fairly reliably. This is a standard quality control procedure in the textile and paint industries. In principle it would seem that application of colour standard matching to water colours should be possible.

Perhaps the best known colour standard system is the Munsell system (Anon 1966). This colour system has three ordinates: Munsell hue (symbol H), value (V), and chroma (C). The Munsell hues are arranged on a scale (actually, a closed circle) divided into 100 hue units. A particular Munsell hue might be designated as 60, or alternatively, and more usefully, as 10 BG (the 10th unit of the blue–green hue range). The Munsell value is an index of the apparent brightness of the colour, and ranges from 0 (black) through grey to 10 (white). The Munsell chroma is an index of colour saturation, and extends from 0 for neutral greys to values of 20 or more for the most saturated (spectral) colours. Munsell standards are designated by a notation indicating hue, value, and chroma in the form H V/C. For example, a relatively bright, low saturation green might be 7.5 G 7/4. Munsell standards are available in a convenient book form that is readily transportable.

Davies-Colley *et al.* (1988) showed that water colour in lakes viewed from a boat through an underwater viewer could be matched to Munsell colour standards with reasonable reliability. Fig. 3.17 shows their comparison of the hues of 14 New Zealand lake waters as measured by spectroradiometer and as matched to standards in the 'Munsell book of colour'. (The reflectance curves and CIE chromaticity plots are given in Figs 2.22 and 2.23.) The agreement between observed and predicted hues is fair, but there was an overall bias, since direct matching gave 'bluer' colour specifications than those 'seen' by the spectroradiometer—for reasons which are obscure. However, the correlation is close, suggesting that hue matching has potential as a simple field technique in water resources survey work.

If Munsell standards are not available direct description in the field should be made of water colour as viewed through an underwater viewer such as that shown in Fig. 3.12. The direction of viewing should be noted. Vertical viewing is to be preferred to horizontal viewing because it accords with the definition of intrinsic water colour given in Section 2.6.1, but this requires that the water be optically deep. In practice, we have found that the hue is essentially identical whether viewed vertically or horizontally, although the vertical colour is typically much darker and somewhat purer (more saturated) than the horizontal colour.

Hue of the water should be assigned to one of the hue classes: blue, blue–green, green, green–yellow, yellow, yellow–red (orange), red. A decision should also be made as to whether the water colour is 'pure' or 'greyish', and whether 'bright' or 'dark'. These subjective descriptions of colour purity and brightness are less reliable than the hue descriptors but may still be useful for later interpretation. Evocative terms such as 'milky green' (more precisely, a bright, greyish green) or 'muddy brown' (a bright, greyish orange) may be ambiguous and should not be used.

Appendix 2, Section A2.11, gives more specific detail about field observation and description of water colours.

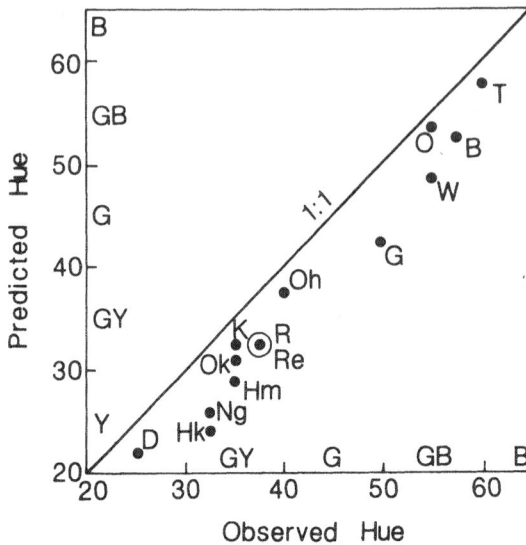

Fig. 3.17. Comparison of Munsell hues calculated from spectroradiometric data (spectral reflectance) with hues matched in the field to the Munsell Book of Color. The numbers are Munsell hue numbers. Y, yellow; GY, green–yellow; G, green; GB, green–blue; B, blue. Point labels are lake codes (as for Figs 2.22 and 2.23). (From Davies-Colley *et al.* (1988), with permission).

3.6 DETERMINANTS OF WATER COLOUR AND CLARITY

An important component of most investigations into colour and clarity problems in natural waters will often be measurement of the concentrations of those constituents contributing to the optical character of the water, especially degraded colour or reduced clarity. After all, it is the light-attenuating constituents of water that are potentially manageable. Measurements of light-attenuating constituents will often need to be carried out on wastewaters as well as the natural water of concern. In this section we consider the measurement of the main natural constituents which contribute to colour and clarity of waters: dissolved yellow substance, mineral suspensoids, phytoplankton and detritus (Table 2.1). Some of these methods can be adapted for quantifying the constituents of optical concern in wastewaters. The optical cross-sections (absorption, scattering, or total beam attenuation per unit concentration) of these constituents are also considered and data presented where available (Table 3.3).

3.6.1 Yellow substance

The only filter-passing constituent of optical importance in waters, as we have seen, is the yellow substance: organic material which is a random polymer of phenolic and aliphatic organic monomers, and is chemically, and often generically, related to soil humic material. 'Aquatic humus' typically constitutes 40–60% of the total dissolved organic carbon (DOC) content of water and 85–100% of the visible absorption of filtrates (Thurman 1985).

The DOC content of waters and wastewaters can be measured using carbon analysis facilities suitable for liquid samples. DOC measurement on wastewaters and very 'dirty' natural waters (DOC >1 g m^{-3}) is fairly straightforward, but it is more difficult to measure DOC at the lower concentrations typical of most 'clearer' natural waters. In any case, the quantity measured is the total DOC present which may be appreciably greater than the DOC of the light-absorbing yellow substance. Natural waters and organic wastewaters contain a very wide variety of dissolved organics, most of which do not absorb visible light significantly, although many absorb strongly in the ultraviolet. Thus yellow substance concentration is only roughly indicated by DOC measurement, in spite of the typical overall correlation of visible light absorption and DOC (e.g. Gjessing 1976). For this reason DOC measurement is not recommended in investigations of optical water quality.

Thurman (1985) has proposed an operational definition of aquatic humus as that fraction of the DOC which can be extracted from natural waters by column chromatography with adsorption on a macro-reticular acrylic resin, XAD-8, under specified standard conditions with the water sample acidified to pH 2.0. The adsorbed material can be recovered by elution from the column of XAD-8 resin with 0.1 M sodium hydroxide solution. Since aquatic humus, so defined, represents most ($>85\%$) of the visible light-absorbing dissolved organics, the recovered DOC can be taken as a measure of the yellow substance DOC. However, there seems to be no compelling reason for making such measurements routinely in water resources management, unless the organic carbon content of the water is of particular interest in its own right, that is, there are wider concerns than optical water quality alone.

Furthermore, the absorption cross-section of yellow substance in terms of organic carbon can be quite variable. Carder et al. (1989) showed that the (variable) fulvic acid fraction of marine aquatic humus has a very much lower absorption cross-section (0.007 m^2 g^{-1} of organic carbon) than the humic acid content making up the remainder of the humic isolate (0.13 m^2 g^{-1}). Thus the absorption cross-section of the whole humic isolate varies greatly depending on the fulvic:humic ratio. This means that measurements of DOC, or even of humic DOC, are of limited value for prediction of optical character of a water.

Here it is recommended that the absorption coefficient at 440 nm (g_{440}, in Kirk's nomenclature) be taken as an index of yellow substance concentration rather than measurement of organic carbon. Thus, rather than relating yellow substance to the DOC or the aquatic humus, yellow substance remains a purely optical concept. The methods discussed in Section 3.2 and detailed in Appendix 2, Sections A2.3–A2.6, should be followed for the measurement of yellow substance absorption coefficients.

3.6.2 Suspended solids

The suspended solids content of water and wastewaters is one of the most common of all water analyses. The suspended solids (SS) analysis is simply performed by filtering a measured volume of well-mixed water sample through a glass fibre filter (e.g. Whatman GF/C) and weighing the filter residue after oven drying at 103–105°C ('non-filterable residue' (APHA 1989)). As with all analytical procedures, it is essential to carry a blank through with each batch of samples to permit correction for analytical artefacts.

As we saw in Section 2.4.2, very often the mass concentration of suspended matter in

Table 3.3. Reported optical cross-sections (PAR-band averages) of natural water constituents

Optical cross-sections (m^2g^{-1})			References	Location, notes
C	B	A		
Mineral suspensoids				
0.4	0.37	0.03	Di Toro (1978)	San Francisco Bay
	0.64 (550 nm)	0.019 (550 nm)	Bukata et al. (1981a)	Lake Ontario
1.2	1.1	0.05	Vant & Davies-Colley (1984)	27 New Zealand lakes
	0.4–2.0		Various studies cited by Di Toro (1978)	Various
0.29–0.48			Peterson (1974)	Marine waters
0.32–0.4			Vant & Davies-Colley (unpublished)	Lakes Rotorua and Karapiro
0.053			Campbell & Spinrad (1987)	Chesapeake Bay
0.025–0.5			Pickrill et al. (1986)	Various (fresh and marine)
0.30	0.29		Vant (1990)	9 New Zealand estuaries
Phytoplankton				
200			Kiefer & Austin (1974)	Marine field studies
	120		Morel (1980)	Marine field studies

			Reference	
110	90		Vant & Davies-Colley (1984)	27 New Zealand lakes
20–130		5–25	Vant & Davies-Colley (unpublished)	12 New Zealand lakes
47–140	44–140	~10	Davies-Colley et al. (1986)	Laboratory cultures of fresh-water algae
90–220	80–220	~10–30	Bricaud et al. (1983)	Laboratory cultures of marine algae
		20–30	Various studies cited by Di Toro (1978)	Various
		7–27	Range from Kirk's (1983) Table 9.1 compilation	Various
		5–14	Atlas & Bannister (1980)	Various field measurements (variation with water colour)
		16	Smith & Baker (1978)	Marine field studies
		30	Di Toro (1978)	San Francisco Bay
180	160	14	Vant (1990)	9 New Zealand estuaries
	30–90 (550 nm)		Privoznik et al. (1978)	Chlorella culture
114 (550 nm)	128 (550 nm)	17 (550 nm); 35 (443 nm)	Bukata et al. (1981a)	Lake Ontario

a water or wastewater is only weakly related to the environmental concern with the sus-pensoids. When, as is typical, the main concern with suspended matter is optical, arguably an optical measurement, such as of light scattering or light beam attenuation, is more appropriate than suspended solids determination. Such optical measurements are also simpler and quicker to perform than SS analysis and can be made continuously. We believe that many routine, but onerous, SS analyses in water management could be avoided, or replaced, by simple optical measurement, if the true objective of the analysis was recognized. However, that said, it is still useful to know SS as well as c or b say, because this permits the average optical cross-section to be calculated—which, in turn, conveys information about the character of the suspensoids.

In Section 2.4.2 we saw that the optical character of mineral solids can be quite differ-ent from that of organic particles. In particular, mineral solids are typically only weakly light absorbing but often account for most of the scattering in natural waters. A simple extension to the suspended solids measurement is often very useful for characterizing SS as regards its organic versus mineral content. The organic fraction of suspended material captured on a glass fibre filter is burnt or volatilized in a furnace. The loss of weight is an index of the organic content known as the *volatile* suspended solids (VSS). APHA (1989) recommends a standard ignition temperature of 550°C, but clay minerals lose structural water at such elevated temperatures (e.g. Mook & Hoskins 1982). A preferred procedure in the presence of clay minerals (which are almost ubiquitous in waters and wastewaters) is to use an ignition temperature of 400°C (and an extended ignition time, e.g. 6 h) to avoid positive bias from loss of structural water.

The residue (ash) remaining after ignition is known as the non-volatile suspended solids (NVSS):

$$SS = NVSS + VSS \qquad (3.31)$$

The non-volatile fraction of the total suspensoids is a useful index of the mineral sus-pended solids content of the water. However, some fraction of the NVSS derives from the ash of phytoplankton and other organic constituents, and ideally the NVSS would be corrected (downwards) to account for this contribution. Unfortunately, such a correction is not straightforward to apply because the inorganic (ash) content of organic material, including phytoplankton, varies widely. The ash content of green and blue–green algae, for example, is mainly the phosphates remaining after ignition—a small fraction, perhaps 5%, of the algal biomass. However, the silica frustules of diatoms can represent 50% or more of the total biomass of these organisms.

The importance of the size of suspended particles as regards their optical properties and environmental behaviour has been emphasized in Section 2.4.2. Unfortunately, mea-surement of the distributions of sizes of particles at environmental concentrations is rather difficult. Standard methods for particle sizing of soils and sediments, such as the pipette method (e.g. Lewis 1984), are suitable only if suspensions of the order of 1000 g m^{-3} can be prepared, and are therefore not generally applicable to natural waters and effluents. Perhaps most suitable for sizing of natural water particles in the clay to silt size range are electrozone particle counters (e.g. Coulter counters) which measure size in terms of the change in resistance depending on the volume displaced by particles as they pass through a narrow orifice (Singer *et al.* 1988). Laser diffraction analysers (e.g. the Malvern range of instruments), which calculate size from the angular distribution of scattering (Singer *et al.* 1988), are useful if samples are sufficiently turbid. Scanning electron microscopy of particles captured on filters can also be used. All these methods require access to expensive instrumentation usually only available in specialist research laboratories, so particle sizing is unlikely to become routine in water management.

Although the optical character of suspended particles in waters and wastewaters varies widely, the attenuation cross-sections of suspensoids are not infinitely variable. There are two main reasons for this. Firstly, grain size distributions of suspensoids in natural waters tend to have modes of the order of 1 µm in size. Coarser particles (e.g. those of coarse silt size or greater, say >10 µm diameter) tend to settle out of the water column. Much smaller particles have huge specific surfaces, S (the total surface area analog of the geometrical cross-section, s; $S = 4s$ for convex particles), and have a strong tendency to agglomerate or flocculate into larger particles under the action of surface electrical forces. Secondly, the attenuation cross-section of a mixture of particle sizes tends to be dominated by those particles in the range of the maximum in Fig. 2.9B (say 0.5–2 µm for mineral particles). The result is that the attenuation cross-section for mineral particles in natural waters varies relatively little, and is typically in the range 0.1–1 $m^2\,g^{-1}$. Some optical cross-sections that have been reported for natural water suspensoids are tabulated in Table 3.3.

3.6.3 Phytoplankton

An ideal measure of phytoplankton biomass would be phytoplankton organic matter or organic carbon—that organic carbon present within algal cells. However, algal cells are always present in natural waters accompanied by other types of organic matter (the particulate detritus and dissolved yellow substance), some of which may be derived from recent algal photosynthesis. Membrane filtration of water samples can be used to remove yellow substance, but the particulate detritus component of the organic carbon cannot be separated easily from the phytoplankton cells. For this reason particulate organic carbon (POC), which can be routinely measured with a carbon analyser, is not often useful as an index of phytoplankton concentration.

If the composition of the phytoplankton is of interest, for example to explain why light attenuation through the year in a particular water body does not correlate well with total algal biomass, microscopic examination may be desirable. Identification of the algal floral composition requires microscopic and taxonomic expertise.

The total phytoplankton biovolume can be calculated if the algal composition data (cell counts for dominant species) are combined with measurements of cell volume. In practice, the dimensions of at least 10 individuals of each species are measured (not including spines and gelatinous envelopes) and volumes are calculated from geometric approximations (Clarke *et al.* 1987). The sizes of planktonic algal cells and cell aggregates vary greatly, so it is perhaps not surprising that the light attenuation by the phytoplankton, expressed per unit biomass, also varies appreciably.

The intracellular concentration of chlorophyll *a* varies appreciably (e.g. more than ten-fold in Table 1 of Morel (1987)), depending on the species and the physiological state of the cells (including their recent light exposure). However, a typical average carbon to chlorophyll *a* ratio is about 40, corresponding roughly to a 1% chlorophyll *a* content in algal biomass (roughly 50% carbon). It is often convenient to regard measurements of chlorophyll *a* as an index of phytoplankton biomass.

Chlorophyll *a* should be measured routinely in investigations of water colour and clarity in waters such as lakes and reservoirs, where the phytoplankton are likely to dominate light attenuation. The main degradation product of chlorophyll *a*, phaeopigment, may be present at significant levels in natural water samples, except under bloom conditions, and should be measured as well as the parent pigment. There are two common, reliable methods for measuring chlorophyll *a*: by spectrophotometer and by fluorometer.

The spectrophotometric method (e.g. Strickland and Parsons 1972) involves the following steps. Firstly, the phytoplankton is captured from the water sample by volumetric filtration onto a glass fibre filter. This should be done as soon as possible after water sample collection; however, chlorophyll on filters can be stored deep frozen and in the dark for a period of months before analysis. The glass fibre filter is then ground in a tissue grinder and extracted with 90% acetone. The acetone extract is clarified by filtration or centrifugation before measurement of the absorbance at 665 nm of the extracted chlorophyll *a* (for which molar absorptivity is well known) in a spectrophotometer. The absorbance is measured again after acidification of the extract, which converts all chlorophyll *a* to phaeopigment. From the absorbance measurements at 665 nm before and after acidification, the chlorophyll *a* concentration and phaeopigment in the original sample are calculated. (Note the shift in the position of the red absorption maximum: 665 nm in acetone compared with the *in vivo* absorption peak at 676 nm.) Recent trends have been towards the use of alcoholic solvents (e.g. ethanol) which are apparently slightly more efficient as extractants than is acetone (Marker et al. 1980).

It is worth noting that direct *in vivo* measurement of chlorophyll *a* by spectrophotometry is possible although it is not as accurate or sensitive as extraction spectrophotometric methods. Direct measurement has the great advantage of obviating the need for solvent extraction and thus is potentially suitable for continuous monitoring. Faust & Norris (1985) reported a method using the second derivative of spectral scans across the *in vivo* red peak of chlorophyll *a* centred on 676 nm. (Using the second derivative avoids interference by other pigments absorbing at 676 nm.)

Fluorometric assay of chlorophyll *a* has some distinct advantages over spectrophotometric methods; in particular a very high sensitivity. This method, which is favoured in marine studies since chlorophyll *a* in marine waters is often low (Nusch 1980), is based upon the fluorescence of chlorophyll *a* at wavelengths around 685 nm when excited with blue light of around 430 nm. Filter fluorometers (e.g. Turner Associates) can be used to measure fluorescence on acetone or alcohol extracts prepared as for spectrophotometry but without any need for clarification. Alternatively, fluorescence can be measured directly on waters (*in vivo* fluorescence) thus permitting continuous monitoring applications or profiling. The *in vivo* fluorescence yield varies with species composition and size, shape and physiological state of the cells, thus correlation of *in vivo* fluorescence with extracted chlorophyll *a* is usually less than perfect. The physiological source of variation can be removed by maximizing fluorescence yield with addition of an electron transport inhibitor, dichlorodiphenyldimethyl urea (DCMU) (Slovacek & Hannan 1977). Fluorometric assay of chlorophyll *a in vivo* is unsatisfactory in the presence of cyanophytes which differ greatly in fluorescence response from other components of the phytoplankton (Vincent 1983a).

The absorption and scattering cross-sections of phytoplankton are of considerable interest for the purpose of predicting change in clarity of waters with change in trophic state. Most of the absorption associated with phytoplankton is by photosynthetic pigments so we expect an overall correlation of absorption with the primary photosynthetic pigment chlorophyll *a*. Most of the available data on the absorption cross-section of phytoplankton are actually in terms of the attenuation coefficient for PAR, that is, an irradiance attenuation cross-section, obtained as the slope of a plot of diffuse attenuation against chlorophyll *a*. (Kirk (1983, Table 9.1) summarizes some reported values.) The irradiance attenuation cross-section is always greater in magnitude than the absorption cross-section because of the influence of scattering which increases the effective path length in water and thus the probability of absorption.

The limited data that have been reported from which absorption cross-sections can be inferred for phytoplankton are mainly for the PAR band (Table 3.3). Some variation is expected with water colour and, as we saw in Section 2.4.3, with variation in pigment composition of the algal flora and in cell or cell aggregate size ('packaging').

3.6.4 Detritus
Detritus is ideally measured as particulate organic carbon (POC), corrected for the organic carbon present in algal cells. The POC in the phytoplankton is typically smaller than the detrital POC in natural waters, except under algal bloom conditions. Thus a constant average value of the carbon to chlorophyll *a* ratio of about 40 can be assumed so that the detrital POC can be estimated:

$$POC_{det} = POC - 40\,Chl\,a \qquad (3.32)$$

Obviously the precision is very poor when the second term on the right is large.

The POC analysis is best carried out using a special-purpose carbon analyser (APHA 1989). The instruments using high temperature catalytic combustion in a stream of oxygen gas, with detection by infrared analyser, are favoured for accuracy and reliability. However, wet combustion by strong oxidizers, such as dichromate, and ultraviolet digestion have also been used to convert organic carbon to CO_2. Other means of measuring the CO_2 produced include gas chromatography. Methods are reviewed by Anon (1979).

If a carbon analyser is unavailable, detrital organic matter can be quantified by loss of weight on ignition of suspensoids in a laboratory furnace (VSS analysis). This measurement is carried out as an extension to the usual suspended solids assay as discussed in Section 3.6.2.

The optical cross-sections (specific absorption and scattering coefficients) of detritus are poorly known and probably rather variable because detrital material occurs in a wide range of physical and chemical states. However, the shape of the absorption spectrum of detritus is known to be similar to that of yellow substance, that is, exponential shaped, but otherwise featureless, and of lower spectral slope (e.g. Bricaud & Stramski 1990). It is difficult to generalize further about the optical properties of detritus but, fortunately, explicit consideration of this component is not necessary for many purposes in water management.

3.7 OPTICAL MONITORING OF WATER QUALITY

If the optical cross-section of a particular constituent of water or wastewater is known to be fairly near constant, at least locally, it may be feasible to monitor the concentration of that constituent optically. The constituents of natural water that can be monitored optically, at least in principle, are the suspended solids, the organic content and the phytoplankton biomass. These constituents can also be measured remotely using optical sensors mounted on satellites or aircraft to detect backscattered natural light ('passive' sensors) or backscattering from their own laser light source ('active' sensors). Whilst the remote sensing of suspended solids has been most successful historically, remote sensing of chlorophyll a is perhaps of greater importance because of its potential for studying spatial and temporal variation in biomass and productivity of phytoplankton over wide areas of the sea or large lakes.

3.7.1 Suspended solids

Since scattering of light in waters is almost entirely attributable to suspended matter, scattering measurements can be used to monitor suspended solids in waters or waste streams. Such measurements can be performed at a small fraction of the cost of performing suspended solids assays. In any case, suspended solids measurements only yield discrete data whereas an optical sensor can measure continuously, thus providing a complete picture of the variation with time, or with space, of the suspended matter.

On-line nephelometers and scattering sensors are made by Partech Electronics Limited, Hach Chemical Company and Sigrist Photometer Limited for example. Some of these instruments require pumping from the water or waste stream and, being consumptive of power, are best for applications where mains electrical power is available as in effluent quality monitoring. Backscattering instruments, which measure light scattered back from the water, have the advantage that no optical windows are in contact with the water, thus avoiding fouling problems. However, backscattering cross-sections are often more variable than forward scattering cross-sections, so these instruments may not be suitable in situations where particulate composition is highly variable.

For suspended solids monitoring in natural water, diffuse transmittance sensors have been developed, such as the Partech 'twin gap' range of sensors. The signal from this sensor is the ratio of diffuse light transmittance over two different path lengths. This ratio is independent of window fouling so long as the transmittance of both paths is equally reduced by the fouling. These instruments effectively measure a diffuse attenuation coefficient:

$$k = \frac{\ln(\Phi_1/\Phi_2)}{r_2 - r_1} \tag{3.33}$$

in which Φ_1 and Φ_2 are detected light fluxes at the ends of paths r_1 and r_2. Because these instruments measure transmittance of diffuse light, k is lower than the beam attenuation coefficient, c. Correlation of k with suspended solids is fair so long as the grain size range and composition of the suspended solids are not too variable. In highly absorbing waters (low b/a ratio) the diffuse attenuation is strongly dependent on absorption as well as scattering, and thus good correlation of k with suspended solids cannot be expected unless the absorption is mostly associated with particulates and covaries with SS.

Measurement of total scattering rather than diffuse transmittance (absorption plus forward scattering) or fixed angle scattering would be preferable for suspended solids monitoring. The scattering efficiency is essentially constant (=2) for particles appreciably larger than the wavelength of light (van de Hulst 1957), which includes most suspended matter in most waters and wastewaters. Thus total scattering in water is proportional to the total projected (cross-sectional) area of particles per unit volume (in $m^2 m^{-3}$). Total scattering can be measured as beam attenuation at wavelengths for which absorption by particulates is low, such as in the near-IR. No optical sensor can accurately monitor suspended solids of very fine and variable grain size, although in principle an X-ray transmittance principle (as is used in the Micromeretics Sedigraph 5000D instrument for measuring grain size distributions of sediments (Stein 1985)) would be a solution. X-rays have much shorter wavelengths than visible light, so the attenuation efficiency factor, Q_c, is constant at 2, even for those particles much smaller than the wavelength of light.

3.7.2 Organic matter

We have seen that of the total amount of organic matter appearing in water (and organic wastewaters) some may be present in algal cells and is best considered separately (linked to the chlorophyll *a* content). A further fraction is non-absorbing (and probably insignificantly scattering) of light, leaving a large but variable fraction which is strongly light absorbing (yellow substance and most of the particulate detritus). In principle the total detrital organic carbon (TOC) of a given water could be monitored by light absorption measurements if the non-absorbing TOC fraction and the phytoplankton fraction are either small or relatively constant. In practice, although a correlation between total absorption and TOC can usually be obtained for a given water, it is often relatively weak ($r^2 < 0.5$) because the non-absorbing fraction may be large or variable. Also, the character, and thus absorption cross-section, of the absorbing fraction of the TOC may vary, contributing further noise.

So-called 'colour monitors' or absorptiometers have been developed to measure continuously total absorption or filterable absorption of water or waste streams. These measure the absorbance of the water or wastewater, sometimes at two different wavelengths, one in the visible (e.g. 400 nm) and one in the near-IR (to correct for scattering). The instrument response is typically used directly as an operational control parameter rather than being converted to TOC (or DOC), although this is feasible where correlation is good.

An example of such an instrument which is commercially available is the Aztec colour monitor developed by the Water Research Centre (WRc) of Medmenham, England (Briers & Wheeler 1986). This instrument measures the absorbance at 405 nm of filtered water sampled on line. Filtration is achieved by pumping through a prefilter and hollow fibre filter cleared by periodic backflushing. The instrument can be used on unfiltered water, in which case diffuse attenuance rather than absorbance is measured. The Aztec colour monitor was designed primarily for raw water colour monitoring and process control in potable water supply.

For the monitoring of TOC in waters and wastewaters, ultraviolet wavelengths are more suitable than visible wavelengths for two main reasons. Firstly, many organic substances do not absorb in the visible region, but most absorb in the UV. Secondly, absorption is much higher in the UV than in the visible, so sensitivity is better at UV wavelengths and scattering is less likely to interfere.

Many workers have suggested the use of UV absorption measurements to monitor organics in both waters (e.g. Banoub 1973; Foster & Morris 1974; Reid *et al.* 1980; De Haan *et al.* 1982) and wastewaters (Dobbs *et al.* 1972; Briggs *et al.* 1976). Usually comparatively short wavelengths (around 250–280 nm) give better correlations with TOC or DOC than longer UV wavelengths. For example De Haan *et al.* (1982) obtained better correlations of DOC with absorbance at 250 nm than at 365 nm on filtered waters from humic-stained Lake Tjeukemeer, Netherlands. Wavelengths shorter than about 240 nm cannot be used, however, because of absorption by inorganic species such as NO_3^- and Br^-. Sometimes correlations of UV absorbance with TOC or DOC in natural waters can be improved by de-seasonalizing the data (Reid *et al.* 1980).

Commercial UV colorimeters developed as detectors for chromatography applications have sometimes been used as on-line monitors of waters and wastewaters. These colorimeters use the 254 nm line, isolated using optical filters from the spectrum of low pressure mercury ultraviolet lamps. Briggs *et al.* (1976) described an instrument designed at the Water Research Centre, England, using such a principle for the on-line monitoring of TOC in sewage and other organic wastewaters. The 510 nm line in the low pressure mercury lamp spectrum was used to correct the 254 nm absorbance for scattering. Biofouling of the optical silica windows of this instrument was not a problem, probably because of the biocidal nature of the UV radiation. Briggs *et al.* (1976) showed that absorbance at 254 nm (corrected for scattering) correlated closely ($r = 0.83$–0.93) with TOC of sewage, but also correlated (less closely and in a non-linear pattern) with biochemical oxygen demand (BOD). BOD, which expresses the oxygen-demanding load on receiving waters, is generally of more concern than TOC of sewage, so even rough correlation with absorption at 254 nm suggests valuable continuous monitoring applications in sewage treatment. We do not know to what extent the WRc instrument or similar UV monitors have been deployed in waste treatment plants and what operating experience has now been accumulated.

In some situations, for example in effluent from waste stabilization ponds, it might be desirable to monitor algal biomass continuously with time. In principle the spectral absorption by phytoplankton could be utilized to estimate chlorophyll *a* concentration roughly. For example the ratio of absorption at 440 nm, near the Soret band absorption peak of chlorophyll *a*, to that at 550 nm could be correlated with chlorophyll *a* content. Alternatively the 676 nm chlorophyll *a* red peak could be used for absorptiometric monitoring. *In situ* measurement of chlorophyll *a* by fluorometry as discussed in Section 3.6 might be the simplest method for continuous monitoring of time variation in algal biomass.

3.7.3 Remote sensing

Where variation over an area, rather than at one point in time, is of paramount importance, remote sensing methods come into their own for the monitoring of phytoplankton biomass and other water quality variables. To date there has been relatively little application of remote sensing to the water quality of inland waters, possibly because most inland waters are small in scale and the spatial resolution (pixel size) of most imagery is too coarse. Also, flyover of satellites is often too infrequent (and cloud cover often degrades images in any case), and special-purpose flights of aircraft are too expensive for the water industry. This situation is changing, however, and in the future we can expect to see significant applications of remote imagery to water resources work, including water quality, as the spectral and spatial resolution of imagery increases, and image processing hardware and software improve. Water managers need to keep a watching brief on developments, particularly in marine studies. In countries such as Canada, problems of access over vast land areas, combined with large numbers of water bodies, have given remote sensing a valuable role in water quality studies. Useful reviews of principles of remote sensing of water features, and applications in water resources work, are given by Gordon & Morel (1983), Kirk (1983) and Hilton (1984). Here we give only the briefest overview.

The only natural water quality constituents which can be remote sensed directly using optical imagery are suspended solids, chlorophyll a in phytoplankton, and organic carbon (yellow substance and detritus). To this list we can add some features of the aquatic environment which may be of interest in water quality studies such as bathymetry, submerged vegetation (macrophytes) and water surface effects (e.g. waves).

In principle, remote sensing can be used to estimate concentrations of all optically significant water constituents if their absorption and backscattering cross-sections (or the wavelength dependence of backscattering) are known, and imagery is available in a sufficient number of different spectral bands (Morel 1980). The general method involves solving a system of simultaneous equations, each giving the ratio of irradiance reflectance in two appropriate bands (as calculated from the ratio of the brightness of the same pixel in two different images) as a function of the backscattering and absorption cross-sections multiplied by the (unknown) concentrations of the constituents. For example, Bukata et al. (1981a,b,c) showed that spectral measurements at 443, 520, 550 and 670 nm (chosen at the centres of bands imaged by the Coastal Zone Colour Scanner—CZCS—on the Nimbus 7 satellite) could be used to estimate four constituents with independently measured optical properties in Lake Ontario: suspended solids, chlorophyll a, detritus (POC), and yellow substance (DOC). However, such theoretically well-founded approaches may not work well in practice for a variety of reasons as discussed by Morel (1980). In particular, real imagery is often degraded by surface reflection, sun glitter, white light scattered from bubbles in the 'white caps' of breaking waves, and atmospheric absorption and scattering between the water and remote sensor. In any case the optical cross-sections of water constituents are typically somewhat variable in time and space (e.g. with change in algal floral composition). For these and other reasons most remote sensing efforts have relied on empirical correlations of pixel brightness with 'ground truth' measurements of the constituent of interest in samples obtained by surface vessels.

Since suspended mineral solids are usually intensely scattering, but do not significantly absorb light, high concentrations of such solids usually render water bright in colour. Sometimes correlations are obtainable between pixel brightness and SS concentration. For example, Munday & Alfoldi (1979) obtained fair empirical correlations of red LANDSAT band radiance ratios and suspended solids in the Gulf of Fundy, Canada. A number of similar studies of suspended solids concentrations are reviewed by Kirk (1983). Even where accurate values of SS cannot be obtained, qualitative patterns of solids dispersal in surface waters are often of interest for water mass tracing and flow visualization.

A promising method for remotely sensing chlorophyll a makes use of the fluorescence of this constituent. As was discussed in Section 3.6.3, phytoplankton (other than blue–greens) fluoresce at about 685 nm when exited at around 440 nm. Usually the natural sunlight-induced fluorescence is largely swamped by backscattered light, but methods have been developed and undergone successfully trials which employ lasers flown on the remote sensing platform to induce fluorescence. For example, Bristow *et al.* (1981) used a pumped dye laser (λ = 470 nm) to induce fluorescence which was sensed by a telescopic radiometer. Both laser and radiometer were flown in a helicopter a few hundred metres above the water in which chlorophyll *a* was to be remote sensed.

Although not strictly water quality variables, bathymetry and biomasses of submerged macrophytes in waters are often of concern in water management. In principle these features can be studied by imaging in spectral bands in which there is sufficient light penetration (clarity) in the water body of interest (usually blue or green light) (Hilton 1984). As an example of such application of remote sensing methods, Kuittinen & Sucksdoroff (1984) used colour aerial photography to estimate the depths of water (up to 2–3 m) available for navigation in Finnish lakes. The method was limited by bottom reflectance and water clarity.

4

Colour and clarity in different aquatic environments

4.1 INTRODUCTION

The science of aquatic optics—the study of the behaviour of light in water bodies—has developed primarily through oceanographic studies. In a special issue of *Limnology and oceanography* devoted to this subject (Spinrad 1989), most of the papers reporting experimental work dealt with marine measurements. The optics of ocean waters is of great scientific interest, and considerable practical importance as regards oceanic primary production and the remote sensing of phytoplankton crops by optical sensors, but it is unlikely that mankind will ever be able to influence greatly the optics of ocean waters. However, the optical character of inland waters and estuaries can be, and often is, affected directly or indirectly by human activity. Thus, while we recognize the all-pervading influence of the optical oceanographers (e.g. Jerlov 1976) on the development of aquatic optical science, this chapter confines itself to discussing the optical character of inland waters: lakes, rivers and estuaries.

Besides ocean waters, one other important category of surface water is omitted, namely springwaters. Freshwater springs are often visually very clear and typically blue–green coloured, that is they are optically pure waters. Optically pure spring waters are represented in New Zealand, the best known being Waikoropupu Springs in the north-west Nelson area of the South Island. This is New Zealand's largest freshwater spring system (total discharge about 11 $m^3 s^{-1}$, reportedly the largest springflow in the Southern Hemisphere (Michaelis 1974)). Like most freshwater springs elsewhere in the world, notably those in Florida (Odum 1957), Waikoropupu Springs are very clear (black body visibility exceeding 60 m) and blue–violet hued, with optical characteristics apparently similar to those of the clearest ocean waters (authors' unpublished data).

Geothermal springs, unlike freshwater springs, often exhibit 'unusual' and variable water appearance, including highly turbid blues, iridescent greens and yellows, and greys (e.g. Howard-Williams & Vincent 1984). These waters are also represented in New Zealand, occurring in the various volcanically active areas of the country, notably the Central Volcanic Zone of the North Island. The optics of these waters has apparently been little studied, although Howard-Williams & Vincent (1984, 1985) reported absorption spectra and light attenuation measurements in some geothermally influenced lakes and speculated that particulate sulphur might be responsible for their high turbidity and bright appearance.

4.1.1 Lakes

As comparatively large standing bodies of water, lakes have tended to be approached with experimental methods adopted from optical oceanography. Since pioneering work began on lakes in Wisconsin (e.g. Birge & Juday 1929; James & Birge 1938) and England (e.g. Pearsall & Ullyott 1933, 1934), certain aspects of aquatic optics have been increasingly studied in lakes. In the main this has arisen from a desire to understand better the underwater light climate for plant growth (e.g. Talling 1971), but considerations of water appearance have also been important (e.g. Smith *et al.* 1973). Section 4.2 of this chapter reviews our understanding of the important optical properties of lakes, and shows how these are affected by the constituents present in lakewaters.

Much of the early work on lake optics was reviewed by Hutchinson (1957, Chapter 6), while Kirk's (1983) book is more up to date and, because of the intervening advances in our understanding of the theory of aquatic optics (e.g. Jerlov 1976), is more rigorous in its approach. The optical properties of New Zealand's many and diverse lakes have been increasingly well studied (e.g. Howard-Williams & Vincent 1984, 1985; Vant & Davies-Colley 1984), so the characteristics of lake waters generally are illustrated by reference to these publications. Note that a practical guide to determining and understanding aspects of the colour and clarity of lakes already exists (Davies-Colley 1987a). This guide, part of a larger work on the management of lakes, complements the material presented here by providing advice on undertaking surveys of lakewater appearance and identifying the causes of poor water clarity.

4.1.2 Rivers

In contrast to lakes, rivers are comparatively shallow water bodies characterized by uni-directional currents. These fundamental differences between standing and flowing waters have some major implications for optical character and, consequently, water colour and clarity. It would probably be fair to say that rivers are optically more complex than lakes. Only in relatively large rivers with highly light-attenuating waters is it possible to ignore the influence of the bed and banks, and perhaps also that of water-marginal features such as riparian vegetation, on the light field in the water. Also, currents in rivers maintain particulate matter in suspension, whereas in lakes suspensoids tend to settle out. This optical complexity was recognized by Westlake (1966) who regarded the light climate of rivers as having some of the features of forests (due to plant shading), some of oceans and lakes (due to attenuation of light in the water column) and some peculiar to rivers such as bed and bank effects.

Part of the charm and attraction of rivers as scenic entities is that the view changes continually as one walks up or down the bank. Indeed, the whole scenic character of river waters typically changes over their length from their headwaters to the sea. Headwater rivers in mountainous terrain are often characterized by whitewaters—white from the spectrally non-selective scattering by air bubbles entrained in rapids or falls. Sometimes the intrinsic hue of river water is enhanced near areas of white water owing to selective absorption in the overlying water of the white light scattered from the bubbles (Plate 8).

In quieter reaches downstream, the surface of river waters is disturbed only in riffle areas. Here the glitter of sunlight from the disturbed water surface tends to mask intrinsic water colour (Section 2.6.1). In runs with undisturbed water the bed features, including sand waves, bedrock shelves, gravel or cobble substrate and macrophytes or filamentous periphyton, may be visible in sufficiently clear and shallow (i.e. 'optically very shallow') water. In pools the water, even in clear streams, may become optically deep and the intrinsic colour of the water can be seen (Plates 9A and 9B). The intrinsic colour may be most noticeable where surface-reflected light is low as in the reflected image of dark overhanging banks and vegetation.

Relatively little work appears to have been carried out on the optical properties of river waters, certainly less than in the sea or lakes. There may be three main reasons for the comparative neglect of the optical properties of rivers. Firstly, because rivers are typically optically shallow, many workers may have considered that light availability (and visual range) in rivers is restricted by limited physical size rather than light attenuation in the water. That is, light transmission in river waters may be regarded as a non-problem. Secondly, it is difficult to make meaningful ambient light measurements when the light field is distorted by the boundaries—as typically occurs in rivers. Often, only instruments with their own light source, such as transmissometers, can be usefully deployed in rivers. Finally, rivers vary greatly with time as regards their water quality in general and optical quality in particular, mainly with state of flow, making it difficult to optically characterize individual river waters.

In New Zealand as elsewhere, measurements related to the optical character and visual quality of river waters have historically been restricted, almost exclusively, to measurements of nephelometric turbidity made as part of general water quality survey. The development of the simple, 'low-tech', black disc method for measuring water clarity (Davies-Colley 1988a) has made possible the routine measurement of visual range, and the estimation of beam attenuation coefficients, in shallow river waters at remote sites (Section 3.4.2, Appendix 2.6). The same information could be obtained by use of suitably designed transmissometers and much more widespread deployment of sensors using a beam transmittance principle is to be expected in future. These sensors have a major advantage over human observations of water clarity in that they are amenable to continuous measurement (including at night) and can be expected to greatly increase knowledge of the optics of river waters.

4.1.3 Estuaries
'To the casual observer, suspended material is one of the most obvious features of estuaries. They tend to be muddy; muddier than the rivers flowing into them or the sea beyond.'   McCave (1979)

Light attenuation in estuaries has been studied much less intensively than in the sea, with some notable exceptions (e.g. Postma 1961; Thompson *et al.* 1979; Pierce *et al.* 1986; Di Toro 1978; McPherson & Miller 1987; Gallegos *et al.* 1990). In New Zealand, some information is available on the optics of several shallow estuaries in the north of the North Island (van Roon 1983; Vant 1990), and for Pelorus Sound, a drowned river valley in Marlborough (Vincent *et al.* 1989). An outline of these results is given in Section 4.4.

Mixing of freshwater and seawater in estuarine basins is usually incomplete.

Furthermore, resuspension of bottom sediments by tidal and wave action in shallow intertidal and subtidal regions of estuaries is common (e.g. Cloern *et al.* 1989). These processes cause marked spatial and temporal variations in the concentrations of light-attenuating suspensoids. In addition, recirculation of suspended matter near the limit of salt intrusion can trap these particulates in a 'turbidity maximum'. As a result optical properties usually vary spatially as well as temporally in estuarine waters, making the characterization of estuarine colour and clarity a complex task.

Section 4.4 considers the sources of the optically dominant mineral suspensoids in estuaries, and the causes of spatial variability of suspended solids concentrations. A brief introduction to some important aspects of estuarine optics, illustrated with the results from New Zealand studies, is given. Relationships describing the effects of suspensoids and phytoplankton on beam attenuation in estuaries are given. Finally the consequences of light attenuation for underwater plant growth in estuaries are examined.

The chapter concludes with a comparison in Section 4.5 of the optical character of lakes, rivers and estuaries. The contrasts in optical quality of these different aquatic environments, and the contrasts in temporal variability, are emphasized.

4.2 LAKES

4.2.1 Sources of light-attenuating constituents

Lakewater colour and clarity are ultimately determined by three factors: catchment type, catchment land use and lake morphometry (e.g. Duarte & Kalff 1989). These regulate the supply to lakes of the various light-attenuating constituents which, directly or indirectly, determine their colour and clarity (Section 2.4, Table 2.1). Table 4.1 lists the relevant constituents, identifying typical sources of them.

Where degraded colour and clarity are encountered in lakewaters, the cause is often the excessive growth of phytoplankton, in response to elevated catchment loads of the plant nutrients nitrogen and phosphorus. Both diffuse runoff from agricultural land and point sources of wastewaters are important. Other constituents of wastewaters, particularly industrial wastewaters, can also cause problems. For example, mineral suspensoids in the wastewaters from open-cast coal mining markedly affected the appearance of Lake Waahi (Kingett 1984, Section 6.3), and highly coloured effluent from a Kraft pulp mill is discharged to Lake Maraetai, an impoundment on the Waikato River (Davies-Colley 1987b).

Catchment soil erosion and bottom sediment resuspension within lakes may also give rise to high levels of mineral suspensoids in lakewaters. Erosion varies widely depending on factors such as rainfall intensity and volume, soil type, vegetation cover and the intensity and nature of land use within the catchment (see Section 5.5). The degree of sediment resuspension within lakes reflects the degree of (submerged) vegetation cover and lake morphometry, particularly depth and area (as an indication of wind fetch). These factors control water movement at the lake bed, and thus the resuspension and transport of fine-grained sediments (e.g. Hoare & Spigel 1987). Resuspension generally becomes less important as lake depth increases but increases with lake surface area and thus wind exposure (e.g. Davies-Colley 1988b).

The specific effects of these various constituents on the optical properties of lakewaters are described later in this section.

Table 4.1. Sources of constituents which affect the colour and clarity of lakes

Constituents (Concentrations encountered in New Zealand lakes)	Sources	Colour	Clarity
Dissolved Yellow substance (g_{440}: <0.1–8.1 m^{-1})	Drainage from peatland and forest catchments; lesser amounts from the decay of aquatic plants.	×	
Particulate (Total suspended solids: 0.2–87 g m^{-3}) Mineral suspensoids (<0.1–36 g m^{-3})	(1) Catchment erosion.		×
	(2) Point discharges (e.g. mining wastewaters).		×
	(3) Resuspension of bottom sediments.		×
	(4) Phytoplankton skeletons (usually trivial amounts).		×
Phytoplankton (chlorophyll *a*: 0.2–236 mg m^{-3})	Produced by growth within lakes (in response to loads of plant nutrients from their catchments).	×	×
Detritus and other organic solids	(1) Produced within lakes by the decay of aquatic plants and other living material (includes bacterial biomass).	×	×
	(2) Derived from catchment soils.	×	×

Where measured, the range of concentrations found in New Zealand lakes is shown (from Howard-Williams & Vincent (1984) and Vant & Davies-Colley (1984)).

4.2.2 Traditionally measured optical properties
Secchi depth
Water transparency as measured with the Secchi disc has long been a classic optical property of lakewaters. It was probably the optical property first formally measured in a lake, is the one which has been most widely measured (in New Zealand and, probably, elsewhere), is simple to measure, and its significance is readily understood by the lay public. Within a few years of the introduction of the Secchi disc to marine studies (see

Tyler (1968) for a discussion), transparency was measured in 1873 in Lake Tahoe, North America (Goldman 1988). Interestingly, in this particularly clear lake, Secchi depths (z_{SD}) measured 100 years later (Smith et al. 1973) were similar to the 33 m originally recorded, but, more recently, transparency has declined (z_{SD} ~ 25 m (Goldman 1988)). Other early records of Secchi depths in lakes were described by Hutchinson (1957).

In the clearest lakes Secchi depths can reach 40–45 m (Hutchinson 1957; Larson 1972; Goldman 1988), rather less than the maximum reported (~80 m in Antarctic seawater (Gieskes 1987)). However, Hutchinson (1957) concluded that lakes in which z_{SD} >30 m are 'very rare'. The absorption and scattering of light by certain dissolved and particulate constituents of lakewaters cause values typically to be rather less than these. In principle there is no lower limit to z_{SD}, although relative measurement errors increase rapidly for z_{SD}< 0.2 m because the disc interferes with the light field (Sections 3.4.1, 3.4.2).

One of the earliest demonstrations that Secchi depths in lakes are related to the concentrations of constituents present in lakewaters was that of Tressler et al. (1940) who found that seasonal variation in the levels of suspended particulates caused Secchi depth to vary. Carlson's (1977) trophic state index for lakes was based on the assumption that these suspensoids are generally mostly phytoplankton related. However, it was quickly pointed out that non-algal light attenuation varies markedly between lakes, and could be considerable (Brezonik 1978; Lorenzen 1980; Megard et al. 1980).

Many Secchi depth measurements have been made in New Zealand lakes. Results are available for most (141) of the 165 lakes included in the 'Inventory of New Zealand lakes' (Livingston et al. 1986). Altogether Secchi depth has been reported at least 1800 times, most often in North Island lakes (86% of measurements). In many (83%) of the 64 South Island lakes for which data were reported, measurements have been made only once or twice. Similarly sparse records were encountered in fewer (36%) of the 77 North Island lakes. Conversely, Secchi depth has been measured on more than 10 occasions in 45% of the North Island lakes and in 5% of those in the South Island. Lakewater clarity is thus better documented in the North Island than the South Island.

The cumulative frequency plots in Fig. 4.1 show the distribution of the mean Secchi depths in the lakes of each island. Secchi depths in South Island lakes are generally greater than those in North Island lakes. The median value for the South Island dataset (5.0 m) is more than twice that for the North Island (2.2 m). The majority of lakes in both islands are of moderate clarity or better (z_{SD} >1.5 m).

Sterner (1990) collated results from more than a thousand lakes in Ontario, showing that Secchi depth increased with increasing lake depth. The same applies to New Zealand lakes (Fig. 4.2). In both islands it is the deeper lakes which tend to be clearer. This is probably because lake size, as broadly measured by maximum depth, is correlated with factors such as slower flushing and thus greater likelihood that particulates can settle out of the water column, less likelihood of resuspension of particulates from the lake bottom, and, in this dataset at least, lower nutrient status so that phytoplankton biomass is low. Deeper lakes are more common in the South Island (Fig. 4.2), which explains in part why lakes there tend to be clearer. The small, turbid lakes in the North Island are often found in the developed lowlands (altitude <80 m above sea level), but this type is uncommon in the South Island (where the median altitude of the 64 lakes is 355 m above sea level).

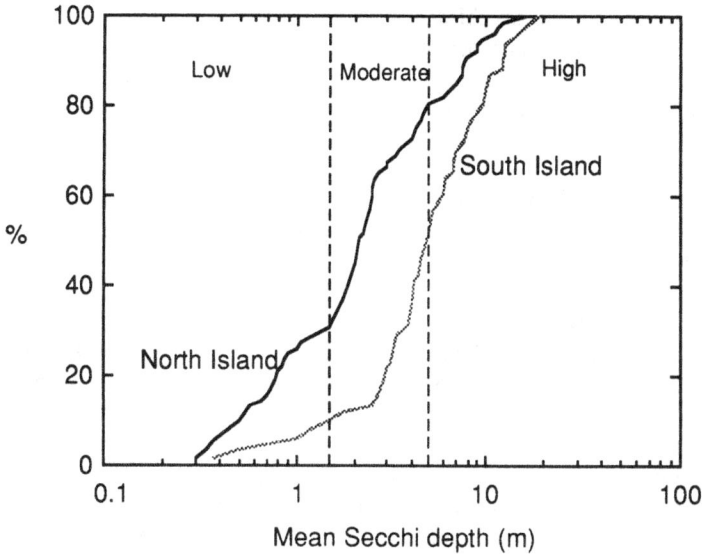

Fig. 4.1. Cumulative frequency distributions of Secchi depths in North and South Island lakes. (Data from Livingston *et al.* (1986).) Boundaries of the water clarity classes suggested by Davies-Colley (1987a) are also shown (at 1.5 and 5 m).

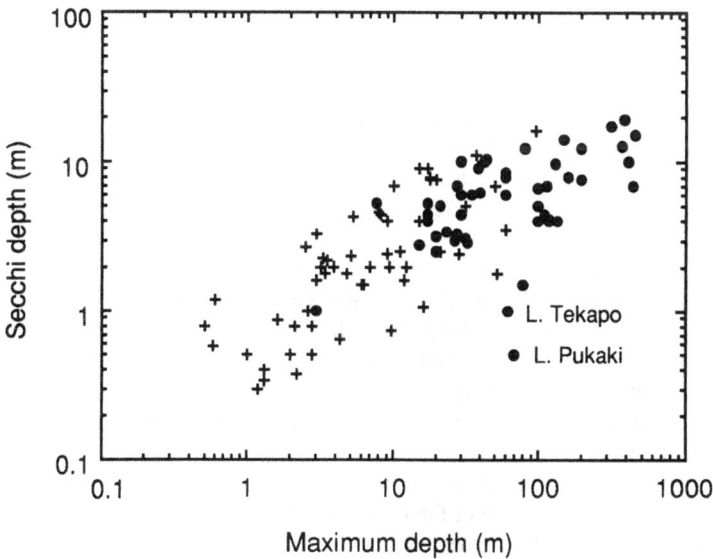

Fig. 4.2. Secchi depths and maximum water depths in North Island (+) and South Island (●) lakes. (Data from Livingston *et al.* (1986).) Lakes Tekapo and Pukaki (Plate 2), two deep glacial lakes with low visual clarity owing to glacial flour, are indicated.

Seasonal variability in Secchi depth can be marked. Values in a large mesotrophic European lake varied by an order of magnitude (1.3–14 m) as phytoplankton biomass fluctuated (Tilzer 1988), although this degree of variability is unusual. By contrast, spatial variability within particular lakes is usually relatively low, because lakes tend to be horizontally well mixed. Occasionally mixing may be incomplete, as in Lake Whangape for example (Davies–Colley & Vant 1988). The dense macrophyte beds then present in this lake (Section 6.3) restricted the mixing of optically distinct water masses following heavy rain, causing Secchi depth to vary by an order of magnitude across the lake (0.13–1.3 m).

Irradiance attenuation
Shortly after Poole & Atkins (1926, 1929) pioneered use of an underwater photometer to measure light attenuation in the sea, similar measurements were made in lakes in Wisconsin (Birge & Juday 1929, 1932) and England (Pearsall & Ullyott 1933, 1934). The latter study demonstrated that attenuation in Lake Windermere was dependent on phytoplankton density. By using a photometer equipped with coloured filters Utterback *et al.* (1942) were able to show how attenuation varied with wavelength in Crater Lake (Plate 1). The subsequent use of spectroradiometers provided more detailed information on the spectral attenuation of diffuse light in this lake (Smith & Tyler 1967; Smith *et al.* 1973). These instruments have since been used increasingly in lakes elsewhere (e.g. Jewson 1977; Kirk 1979; Kishino *et al.* 1984; Bowling *et al.* 1986; Weidemann & Bannister 1986).

After Secchi depth, irradiance attenuation is the next most widely measured optical property of New Zealand lakes (and, probably, of lakes in general). Attenuation was measured in a total of 25 lakes prior to 1975, generally using yellow–red-sensitive photocells (e.g. Green 1975). During 1980–1986 the attenuation of irradiance was measured with sensors of photosynthetically available radiation (PAR) in 59 lakes, including 14 of those in which measurements had been made previously (Howard-Williams & Vincent 1984; Vant & Davies-Colley 1984, 1988; Vant *et al.* 1986; Davies-Colley & Vant 1988). Attenuation was generally measured only once in each lake, except for 12 North Island lakes where conditions were monitored at monthly intervals for a year (Vant *et al.* 1986). Overall this means that irradiance attenuation has been measured in about 70 New Zealand lakes, three-quarters of which are located in the North Island.

In general, irradiance attenuation is lower in the deeper New Zealand lakes (Fig. 4.3), as has been inferred for Ontario lakes from their Secchi depths (Sterner 1990). Note, however, that for any given lake size (i.e. maximum depth), K_d can vary by at least an order of magnitude. The association between size and irradiance attenuation in Fig. 4.3 is only weak, being moderated by the fact that constituent composition of the lakewaters can vary markedly between similarly sized lakes. For example Lakes Haupiri and Kaiiwi (see Fig. 1.1 for locations) are of similar depth (16–17 m), but K_d in Haupiri is 12 times that in Kaiiwi (Fig. 4.3) because of the high concentrations of light-absorbing dissolved organic matter present in the former lake, whose catchment includes swamp and dense forest.

As well as irradiance attenuation, constituent composition and irradiance reflectance were also measured in the 59 lakes studied during 1980–1986. Fig. 4.4 shows how K_d tends to increase with increasing phytoplankton biomass (as chlorophyll *a*). The extent to

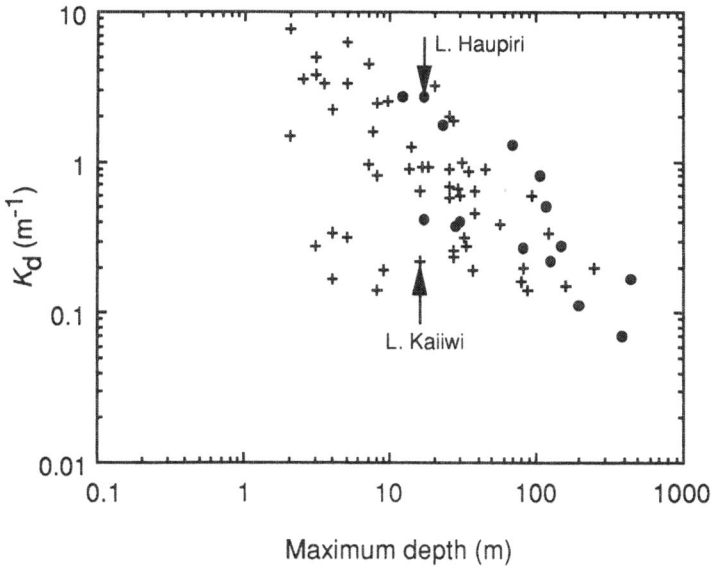

Fig. 4.3. Irradiance attenuation coefficient (K_d) and maximum depth in North Island (+) and South Island (●) lakes. Lakes Haupiri and Kaiiwi (compared in the text) are identified. See text for data sources.

which the points fall above the line indicates the relative size of the contribution by other constituents, particularly mineral suspensoids ('turbid' lakes) and dissolved yellow substance (humic lakes). It is interesting to note that the phytoplankton effect is non-linear, corresponding to $K_d \propto$ [chlorophyll a]$^{0.64}$. This non-linearity, which has also been observed in the sea (Morel 1988), probably arises because phytoplankton in eutrophic waters tend to be larger and less efficiently light attenuating than those in oligotrophic waters, whilst small particles of detritus and planktonic bacteria are relatively more important in oligotrophic waters.

Where phytoplankton dominated attenuation in the data of Vant & Davies-Colley (1984, 1988) and Howard-Williams & Vincent (1984), irradiance reflectance was generally moderate (mean 5%, $N = 36$). In turbid lakes where phytoplankton were only a minor component of the total suspensoids, irradiance reflectance was higher (5–36%, $N = 5$) because of the greater scattering (Kirk 1981b). By contrast in the highly absorbing humic lakes (chlorophyll a <2 mg m^{-3}, $g_{440} > 0.5$ m^{-1}; $N = 5$), reflectance was low (0.4–1.0%), although it increased (to 2–10%) with higher levels of phytoplankton (chlorophyll a >10 mg m^{-3}; $N = 6$).

These studies involved measurement of broad waveband (e.g. 400–700 nm) light. As is well known, however, the spectral composition of underwater light changes with depth as certain wavelengths are attenuated more rapidly than others (see Fig. 3.6). As a result K_d for broad waveband light is often found to decrease with depth (e.g. Kirk 1977), as the more rapidly attenuated wavelengths are progressively removed. The effect is particularly noticeable in clearer waters (e.g. Fig. 2.4). Howard-Williams & Vincent (1985) demonstrated this non-linearity in a group of 34 New Zealand lakes, showing that the

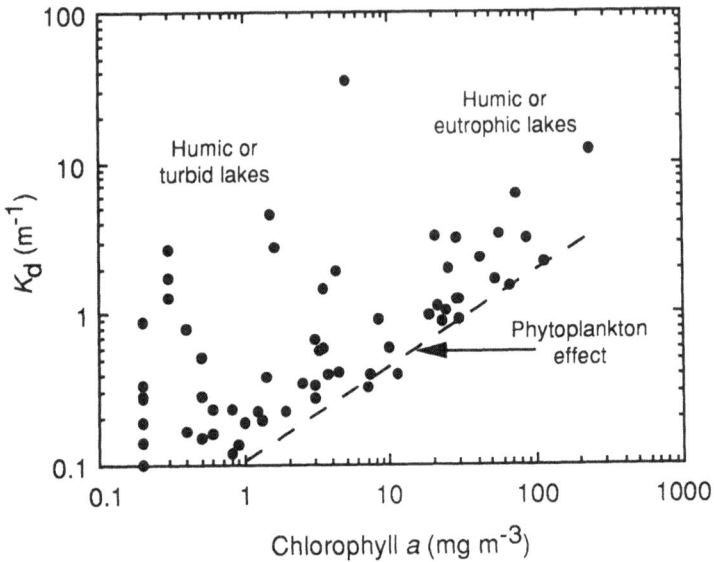

Fig. 4.4. Irradiance attenuation coefficient (K_d) and chlorophyll a in New Zealand lakes. (Data from Howard-Williams & Vincent (1984) and Vant & Davies-Colley (1984, 1988).) The equation of the line is $K_d = 0.1[\text{chlorophyll } a]^{0.64}$.

most penetrating (i.e. least attenuated) waveband varied with lake type. In oligotrophic lakes the most penetrating waveband is blue–green because the absorption of red wavelengths by water itself is dominant in these lakewaters containing few light-attenuating constituents. In humic lakes, by contrast, absorption of blue wavelengths by dissolved organic matter is dominant, so the most penetrating waveband is yellow–orange.

Partitioning irradiance attenuation

Identification of the main causes of irradiance attenuation in waters has been a major focus of aquatic optics. This is largely due to a widespread interest in underwater plant growth and, in particular, in the extent to which attenuation of light by water constituents reduces the amount of light available for photosynthesis (e.g. Bannister 1974; Kirk 1983). In most cases the contributions of water constituents to diffuse light attenuation have been considered. As Kirk (1983) observed, the fact that K_d is an apparent optical property means that such attempts should be regarded only as first approximations. The assumption that a constituent's contribution to diffuse light attenuation is linearly related to its concentration is robust only in waters in which the attenuation of light by scattering is relatively low (e.g. ocean waters (Morel 1988)). Even so, Baker & Smith (1982) defended the partitioning of K_d, arguing (1) that diffuse light attenuation was directly relevant to underwater photosynthesis, (2) that it was tedious to measure an apparent property (K_d), to convert it to inherent properties (absorption and scattering), to partition these and then to re-convert back to partitioned K_d, and (3) that K_d exhibited 'quasi-inherent' properties anyway (e.g. as shown for ocean waters by Gordon (1989)).

Most workers have chosen to simply distinguish 'algal' and 'non-algal' effects on K_d, with the latter term encompassing the contributions from mineral suspensoids, detritus, dissolved yellow substance and water itself. Having assumed that K_d varies linearly with phytoplankton biomass, and using chlorophyll a as the measure of biomass, the slope of a plot of K_d (m^{-1}) against chlorophyll a (mg m^{-3}) provides an estimate of k_c, the specific vertical attenuation coefficient for phytoplankton (in m^2 mg^{-1}). The intercept on the vertical axis — the value of Kd when the chlorophyll a concentration is zero — is k_x, the non-algal attenuation. The contributions of phytoplankton (i.e. 'self-shading') and the non-algal components to diffuse attenuation are therefore k_c[chlorophyll a]/K_d and k_x/K_d respectively.

Values of k_c do not vary widely, generally being in the range 0.01–0.04 m^2 mg^{-1} depending mainly on the composition of the phytoplankton community (e.g. Talling 1971; Jewson 1977; Tilzer 1983; Schanz 1985). Values of k_x are more variable, however, depending on the nature and concentration of the other light-attenuating constituents. If concentrations are very low then k_x will be similar to that of water itself, ~0.03 m^{-1} as in clear ocean water (Smith & Baker 1981), with values 10–100 times greater than this being encountered in productive, stained or turbid lakes (e.g. Megard *et al.* (1980), and the lakes described by Vant & Davies-Colley (1988)).

Kirk (1983, Table 10.2) showed how different combinations of phytoplankton biomass and k_x caused marked differences in the proportion of the underwater light attenuated by the phytoplankton population in some hypothetical lakes. In a relatively clear, unproductive lake phytoplankton only attenuated a small fraction (10–30%) of the light, while more (over half) could be attenuated in productive waters, even though k_x was greater in these. If their density is high enough, phytoplankton can absorb a large proportion of the available light regardless of high levels of non-algal attenuation. In two productive North Island lakes about half of the attenuation was attributable to phytoplankton (mean chlorophyll a 90–170 mg m^{-3}), even though non-algal attenuation was high (k_x = 2.6–3.6 m^{-1} (Vant & Davies-Colley 1988)).

4.2.3 Absorption and scattering
Absorption
Lakewater absorption spectra measured in the laboratory have been reported since 1903 (Hutchinson 1957), but, because suspended particles scatter light out of the spectrophotometer beam (Section 3.3.2), early studies on the absorption of unfiltered lakewaters are of somewhat dubious value. James & Birge's (1938) work with samples from the Wisconsin lakes did, however, include the examination of filtered samples, and showed that lakes differ considerably in the concentration of dissolved coloured organic matter (i.e. yellow substance). Kirk (1980) used an integrating sphere attachment (see Section 3.3.2) to overcome the problem of scattering from a concentrated sample of lakewater particulates, in order to measure the contribution to spectral absorption of the particulates present in his lakewater samples. The relative contributions of yellow substance, phytoplankton and non-algal particulates to total absorption in different lakes depended on the constituent composition of the waters. Davies-Colley (1983) and Weidemann & Bannister (1986) also used this technique to distinguish the contributions of dissolved organic matter and particulates to absorption in lakewaters. For example, Fig. 3.9 shows the results for hypertrophic Lake Hakanoa. The contribution of water itself was only

significant at long wavelengths (>600 nm) in this very highly light-absorbing water. Dissolved yellow substance was only significant at short wavelengths (becoming dominant in the UV, i.e. <400 nm). Throughout the visible spectrum the dominant effect was that of the particulate constituents, mainly phytoplankton (the chlorophyll peak at 660–670 nm is apparent in Fig. 3.9).

Kishino *et al.* (1984) measured particulate absorption spectra on filter residues from two turbid Japanese lakes, examining the particulate fraction further by distinguishing the effects of phytoplankton and detritus on absorption in the region 400–700 nm. Detritus was an important absorbing constituent in both lakes (40–50% of total absorption), while phytoplankton was similarly important in the more turbid of the two where chlorophyll *a* concentrations were higher (by an order of magnitude).

Absorption by phytoplankton in lakewaters is reasonably well known. As with the absorption by yellow substance and water itself, phytoplankton absorption is markedly wavelength dependent (e.g. see Figs. 2.10 and 2.11). Using the integrating sphere technique to measure the spectral absorption of concentrates of reasonably pure phytoplankton cultures, Davies-Colley *et al.* (1986) found that PAR-band average absorption was generally about 10 m^2 (g chlorophyll $a)^{-1}$. Similar values have been obtained from measurements of absorption and phytoplankton biomass in a number of lakes (e.g. an average of 9 $m^2 g^{-1}$ for 27 New Zealand lakes (Vant & Davies-Colley 1984) and 8–10 $m^2 g^{-1}$ in a eutrophic North American lake (Weidemann & Bannister 1986)).

Scattering
Underwater light scattering is difficult to measure directly. However, Kirk (1981b) devised a simple technique allowing scattering to be estimated from measurements of up- and downwelling irradiance (i.e. from the attenuation and reflectance of diffuse light—see Section 3.3.4). Estimates using this or similar techniques have now been made in a variety of lakes (e.g. Bukata *et al.* 1981a; Kirk 1981b; Effler *et al.* 1985; Weidemann *et al.* 1985; Weidemann & Bannister 1986; Bowling *et al.* 1986; Effler 1988), including a number in New Zealand (Vant & Davies-Colley 1984, 1986, 1988). Results from several of these studies have shown that the scattering coefficient in lakewaters is often numerically similar to the nephelometric turbidity, the ratio of the two measures generally being in the range 0.8–1.2 (e.g. Fig. 4.5). The estimated scattering coefficients are derived from PAR-band measurements of attenuation and reflectance and are thus PAR-band averages. However, while light absorption is highly wavelength dependent, scattering is much less so (Kirk 1981a,b), so the PAR-band average scattering coefficients are likely to be comparatively robust.

The contributions of the constituents of lakewaters to scattering have been determined in a few cases. Laboratory measurements of the scattering caused by cultures of six phytoplankton species isolated from New Zealand lakes were made by Davies-Colley *et al.* (1986). Scattering cross-sections were 40–140 $m^2 g^{-1}$, depending mainly on the size and chlorophyll *a* content of the cells or cell aggregates. Similar values (50–150 $m^2 g^{-1}$) were reported from natural populations in a eutrophic North American lake (Weidemann & Bannister 1986), with the higher values being associated with blue–green algae in which scattering is enhanced by the presence of gas vacuoles. An average value of 90 $m^2 g^{-1}$ was obtained from measurements in 27 diverse New Zealand lakes (Vant & Davies-Colley 1984).

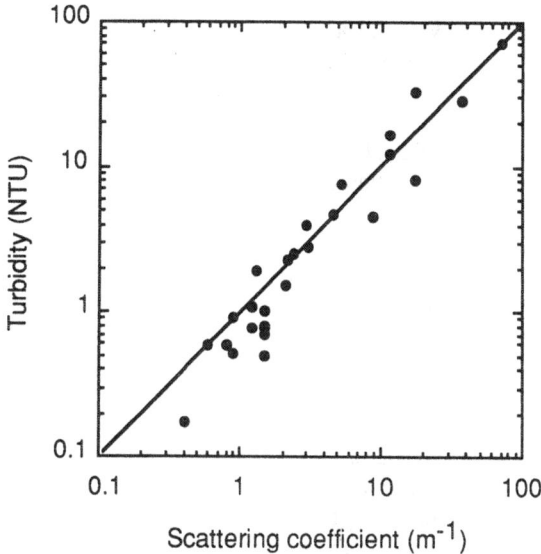

Fig. 4.5. Nephelometric turbidity and scattering coefficient in 27 New Zealand lakes. (Data from Vant & Davies-Colley (1984).) Numerically equal values fall on the line.

The other major cause of scattering in lakewaters is mineral suspensoids. Scattering cross-sections depend on the refractive index and size of the particles (Fig. 2.9, Section 2.4.2). An average value of 1.1 $m^2 g^{-1}$ was obtained for 27 New Zealand lakes (Vant & Davies-Colley 1984), but this value, obtained from a statistical analysis dominated by data from lakes in which the mineral suspensoids were mainly clay minerals (diameters mostly <2–5 μm (Vant 1990)), may be higher than typical (see Section 6.4).

4.2.4 Lakewater appearance

The aspects of water appearance which are most relevant to human observers are visual clarity and two aspects of colour: brightness and hue (Sections 1.1.1 and 2.6.1). Beam attenuation is a useful measure of visual clarity, reflecting the fact that both absorption and scattering contribute to the loss of clarity (Section 2.5.1). Visual clarity can be similarly poor in humic stained lakes where absorption is high, and in turbid lakes where scattering is dominant (e.g. Lakes D and Tekapo respectively (Vant & Davies-Colley 1984)). Clarity measured as Secchi depth usually covaries strongly with beam attenuation (Fig. 2.5A). A possible refinement (Kirk 1982) is to include also the irradiance attenuation coefficient in the description of visual clarity, resulting in a 'contrast' attenuation coefficient, K_c (where $K_c = c + K_d$). However, as c and K_d usually covary strongly, so too do c and K_c, so little is gained by including the extra term.

The brightness of a water body is related to the irradiance reflectance (the proportion of the downwelling irradiance which is reflected back out of the water column), which, in turn, is largely a function of the ratio of scattering to absorption (Section 2.6.2). The reflectance of lake waters is typically around 5% but can vary markedly (e.g. Davies-Colley & Vant 1988). While b was always greater than a in 27 New Zealand lakes, the ratio b/a varied markedly (Vant & Davies-Colley 1984), as did the reflectance, and therefore, by inference, the brightness. The ratio b/a was as low as 1.5 in a humic–eutrophic lake (Lake D), but was >30 in Lake Pukaki (Plate 2) where levels of phytoplankton and dissolved organic matter are negligible, most of the attenuation being due to the highly scattering and weakly absorbing glacial 'flour' in this lake.

Lakewater colour is the sensation produced by the light scattered back from the water column (defined as intrinsic colour in Section 2.6.1). Both yellow substance and particulate constituents, as well as water itself, contribute to hue of lakewaters by the selective absorption of light. The hue of lakewaters has long been matched against colour standards—Hutchinson (1957) quotes a number of observations reported during the period 1897–1933. Traditionally the 21 Forel–Ule standards, which vary from blue through to brown, have been used, but colours have also been described in terms of the Munsell colour standards (Smith *et al.* 1973; Davies-Colley *et al.* 1988). Lake colour has also been measured objectively, although only rarely. Smith *et al.* (1973) measured up- and downwelling spectral irradiance in two North American lakes and used chromaticity analysis (Chapter 2) to calculate lake colour. The technique was also used to determine the colour of 14 North Island lakes (Davies-Colley *et al.* 1988; see Fig. 2.23). The hues measured in these were similar to those observed directly and matched in the field to Munsell standards (Fig. 3.17).

Davies-Colley (1983) applied chromaticity analysis to 'relative reflectance' spectra (the ratio of spectral scattering to spectral absorption—Section 3.5.3) measured on eutrophic lake waters (Fig. 4.6). The structure observed in the reflectance spectra is mainly due to that in $a(\lambda)$, because $b(\lambda)$ varies weakly with wavelength, typically inversely. Those constituents which are most spectrally selective in their absorption of light (phytoplankton, detritus and yellow substance) are thus the main contributors to lakewater hue. However, absorption by water itself was the main cause of the blue hue of Lake Coleridge in the South Island of New Zealand (Biggs & Davies-Colley, 1990; Section 6.2). Concentrations of both dissolved and particulate organic materials were too

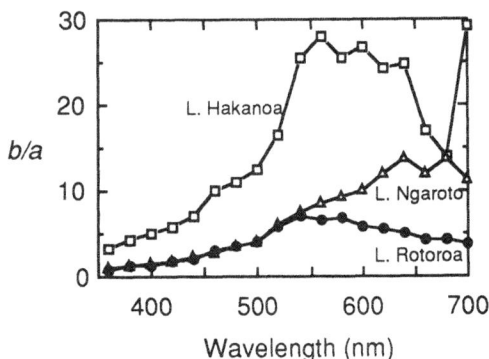

Fig. 4.6. 'Relative reflectance', the ratio of spectral scattering to absorption, in three Waikato lakes. (Redrawn from the data of Davies-Colley (1983) with permission.)

low to contribute to the hue in this lake, while the mineral suspensoids present at low concentrations (mean <1 g m^{-3}) merely increased the reflectance, making the waters appear brighter than optically pure water.

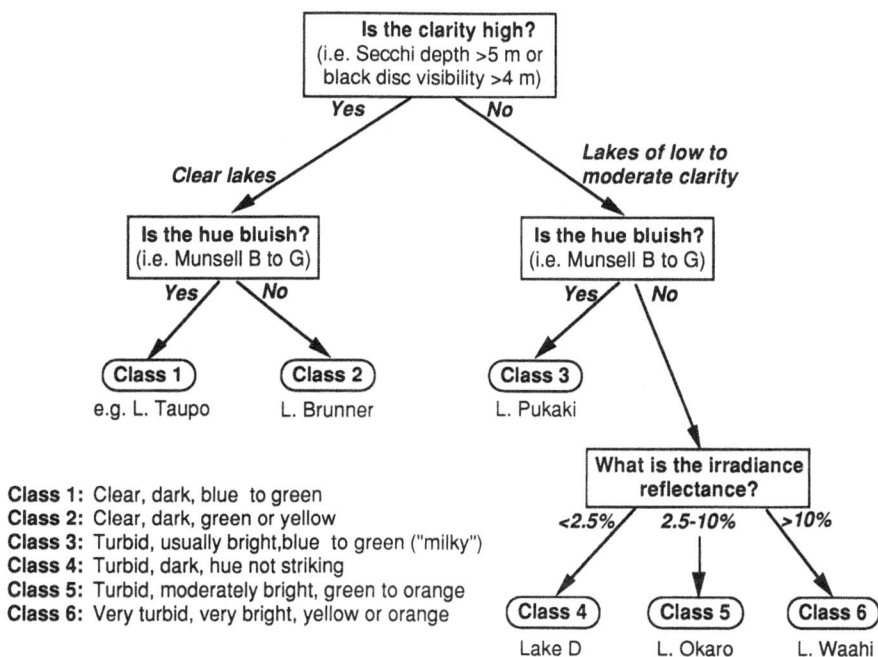

Class 1: Clear, dark, blue to green
Class 2: Clear, dark, green or yellow
Class 3: Turbid, usually bright,blue to green ("milky")
Class 4: Turbid, dark, hue not striking
Class 5: Turbid, moderately bright, green to orange
Class 6: Very turbid, very bright, yellow or orange

Examples

Lake	Secchi depth (m)	Munsell hue	Reflectance (%)
Taupo	15[a]	7.7BG[a]	1.5[a]
Brunner	5.5[b]	(Yellow)[b]	0.7[c]
Pukaki	0.45[d]	(Blue-green)[d]	36[d]
Lake D	0.9[e]	2Y[b]	1.8[e]
Okaro	2.1[e]	1.5GY[b]	3.6[e]
Waahi	0.3[f]	5Y[f]	10[f]

[a] Davies-Colley et al. (1988).
[b] Paerl et al. (1979).
[c] Howard-Williams & Vincent (1984).
[d] Vant & Davies-Colley (1984).
[e] Vant & Davies-Colley (1988).
[f] Authors' unpublished data.

Fig. 4.7. A scheme for classifying lake water appearance. Examples of each lake class are given and the data used for classifying these lakes are tabulated.

Water appearance of lakes varies widely. For example, in New Zealand lakes, beam attenuation coefficients (and Secchi depths) vary over more than two orders of magnitude (see Figs 4.1, 4.2 and 4.3), irradiance reflectances range from <1% (very dark) to >30% (very bright) (Table 3.2), and hues range from blue to orange (Vant & Davies-Colley 1984, 1988; Davies-Colley et al. 1988). Given this diversity it would be useful to have a system for classifying lake appearance.

Fig. 4.7 outlines such a classification scheme based on water clarity, hue and brightness. Not all possible combinations are encountered. Clear lakewaters (classes 1 and 2) are dark since the scarcity of scattering particles means that b/a and, therefore, reflectances are low to moderate. Whether these clear lake waters are blue to green (class 1, e.g. Lake Taupo) or green to yellow (class 2, e.g. Lake Brunner, also Lake Rotomanuka, Plate 4) depends primarily on the concentration of yellow substance. Similarly, if lakes are blue and turbid they must also be bright (class 3, e.g. Lake Pukaki: Plate 2). Since there is little shift in hue from the blue of optically pure waters in these lakes, then suspensoids causing the turbidity must be weakly absorbing so that the ratio b/a, and thus the reflectance, is high.

The turbid lakes, apart from those in class 4 (turbid and dark lakes), are moderately bright to very bright (classes 3, 5 and 6). In class 4 lakes, absorption by dissolved or particulate organic matter minimizes the amount of light scattered back from the water column, thereby producing a dark appearance. Class 4 lakes represent an exception to the generalization that dark lakewaters tend to be visually clear (as in classes 1 and 2).

Some generalizations can be made as to the constituents responsible for the appearance for particular classes. Class 5 lakes are typically eutrophic with abundant levels of phytoplankton (e.g. Lake Okaro). If yellow substance is also abundant, then a eutrophic lake falls into class 4. Class 2 lakes also have fairly high concentrations of yellow substance, but suspensoid concentrations are low. In principle, suspended particulates other than phytoplankton could contribute to the turbidity of class 4 lakes, but such lakes are outside our experience. Mineral suspensoids are usually responsible for turbidity in classes 3 and 6. If little organic matter is associated with these suspensoids the hue shift is small, resulting in class 3 lakes. Greater hue shift due to higher levels of organic matter (either yellow substance or detritus) results in class 6 lakes (e.g. Lake Waahi, Plate 5).

4.2.5 Underwater light for plant growth

It has long been recognized that the attenuation of underwater light is a major factor limiting the growth of plants in lakes (e.g. Talling 1971). Hutchinson (1967) cites several studies from the late nineteenth century which indicate that workers at that time realized that the upsurge in phytoplankton growth observed in European and North American lakes each spring must have been caused by the marked increase in underwater light occurring then (water temperatures having increased only slightly). Similarly Spence (1976) and Hutchinson (1975) cite pioneering studies published during 1926–1955 which described how the depth to which submerged benthic plants grew was limited by the extent to which water constituents—particularly phytoplankton—attenuated the available light.

This controlling role of the underwater light climate reflects the fact that the amount of light available for plant growth decreases with depth. At a certain minimum light intensity—the 'compensation' irradiance—photosynthesis and respiration will be in balance so net growth cannot occur (see Section 2.5.2). At the lower irradiances found at greater depths, growth is not possible. Maximum depths at which plants can grow—'critical' depths—can thus be identified. These vary from lake to lake depending on the irradiance attenuation, and in very clear waters can be >100 m (Spence 1982). For benthic plants, K_d alone is sufficient to characterize the light-dependent maximum depth in lakes at similar latitudes (i.e. receiving similar incident radiation). However, the growth of freely circulating phytoplankton is also dependent on the amount of time the cells spend at depths greater than the euphotic depth (i.e. in the 'dark'), so the lake mixing regime is also relevant to the growth of these plants.

Macrophytes
The light climate of macrophytes, and also benthic algae, depends only on irradiance attenuation and the diurnal and seasonal cycles of sunlight. Vant *et al.* (1986) showed how the maximum depth to which macrophytes grow in some North Island lakes was a simple function of annual average K_d (Fig. 2.15), indicating that the macrophyte communities in each lake all had broadly similar light requirements. Similar relationships have been determined for macrophytes in lakes elsewhere (e.g. Canfield *et al.* 1985; Chambers & Kalff 1985), although these have often been based on instantaneous measurements (usually Secchi depth) rather than the annual average irradiance attenuation—which better defines the light climate.

Macrophyte communities have collapsed and largely disappeared from an increasing number of shallow New Zealand lakes in the past 20 years. Examples include Lakes Ellesmere (Hughes *et al.* 1974; Gerbeaux and Ward 1986) and Omapere (Judd & Kokich 1986) and several lowland Waikato lakes including Waahi and Waikare (Kingett 1984; Wells *et al.* 1988, Section 6.3). In certain cases, increased light attenuation by suspensoids is thought to have contributed to the decline (Vant 1987b; Wells *et al.* 1988). Regardless of the causes of particular declines, the poor clarity prevailing after macrophyte collapse has often prevented successful regrowth of the plant communities. With water movement no longer dampened by abundant plant biomass, wind waves are generated which resuspend sediments from the now-unprotected bottom (Section 6.3). The increased suspended solids concentrations cause light levels over the deeper regions of the lake to be too low for macrophytes to re-establish. Even though average irradiance may be adequate for plant growth in very shallow regions (e.g. <0.5 m deep), wave action in this zone means plants cannot grow here either.

The effect of highly turbid floodwaters on the macrophytes in Lake Tutira was described by Johnstone & Robinson (1987). The floodwaters entering this lake following a major storm increased mid-lake irradiance attenuation 5-fold. One to three months later, a marked reduction in the maximum depth to which macrophytes grew in the lake was observed, broadly in line with Fig. 2.15. Fortunately, in this case the macrophyte community was eventually re-established from surviving rhizomes when the lakewaters clarified.

Phytoplankton
The light climate of freely circulating phytoplankton depends on depth of mixing as well as light penetration into the water. Kirk (1983, pp. 254–258) gives a discussion of the critical depth of mixing, which depends most strongly on the irradiance attenuation coefficient, as expected, but also on the ambient irradiance and certain parameters characterizing

production by the phytoplankton as a function of irradiance (the respiration rate, light-saturated production rate and compensation irradiance—Fig. 2.14). However, a useful rule of thumb is that the critical depth is about five times the euphotic depth: if the mixing depth is more than five times the euphotic depth, growth is precluded by insufficient light. Where the mixing depth is less than the euphotic depth, phytoplankton are light saturated, and at intermediate mixing depths phytoplankton are expected to experience some light limitation (e.g. Talling 1971; Grobbelaar 1985). The average irradiance in the water column mixed to depth z_{mix} is given by

$$E(z_{mix}) = \frac{E(0-)}{K_d z_{mix}}\left[1 - \exp(-K_d z_{mix})\right]$$

where $E(0-)$ is the irradiance just below the surface. Using this equation we can calculate that when $z_{mix} = z_{eu}$ the average irradiance of the mixed layer is about 22% of the surface value (i.e. $0.22E(0-)$), and when $z_{mix} = 5z_{eu}$ the average irradiance is about 4% of $E(0-)$.

In isothermal lakes the maximum depth is obviously the upper limit to the mixed depth, but in thermally stratified lakes the mixed depth is the depth of the epilimnion, which is roughly predictable for a given region (climate) from lake surface area (an index of wind exposure). Davies-Colley (1988b) showed that mid-summer (minimum) epilimnion depth in New Zealand lakes is about $7.69 \times area^{0.232}$. Fig. 4.8 shows the irradiance attenuation data of Fig. 4.3 for New Zealand lakes replotted as euphotic depth

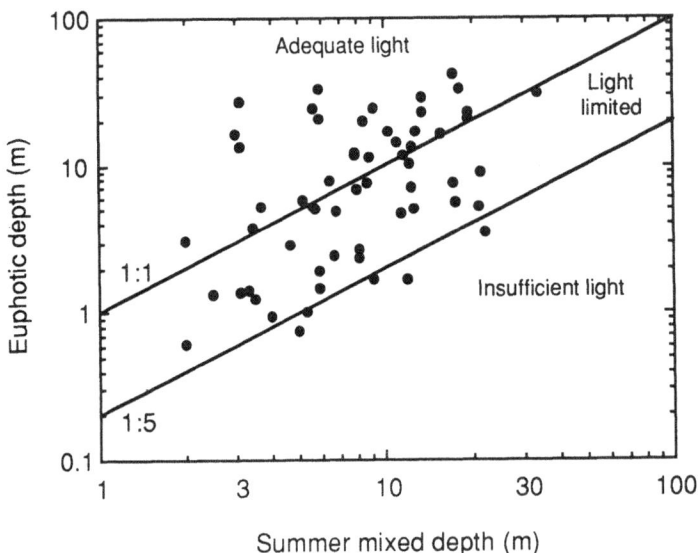

Fig. 4.8. Euphotic depth versus estimated mid-summer mixed depth (see text) in some New Zealand lakes. The lines shown represent ratios of euphotic depth: mixed depth of 1:1 and 1:5, separating adequately lit, light-limited, and insufficiently lit mixed layers.

$(4.6/K_d)$ against the smaller of the maximum depth or the mid-summer epilimnion depth. The mixed layer lighting guidelines are overlain on the figure. Phytoplankton growth is expected to be minimal in a few of the lakes—those (usually peat stained or turbid) represented by points below the line for $z_{mix} = 5z_{eu}$. Some degree of light limitation appears likely in about half of the other lakes, while the remainder, usually clear lakes with a shallow mixed depth (i.e. small surface area), have adequate light.

Note that parts of a given lake will be shallower than the mid-summer mixed depth as defined above, which is therefore the upper limit to the average depth of mixing. On the other hand, the mid-summer mixed depth represents a minimum in stratifying lakes— these lakes generally mix more deeply—and to the bottom after overturn. This fact, together with the seasonal cycle of incident solar radiation, means that lighting of the mixed layer is generally maximal in mid-summer and is lower at other times of the year. For example, Viner & White (1987) calculated that, in certain stratifying North Island, New Zealand, lakes, the average irradiance in the mixed layer in summer can be up to 12 times that in the winter. (As an aside, it can be noted that, despite the lowered light levels, phytoplankton growth is often higher during the winter in some central North Island lakes because nutrient availability is greatest then (e.g. Vincent 1983b).) We may conclude that light limitation in New Zealand lakes is probably more important than Fig. 4.8 indicates, and light limitation in lakes generally may be more widespread than is commonly recognized (Kirk 1983, pp. 256–258).

In stratified lakes, the available nutrients (nitrogen and phosphorus) in the mixed layer may become depleted during the typical spring phytoplankton bloom so that the phytoplankton biomass declines in summer. However, available nutrients are present in abundance below the mixed layer at this time. If the overlying water is sufficiently clear such that sufficient light for photosynthesis penetrates deeper than the mixed depth (as a rule of thumb, if the mid-summer mixing depth is less than the euphotic depth) a population of algae may develop in the thermocline. The resulting peak of chlorophyll in the water column is often termed a *deep chlorophyll maximum* and is a common phenomenon in the sea, but much less common in lakes (e.g. Kirk 1983, pp. 258–260). Deep chlorophyll maxima have been reported from some very clear lakes such as Crater Lake (Larsen 1972), and in clear, oligotrophic New Zealand lakes, such as Waikaremoana (Howard-Williams *et al.* 1986).

4.3 RIVERS

4.3.1 Light-attenuating constituents of river waters

The optical character of rivers has not been much studied, but the available information shows that rivers contrast appreciably with lakes. Fig. 4.9 shows distribution curves for the concentrations of light-attenuating constituents in New Zealand river waters. The data (from Davies-Colley & Close 1990) are for 190 river water samples from 96 different rivers, sampled, on average, twice each under baseflow (operationally <median flow) conditions.[*]

Phytoplankton, as we have seen (Section 4.2) are an important light-attenuating constituent of lake waters, but these are seldom present at high concentrations in rivers, except when lake fed, or impounded (e.g. the Waikato River (Davies-Colley 1979, 1987b)). Much of the chlorophyll *a* suspended in the water column of rivers may be that of

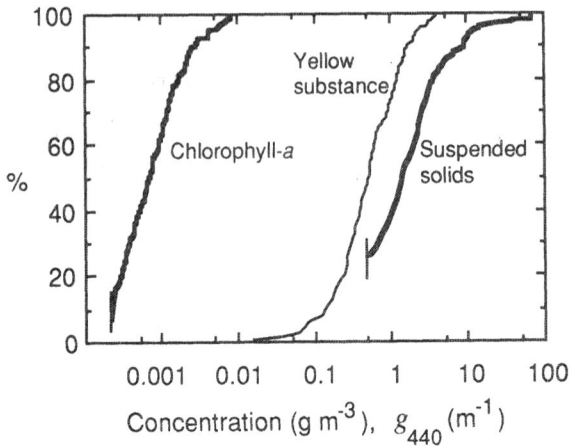

Fig. 4.9. Distribution curves for light-attenuating constituents (chlorophyll a, yellow substance (as g_{440}), suspended solids) in 96 New Zealand rivers. (Data from Davies-Colley & Close (1990).) The detection limits (determined by necessarily limited sample volumes) for SS and chlorophyll a were 0.5 g m^{-3} and 0.2 mg m^{-3} respectively, which explains the truncation of the distribution curves at these concentrations.

sloughed periphyton rather than true phytoplankton, except in lake-fed rivers. The chlorophyll a content of New Zealand rivers is quite low (median = 0.62 mg m^{-3} in Fig. 4.9), and more nearly comparable with concentrations in ocean waters than in lakes. Even the highest concentrations encountered (associated with high periphyton biomasses) would be considered indicative of unproductive conditions in lakes (e.g. Fig. 4.4). Davies-Colley & Close (1990) found that algal cells in the water column were responsible for a negligible fraction of total light beam attenuation in New Zealand rivers, and that colour (hue) was negligibly affected by chlorophyll a concentration.

Other types of suspensoids are typically far more important in rivers than phytoplankton because, in flowing waters, particles tend to be entrained from erosion of the river bed and banks, and then maintained in suspension by turbulence. The median suspended solids (SS) concentration in 96 New Zealand rivers in baseflow was rather low at 1.3 g m^{-3} (Fig. 4.9), but median SS values would be expected to be appreciably higher for data over all states of flow.

Davies-Colley & Close (1990) found that, on average, 87% of the total light beam attenuation in 96 New Zealand rivers was caused by suspended particles. In spite of this high contribution of SS to c, the correlation of these two variables was rather poor because of wide variation in the attenuation cross-section of the suspensoids (Fig. 4.10). As we saw in Section 2.4.2, organic detritus generally attenuates light more strongly than mineral particles, and is also strongly light absorbing. However, fine grained mineral particles such as fine silts or clays have higher attenuation cross-sections (of the order of 1 m^2 g^{-1}) than any organic suspensoids, and tend to dominate the overall light beam attenuation in river waters. This may explain why the modal attenuation cross-section in Fig. 4.10 is about 1 m^2 g^{-1}. The suspended matter in rivers at baseflow could conceivably

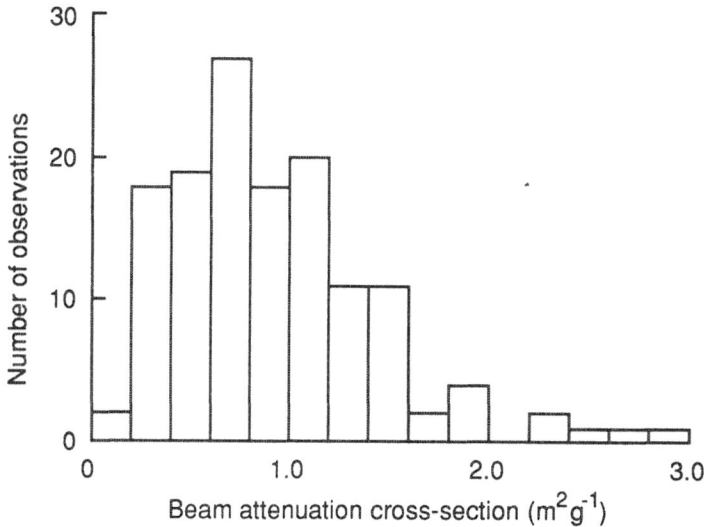

Fig. 4.10. Distribution of the beam attenuation cross-section for suspensoids in 96 New Zealand rivers at baseflow. (Data from Davies-Colley & Close (1990).)

range from large, low density organic detrital particles with low attenuation cross-sections to efficiently attenuating clay mineral particles of the order of 1 μm diameter. Droppo & Ongley (1992) have shown that many particles in river waters occur as fairly small (of order 10 μm) and stable (to shear stresses) flocs of primary particles such as silt and clays which are probably aggregated by sticky 'biopolyelectrolytes' produced by bacteria. Thus the attenuation cross-section of particles in river waters depends on the extent of aggregation as well as the size and composition of primary particles. The range of average attenuation cross-section illustrated in Fig. 4.10 can probably be accounted for by the expected variation in composition and size of particles and particle aggregates in river waters.

Yellow substance can be an important light-attenuating constituent in rivers, as it is in lakes, although there are reasons for thinking that the chemical nature of yellow substance in rivers may be different in some respects from that in lakes. In lakes, a fraction of the yellow substance derives from autochthonous phytoplankton production (e.g. Davies-Colley & Vant 1987), with the remainder being derived from terrestrial production in the catchment. In rivers, catchment sources probably dominate, although production by benthic algae or macrophytes is expected to contribute to the pool of dissolved organic carbon. Typically yellow substance is the main contributor to light absorption by river waters except during floods when turbidity is high and absorption by the particulates, notably eroded soil particles, is correspondingly high.

As we saw in Section 3.6.1, the absorption coefficient of filtrates at 440 nm (g_{440}, Kirk 1976) is a useful index of yellow substance concentration. Median g_{440} in the 190 baseflow river water samples analysed by Davies-Colley & Close (1990) was 0.52 m^{-1} (Fig 4.9). About 26% of the river waters had 'high' concentrations of yellow substance

$(g_{440} > 1 \text{ m}^{-1})$ and 7% had 'low' concentrations $(g_{440} < 0.1 \text{ m}^{-1})$. In New Zealand's National Water Quality Network, the median of all g_{440} data for 77 river sites sampled at all states of flow was 0.71 m^{-1} (30% of samples were 'high' and <1% were 'low' in yellow substance).

4.3.2 Visual clarity
It is generally well known that the visual clarity of river waters ranges widely, both between different water bodies with catchments of different character, and in the same river at different times. Davies-Colley & Close (1990) found that 190 black disc visibility measurements in 96 New Zealand rivers sampled under baseflow conditions (Fig. 4.11) varied more than 40-fold (from 0.25 m to just over 10 m). Median clarity was 3.2 m and 32% of the observations were of 'high' clarity $(y_{BD} > 4 \text{ m})$ and 7% of 'low' clarity $(y_{BD} < 1.2 \text{ m})$.

At higher states of flow, visual clarity is expected to be lower than at baseflow. Visual water clarity has been routinely measured by the black disc method at 77 river sites in New Zealand's National Water Quality Network (Smith & McBride 1990) since January 1989, and the resulting dataset includes measurements over a wide range of states of flow. Fig. 4.11 gives the distribution of median black disc visibilities characterizing the 77 river sites (i.e. the distribution of clarity characteristic of rivers) as well as the distribution of all clarity observations (i.e. the distribution of clarity of river waters; $n = 1846$ up to December 1990). As expected, the median of the two distributions is nearly identical (1.30 m), but the distribution of all observations is somewhat more variable (lower curve slope) because measurements of low clarity at high states of flow give rise to a longer low-clarity tail in the distribution of all observations. As expected, visual water

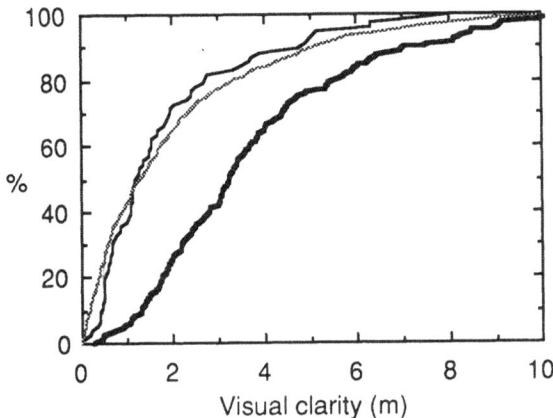

Fig. 4.11. Distribution of visual water clarity in New Zealand rivers. ———, distribution of 190 observations of black disc visibility in 96 rivers sampled at baseflow (from Davies-Colley & Close (1990)). ········ , distribution of 1846 unpublished observations of black disc visibility at 77 river sites at all states of flow (monthly samplings, at approximately monthly intervals, January 1989 to December 1990) in New Zealand's National Water Quality Network (Smith & McBride 1990). ———, corresponding distribution of the median black disc visibility at the 77 river sites.

clarity tends to be higher at baseflow than at higher states of flow in New Zealand rivers.

There were indications from the dataset of Davies-Colley & Close (1990) that rivers in hard rock areas were generally clearer than those in soft, erodible rock catchments—as could be expected. The lowest baseflow clarity value (y_{BD} = 0.25 m) was measured in the Waipaoa River, near Gisborne, a river that drains a rapidly eroding catchment with the highest sediment yield in New Zealand (Griffiths & Glasby 1985) and one of the highest in the world. At the other extreme, visual ranges of the order of 10 m were observed in a number of rivers (e.g. Tongariro, Rangitikei and Maraewhenua Rivers), most being lake fed or spring fed, or with catchments dominated by hard, erosion-resistant rocks such as schist, indurated sandstones and igneous rocks under native forest cover. Land use was apparently less significant than lithology and hydrology, but still important in controlling sediment yield and river water clarity at baseflow. Rivers draining near-pristine catchments with native forest cover may be expected to have higher clarity waters on a given rock type than rivers draining catchments developed for pastoral agriculture (Duncan 1987; Davies-Colley 1990a), but this has yet to be confirmed.

Most people are probably aware that rivers tend to be clear at low flow and discoloured or 'muddy' at high flow. This perception of an inverse relationship between clarity and flow is confirmed by studies on rivers over a range of states of flow. For example Fig. 4.12 shows visual clarity plotted against flow in the Grey River at Stillwater, South Island, New Zealand, on both linear and log scales. The log-transformed data are fairly well fitted by a straight line with a negative slope. This type of relationship was suggested originally by the typical log–log relationship of SS and flow in rivers (e.g. Rieger *et al.* 1988), together with the inverse correlation of clarity and SS (e.g. Davies-Colley & Close 1990).

Davies-Colley (1990a) studied the visual water clarity of twelve New Zealand rivers

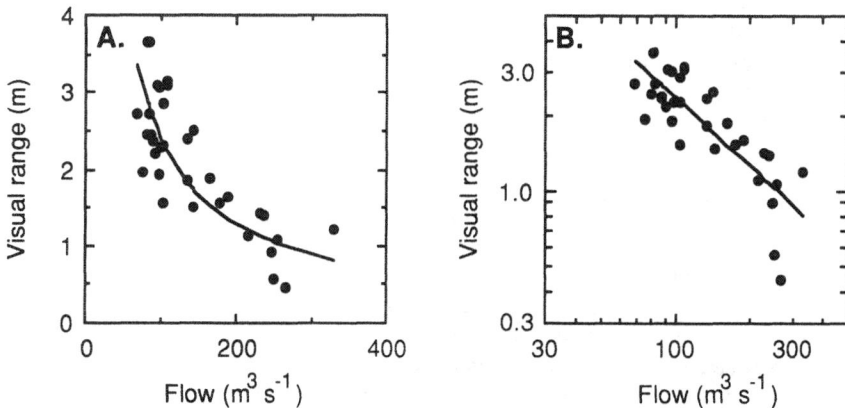

Fig. 4.12. Relationship between visual water clarity and flow in the Grey River at Stillwater, New Zealand (mean annual flow 300 m³ s⁻¹). (A) Linear scales. (B) The same data plotted on log scales. The line shown fitting the data is the linear regression equation for log-transformed variables: y_{BD} = $147Q^{-0.90}$. (Authors' unpublished data.)

at four-weekly intervals over a year. In all but one of these rivers the log of visual water clarity as measured by the black disc method was fairly closely related to the log of flow, and r^2 values ranged from 0.41 to 0.92 and averaged 0.71 (i.e. 71% of the variance on average was explained). In the Maraewhenua River, South Island, which drains very erosion-resistant schist rock, the clarity remained high over the range of flow encountered and was only weakly related to flow.

The typical inverse linear relationship between log transforms of clarity (y_{BD}) and flow (Q) suggests that a power law model with a negative exponent may be generally applicable:

$$y_{BD} = \alpha Q^{-\beta}$$

where α and β are parameters estimated from the empirical data fit. There is no particular interest in the coefficient, α, since this reflects the size (flow range) of the river as much as clarity range. However, the exponent, β, determines the shape of the clarity–flow relationship and is thus of considerable interest, being characteristic of an individual river. Davies-Colley (1990a) suggested that high values of β(>1.0), implying very strong dependence of visual clarity on flow, are to be expected in rivers draining soft erodible rock catchments, whereas much lower values (approaching zero), implying weak dependence of clarity on flow, may be characteristic of rivers draining hard rock catchments.

As yet there is little information available on the sources of the residual variation not accounted for by the power law function relating visual clarity to flow in individual rivers. However, by analogy with the much more-studied suspended solids–flow relationships we may expect that some of the residual variation will be explicable eventually. It is often found that SS in rivers during storms is not uniquely related to Q, and tracks differently in different events as illustrated in Fig. 4.13. Simple clockwise hysteresis curves for storms appear to be most common, whereby SS is generally higher at a given flow on the rising limb of the storm hydrograph than on the falling limb, although anticlockwise hysteresis has also been reported. Clockwise hysteresis is related to the so-called 'first-flush' effect in which sources of suspendable materials erodible by flows of a given magnitude are progressively depleted during individual storms (e.g. Van Sickle & Beschta 1983) and during a storm season. The first increase in flow on the rising limb of the first storm of the wet season yields more suspended matter than the falling limb of the same storm, and more than later storms. Similar hysteresis is to be expected in variables related fairly closely to SS, notably visual water clarity.

As well as varying with time within storm events, the relationship between clarity and flow may be expected to vary with season, again reflecting SS–Q relationships (as well as trends in attenuation cross-section). For example, Walling (1977) found that the relationship between SS and flow in a small catchment in England changed significantly with season, such that during the winter SS was generally lower for a given flow than during the drier summer season.

The finding that most of the variation in visual water clarity encountered in rivers is related to variation in flow is of great interest in itself, but is also of immediate practical importance. To characterize the clarity at a particular site on an individual river, we require a frequency distribution curve for visual range. Such a distribution curve can be obtained

Single Rise

Simple relationship | SS lead | SS lag

SS

Q   Q   Q

Multiple Rise

SS lead | SS lag | SS lead-lag

SS

Q   Q   Q

Complex
(no recognisable pattern)

SS

Q

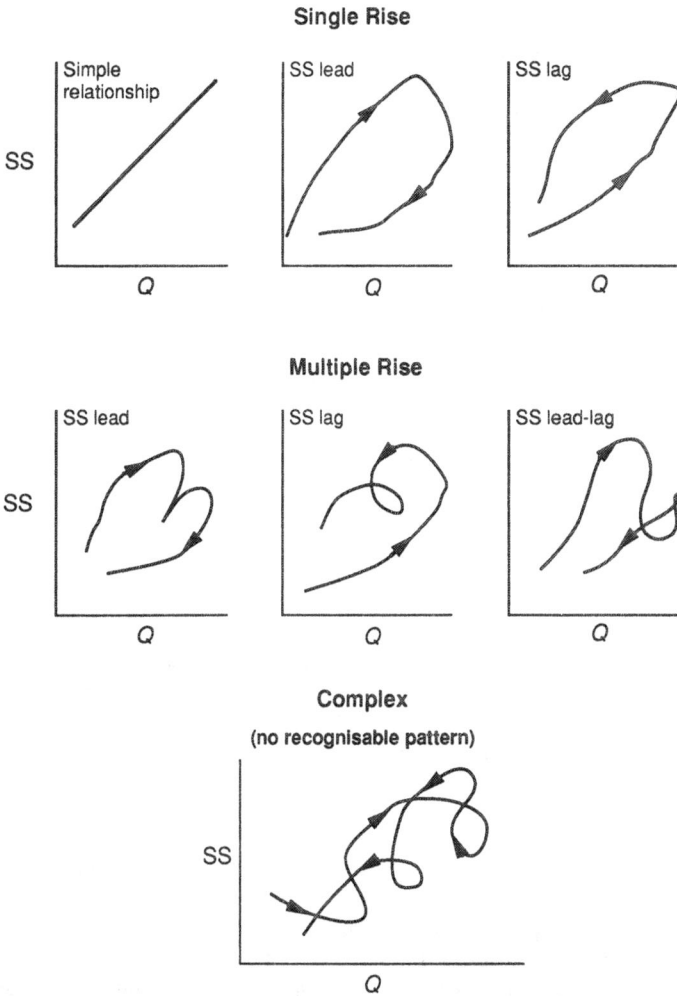

Fig. 4.13. Types of relationships between suspended solids and flow during storms. (After Rieger *et al.*
(1988), with permission.)

by sufficient independent clarity measurements, say >100, but this is a major resource
commitment if appreciable travel is involved. However, with far less effort we can estimate the frequency distribution of clarity by combining the clarity–flow relationship with
the flow–frequency distribution based on rated long-term water level recording in rivers.

Fig. 4.14 illustrates the method. A particular percentile flow obtained from the
flow–frequency distribution is interpolated (e.g. the 80 percentile). This flow value is
then used in the empirical power law expression obtained for that river to estimate the
corresponding 80 percentile clarity. Repeating this process for sufficient percentiles
across a wide range permits definition of the whole clarity distribution curve.

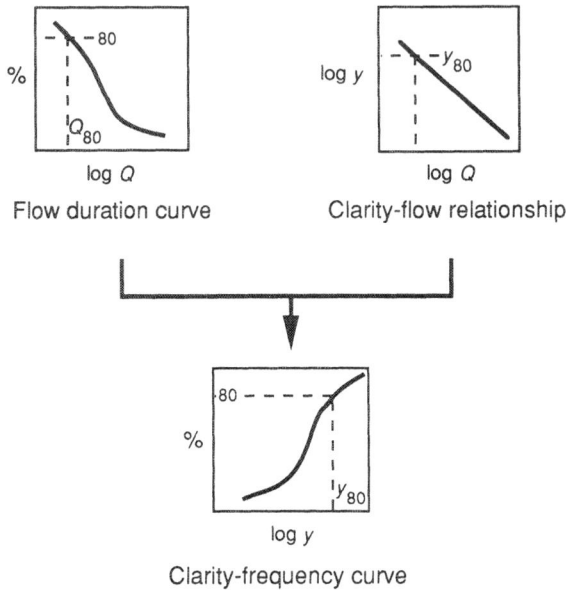

Fig. 4.14. Schematic illustrating the calculation of a cumulative frequency distribution curve for visual clarity (y) in a river by combining the cumulative frequency distribution curve for flow (Q) (often known as the flow duration curve or flow exceedance curve) with the clarity–flow relationship. Note that the cumulative frequency of flow is given according to common convention as percentage of time flow is greater than a given value, whereas the cumulative frequency of clarity is given as percentage of time black disc visibility is less than a given value.

This method for estimating the visual clarity distribution curve is expected to provide robust estimates of the central tendency, e.g. median clarity, but may cause bias at the extremes (e.g. 5 and 95 percentiles) because that variation in clarity unrelated to flow is not accounted for (Davies-Colley 1990a). This unexplained variation makes extreme clarities, at both the high and low ends of the distribution curve, more probable than predicted. Therefore the true clarity distribution curve, which could be estimated from sufficient independent samplings of clarity, would be expected to have a more pronounced S-shape and lower slope (implying greater overall variability), to an extent dependent on the magnitude of the regression error.

A preferred method for obtaining the distribution curve for visual water clarity in a river that is not subject to bias would be to measure continuously some optical property strongly related to black disc visibility. Ideal for this purpose would be a recording transmissometer, the transmittance record from which could be used to calculate beam attenuation and thence black disc range in the river water. Alternatively, a continuously recording turbidimeter could be used to develop a clarity distribution curve for the river once a (locally applicable) turbidity–clarity calibration had been developed.

Casual observation suggests that many large rivers suffer progressive degradation of visual clarity as the water moves downstream. Such a trend has been demonstrated on the

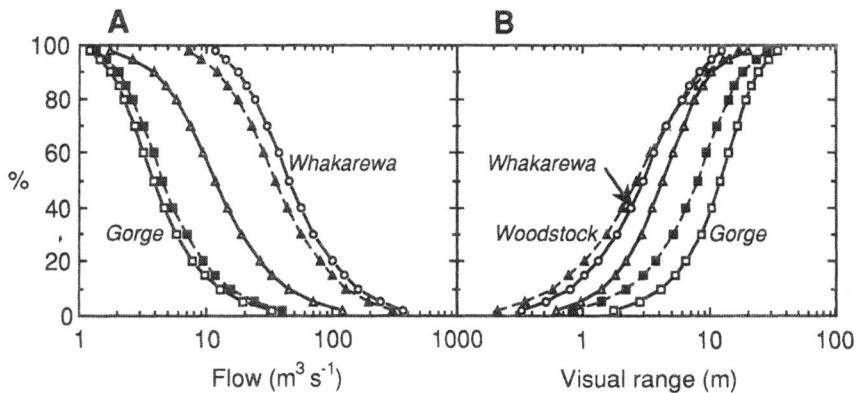

Fig. 4.15. Cumulative frequency distribution curves for flow and clarity in the Motueka River. (A) Flow–frequency distributions (flow duration curves)—percentage of time a given flow is exceeded. (B) Visual clarity frequency distributions—percentage of time visibility is lower than a given value. (From Davies-Colley (1990a), with permission.)

Waikato River, North Island (Davies-Colley 1987b) and on the Motueka River near Nelson in the South Island of New Zealand (Davies-Colley 1990a). Fig. 4.15A shows the flow duration curves at five different sites on the Motueka River. Median flow increased progressively down-river from 3.8 m^3 s^{-1} at the gorge to 35 m^3 s^{-1} at Whakarewa near the river mouth. Median clarity, in contrast, decreased progressively going downstream (Fig. 4.15B), except for the reversal of order at the two lowest sites (Woodstock and Whakarewa). At the gorge the Motueka River is a very clear water body with a median clarity of 12.1 m but clarity is unexceptional in the lower reaches (e.g. median = 2.7 m at Woodstock) compared with New Zealand rivers in general (median 1.3 m, Fig. 4.11).

The typical trend in clarity in the downstream direction undoubtedly reflects the change in catchment character in rivers. In their headwaters, many rivers drain near-pristine mountain lands or climax forest, or else, like the Waikato River in New Zealand, have their source in clear lakes. However, in lower reaches these same rivers typically receive muddy runoff from agricultural lands, swampland drainage and turbid effluents and urban stormwaters. The substrate changes from bedrock, boulders and cobbles in the headwaters to easily entrained sands and highly light-attenuating silt and clay in the lower reaches.

How clear are the clearest river waters? In principle the optics of flowing waters is governed by the same phenomena as that of standing waters, so that the the visual ranges of the clearest river waters might be expected to approach those of the clearest lakes in which attenuation of image-forming light by water itself becomes predominant. However, precisely because river water is flowing, and is therefore able to entrain any erodible particulate material and maintain it in suspension, visual clarities approaching those in the clearest lakes (e.g. Secchi depths of 40–45 m, Section 4.2.2, corresponding to a black disc range of roughly 35 m) are probably very rare. The clearest river waters are likely to be those in reaches immediately downstream of very clear lakes and in reaches fed by large spring systems. The waters of the Waikoropupu River fed by

Waikoropupu Springs, in the Nelson Region of New Zealand (reputedly having the largest flow of any freshwater springs in the Southern Hemisphere), are extremely clear ($y_{BD} \sim 50$ m (authors' unpublished data)). These springs probably rank as the clearest inland waters in New Zealand and are among the clearest natural waters in the world.

As for the other extreme, visual clarity of river waters during extreme floods may be very low (of the order of centimetres)—and dangerous as well as difficult to measure directly. The remarkably muddy Waipaoa River near Gisborne frequently has clarities lower than 100 mm in modest floods, and clarities lower than 25 mm in large events (unpublished data from the National Water Quality Network). The overall range of visual water clarity in rivers is expected to exceed three orders of magnitude, from less than 25 mm to perhaps 25 m or more in the clearest river waters.

4.3.3 Light penetration

Sand-Jensen *et al.* (1988) have maintained that nutrient limitation of the growth of periphyton and macrophytes in streams has been overemphasized, since physical factors, notably light climate and water velocity, are often overriding. Penetration into river water of diffuse light for plant photosynthesis and underwater illumination may be comparable in importance with visual clarity as regards aquatic life, but light penetration into river waters has been little studied except in certain highly light-attenuating systems. Many workers may have assumed that light attenuation by the water column in rivers is generally less significant than shading by riparian trees and overhanging banks, and have therefore ignored light attenuation in the water except in large and 'dirty' river waters. Pioneering work was done by Westlake (1966) using a selenium photocell and optical filters in rather highly light-attenuating rivers in southern England.

A useful optical parameter in discussion of light climate is the dimensionless *attenuation depth*, $\zeta = zK_d$, where K_d is the average diffuse attenuation coefficient from the water surface down to depth z. The euphotic depth, defined as the depth at which light is reduced to 1% of the surface value (Kirk 1983), corresponds to $\zeta = 4.6$ and the mid-level of the photic zone (10% of surface light) to $\zeta = 2.3$. No benthic plant growth (excepting shade-tolerant bryophytes) would be expected in rivers at depths for which $\zeta > 4.6$. Plant growth may be slowed, but not extinguished, because of light limitation at shallower depths.

It is of interest to consider the distribution of ζ (and thus the proportion of surface lighting) at the bed in New Zealand rivers. Only a rough indicative calculation can be made because K_d varies with time (flow condition) and z with space (along and across the channel) as well as with time. The 70 New Zealand river sites that were surveyed for depth in the '100 rivers' study (Biggs *et al.* 1990) averaged about 0.5 m depth at median flow, and very few rivers (about 5%) had average depths >1m, suggesting that variation in K_d rather than z may be the more important contribution to variation in the parameter ζ.

K_d was not measured directly by Davies-Colley & Close (1990) in their study of New Zealand rivers, nor is it measured in the National Water Quality Network (NWQN) (Smith & McBride 1990), but this quantity can be estimated from the beam attenuation coefficient with which it is very roughly correlated ($c \sim 3K_d$ in darker waters (e.g. Højerslev 1986), and is more typically about $5K_d$–$10K_d$: we can therefore assume $c > 3K_d$). In 'typical' New Zealand rivers of about 0.5 m average depth, $\zeta = 0.5\ K_d$ Therefore $\zeta < 0.5c/3 = c/6$. The distribution of this last quantity (termed ζ_{max} and obtained from the distribution of c from the NWQN data given in Fig. 4.11) has a median value of 0.62 corresponding to 54% of surface light at the bed. The 75 percentile of $\zeta_{max} = 1.7$ (18% of surface light).

Evidently light attenuation is rather slight in typically shallow and clear New Zealand rivers, but may be relatively more important in countries with generally more light-attenuating river waters. Shading by riparian vegetation and high banks can sometimes result in <10% of available sunlight reaching the river surface (equivalent to water column attenuation with $\zeta > 2.3$). Thus shading may often reduce lighting of the river bed more than attenuation by the water column. However, we should not dismiss as unimportant the shading of the river bed by the water column in rivers, especially deep or 'dirty' waters typical of rivers in their lower reaches. Periphyton communities in rivers are usually to some degree self-shading (e.g. Jorgensen & Des Marais (1988) demonstrated very strong light attenuation within periphyton mats using a fibre optic microprobe), and macrophytes are typically severely self-shading (e.g. Westlake 1966) so that subsaturating (i.e. limiting) irradiances of benthic plant stands can be much higher than for phytoplankton. Primary production measurements in rivers show that even midday irradiances in unshaded reaches under clear skies (around 2000 μmol m^{-2} s^{-1}) can be subsaturating for thick and severely self-shading periphyton mats (C. W. Hickey and J. M. Quinn, Water Quality Centre, unpublished data). Such growths are commonly absent from shaded reaches of rivers suffering nuisance biomasses in unshaded reaches. In streams subjected to discharge of clays from alluvial gold mining operations (Plates 10A and 10B), Davies-Colley et al. (1992) showed that periphyton biomass and productivity fell almost in proportion to the reduction in bed lighting. Thus light limitation due to a combination of riparian shading (where present), attenuation in the water column, and self-attenuation by plants may commonly restrict growth of benthic plants in rivers. This hypothesis warrants further scientific investigation.

4.3.4 Colour
Colour, as we have seen in Section 2.6.1, is generally difficult to assess because reflections of light from the water surface often represent a stronger signal than the light scattered back from within the bulk of the water. The colour of river waters is sometimes more difficult to observe than the colour of (optically) deep water bodies because of the confusion caused by reflections from bed features, as well as reflections of riparian features and sky colour from the water surface. The intrinsic colour of water in rivers during baseflow conditions is likely to be most noticeable where the water can be viewed at an acute angle so that the surface-reflected signal is small, or in the dark shadow of riparian vegetation (Plate 11). However, the much brighter 'muddy' colours typical of flood flows are well-known features of rivers. Fishermen, canoeists and other recreational users often speak of rivers at such times as being 'discoloured'.

Hue is the main aspect of water colour which has been studied in New Zealand rivers. Davies-Colley (1987b) reported direct Munsell hue matches to water colour observed in the field through a viewer at five stations on the 330 km long Waikato River (North Island) from its source in Lake Taupo to near the sea at Tuakau (Fig. 4.16). The optical properties of this water body are not typical of rivers in general because a chain of eight hydro-electric reservoirs impounds the waters, permitting phytoplankton growth to eutrophic levels. Hue changes monotonically in moving downstream in the Waikato

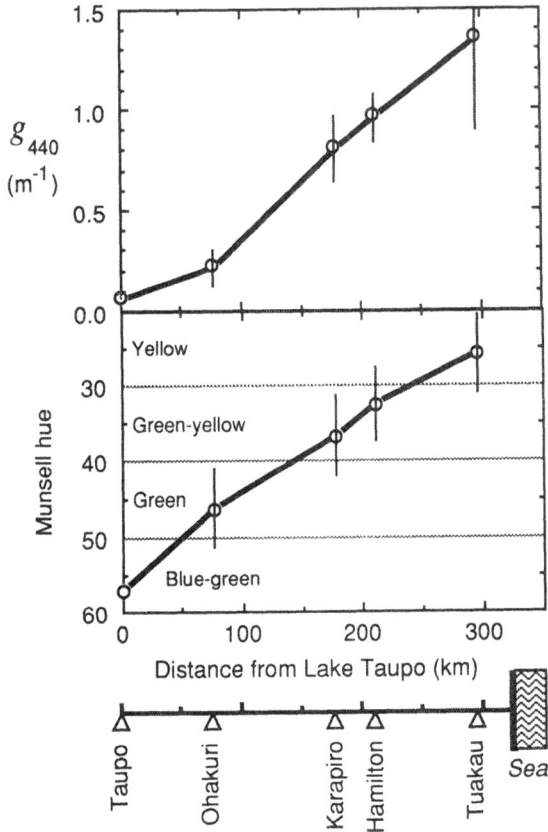

Fig. 4.16. Longitudinal profiles of yellow substance (absorption at 440 nm) and of Munsell hue in the Waikato River. (Adapted from Davies-Colley (1987b), with permission.) Measurements were made at 5 sites on the river. The source of the river is Lake Taupo at the town of Taupo. The first major hydro-electric impoundment is Ohakuri and there are six other large reservoirs downstream to and including Karapiro which re-regulates flows in the lower river. Measurements in the lower river were made at Hamilton and at Tuakau near the mouth of the river where it meets the Tasman Sea. The hue of the river is affected by the discharge of Kraft pulp mill effluent to Lake Maraetai, a reservoir about midway between Ohakuri and Karapiro.

River, consistent with an increasing trend in both total and filterable absorption. The blue–green water leaving oligotrophic Lake Taupo becomes green at Lake Ohakuri owing to the growth of phytoplankton in this eutrophic hydro-electric reservoir. A further hue shift occurs in Lake Maraetai (about halfway to the next monitoring site at Karapiro) owing to the discharge of highly light-absorbing Kraft pulp mill effluent, which has an absorption spectrum similar to natural yellow substance. At Karapiro, the furthest down-stream hydro-electric reservoir, the hue averages a green–yellow. Further changes in hue occur in the uncontrolled lower reaches of the river downstream of Karapiro owing to the discharge of tributaries draining large wetland areas (Davies-Colley 1979). At Tuakau the Waikato River averages a yellow hue, but the colour at this site is rather variable, depending on state of flow.

Davies-Colley & Close (1990) reported analyses of horizontal observations of water hue that were made at baseflow on 96 rivers throughout New Zealand using the black disc viewer (Davies-Colley 1988a) at times of visual water clarity measurement. The field observations were simple subjective descriptions rather than matches to standards since hue standards, such as those in the Munsell system, were not available to the 14 field teams who made the nationwide observations. However, validation studies of this method of hue description, reported by Davies-Colley & Close (1990), suggested that accuracy was fair (±4 Munsell units) and operator bias was absent. Fig. 4.17 shows the distribution of hues described on 175 occasions at 92 river sites as assigned to Munsell hue numbers. The straight line shape of the cumulative distribution curve shows that the distribution of hues was uniform except at the yellow–red extreme. Appreciable numbers of rivers were blue coloured at one extreme while others were yellow coloured, but a majority (58%) were in the green range of hues (green–yellow to blue–green).

The hues observed in the 92 rivers were weakly correlated ($r = -0.48$) with $\log(g_{440})$. The curvilinear (linear–log) relationship between hue and yellow substance concentration is consistent with the results of water colour modelling by Bukata *et al.* (1983). In blue-coloured waters, even small increments of yellow substance produce a marked hue shift towards green, but in yellow-coloured waters relatively large increases in yellow substance are required to produce any appreciable further hue shift towards red. Overall, a relative change in g_{440} of a given magnitude (e.g. 20%) produces approximately the same hue shift irrespective of hue.

Yellow substance accounted for only about one-quarter of the total variation in hue of the New Zealand river waters studied by Davies-Colley & Close (1990), so it is misleading to refer to yellow substance content (or, equivalently, platinum–cobalt measure-

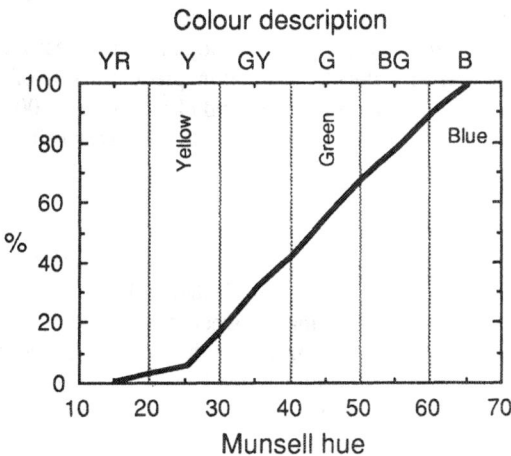

Fig. 4.17. Cumulative distribution of hue descriptions on the Munsell scale for 92 New Zealand rivers sampled at least once in 'baseflow' (< median flow) conditions (*n* = 175). YR, yellow–red; Y, yellow; GY, green–yellow; G, green; BG, blue–green; B, blue. (After Davies-Colley & Close (1990), with permission.)

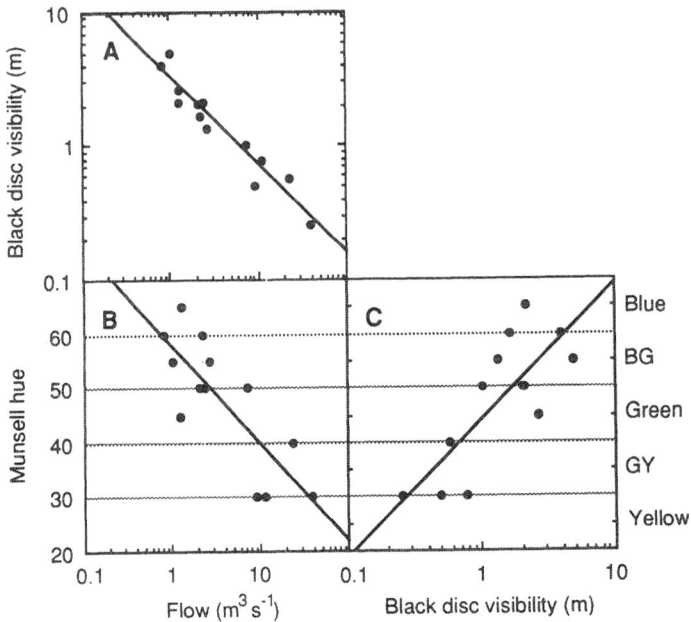

Fig. 4.18. Mutual relationships between colour (hue), visual clarity and flow in the Moawhango River, Central North Island, New Zealand. (A) Black disc visibility versus flow. (B) Hue versus flow. (C) Hue versus black disc visibility. (Plotted from data in Davies-Colley (1990a) and unpublished data.)

ments) as 'colour'. Colour (hue) of river waters also depends on the pigmented particles. Chlorophyll *a* is typically present only at low concentrations in river waters, and contributes insignificantly to their hue, but detrital organic particulate matter is expected to significantly shift hue in river waters by absorption, predominantly of blue light.

At 12 of the river sites sampled for colour and clarity in the '100 rivers' study, measurements were carried out at four-weekly intervals for a year using the same methods. The clarity observations have been published by Davies-Colley (1990a), but not the colour observations, hitherto. The results suggest that colour (hue) in river waters may be generally related to flow in accordance with common experience, in a similar way to clarity. For example, Fig. 4.18 shows the relationships between colour, clarity and flow in the Moawhango River, North Island, New Zealand. Overall the relationship between colour and flow was not as close as that between clarity and flow, in part, no doubt, because of larger operator error affecting colour than clarity. However, the relationship was sufficiently good to justify calculation of the hue-frequency distributions for some of the 12 rivers. Median hues are characteristic of rivers and permit their classification into 'blue', 'green' and 'yellow' water bodies.

Does a general relationship between colour and clarity exist for rivers? Fig. 4.19 shows Munsell hue number plotted against visual range (log scale) for 175 observations in 96 New Zealand rivers at baseflow. A weak ($r = 0.38$) but highly significant relationship is evident: as clarity increases from low values of a few hundred mm towards high values

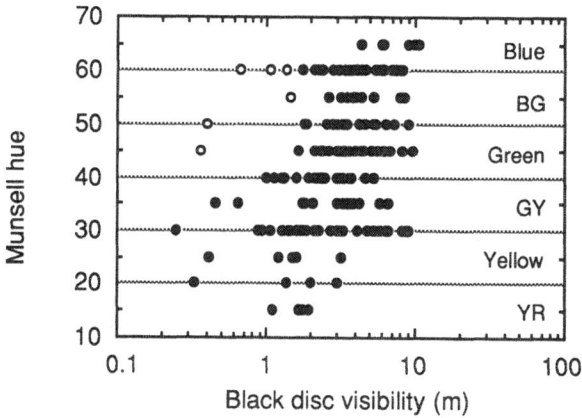

Fig. 4.19. Observations of water hue (as Munsell hue number) plotted against black disc visibility in 92 New Zealand rivers (up to three observations were made in each river). ○, observations in blue-coloured, but turbid, rivers draining the Southern Alps in the South Island of New Zealand, such as the Shotover River near Queenstown. (After Davies-Colley (1990a), with permission.)

around 10 m or more, there is a tendency for colour to change from orange or yellow towards blue–green or blue. That is, clear river waters tend to be blue, and turbid rivers tend to be yellow coloured (muddy in appearance). Some rivers rising in the Southern Alps of the South Island (e.g. the Shotover River) were exceptional in having green or blue water colours yet rather low water clarities (Fig. 4.19). These rivers have catchments dominated by physical weathering processes (i.e. glaciation) rather than chemical weathering involving organic matter, and have relatively high loads of suspended glacial flour and low organic matter concentrations.

Some rivers can have yellowish colours, yet fairly clear water. There were no obvious examples among the rivers studied by Davies-Colley & Close (1990) (which would have plotted on the lower right in Fig. 4.19), possibly because most sampling sites were near headwaters which are generally less affected by coloured wetland drainage than are lower reaches of rivers. As we have seen, a general trend towards hues at the orange end of the spectrum is often noticed in moving down-river from headwaters to the sea. However, on the humid west coast of the South Island there are numerous rivers with high contents of yellow substance (e.g. Collier 1987) that shifts the hue to yellow or even orange, depending on concentration. The yellow substance does not greatly attenuate image-forming light so that visual range is fairly high (>2 m) at times of low suspended solids concentrations during baseflows, even in orange-hued waters (Plates 7 and 10B).

4.4 ESTUARIES

Estuaries are semi-enclosed coastal waterbodies within which seawater is measurably diluted by inflowing freshwater (Pritchard 1967; Dyer 1973). The various ways in which saline and fresh water can mix within estuaries (e.g. Fischer *et al.* 1979) have a bearing on estuarine optics, particularly the spatial variability of scattering. In particular, under certain circumstances a circulation pattern arises which tends to concentrate suspensoids near the head of the estuary in a conspicuous 'turbidity maximum'. The turbulence generated by various water motions in estuaries serves to maintain particles in suspension (with a noticeable clarification near slack water when tidal currents are low (e.g. Postma 1961)), as well as resuspending sediments from the estuary bottom.

4.4.1 Suspended solids and the turbidity maximum
Suspended solids are transported to estuaries from the sea and, usually more importantly, the inflowing rivers. Biological production within estuaries is a further source of suspended particulates. Suspensoids are lost to the sea as well as to the estuary bottom, from whence they may be resuspended. Factors such as rainfall, catchment area and erodibility, estuarine morphometry and bottom sediment erodibility, wind strength and direction, and the tidal current regime all contribute to the external or internal supply of suspended solids to estuaries (McCave 1979; Dyer 1986, 1988), and are thus important in determining the colour and clarity of estuarine waters.

Resuspension of bottom sediments is often the main source of the suspensoids in estuarine waters (e.g. Gabrielson & Lukatelich 1985; Dyer 1986). Both wind waves and tidal currents are important eroding forces. For example, wind-induced resuspension is a major source of suspended solids to the Peel-Harvey estuary (Gabrielson & Lukatelich 1985), to Chesapeake Bay (Ward 1985) and to a Danish North Sea estuary (Pejrup 1986). Cloern et al. (1989) observed that resuspension by tidal currents was greater at spring tides than neap, with spring tides causing suspended solids concentrations to vary by an order of magnitude over a tidal cycle in the Tagus estuary (Vale & Sundby 1987). Sediment resuspension by tidal currents, and by wind waves, caused marked decreases in water clarity of a small inlet of Manukau Harbour, New Zealand (Vant 1991).

Where a substantial inflow of freshwater occurs, a zone of high turbidity is often present in estuarine waters. In this zone, located in regions of low salinity (1–5 salinity units), suspended solids concentrations are higher than in either the upstream river or the increasingly saline water towards the sea (salinity ~35). For example, Postma (1960) observed suspended solids concentrations in the range 150–200 g m^{-3} in the turbidity maximum of the Ems-Dollard estuary (at salinities of 1–2). By contrast, in the fresher, upstream waters concentrations were 30–50 g m^{-3}, with similarly low values (10–50 g m^{-3}) in the more saline reaches (salinity >10) towards the sea. Such a turbidity maximum has been recorded in many other partially mixed estuaries (e.g. the Thames (Inglis & Allen 1957), Chesapeake Bay (Schubel 1968) and the Loire (Gallene 1974)).

The turbidity maximum is thought to arise as follows (see reviews by Dyer 1986, 1988). Catchment-derived particles are carried downstream in the light, fresh surface waters until decreasing current velocities and saline-induced flocculation in the middle of the estuary cause them to settle into the denser, more saline, bottom waters. These bottom waters then carry the particles back upstream. Suspended solids are thus recycled and concentrated in a conspicuous zone of turbidity near the head of the estuary. This zone moves seaward with the ebbing tide, and also with increases in the freshwater inflow.

4.4.2 Light absorption in estuaries
Studies of estuarine optics have generally tended to focus more on the effects of light scattering than on absorption. Thus less is known about water colour in estuaries, for

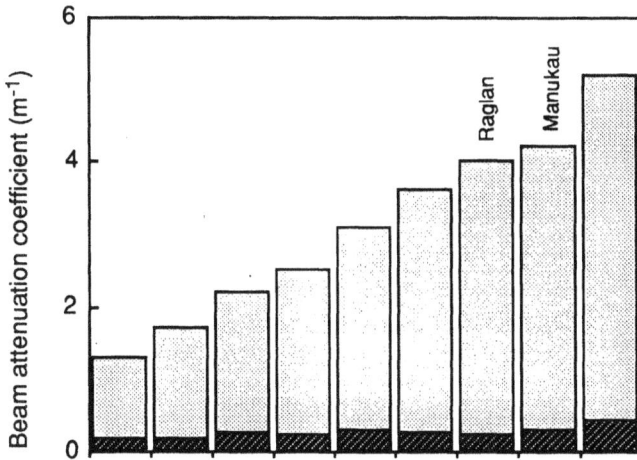

Fig. 4.20. The contributions of absorption (▨) and scattering (▦) to PAR waveband-average beam attenuation in nine northern New Zealand estuaries, summer 1988–1989 (after Vant (1990), with permission). The bar graphs for Manukau Harbour and Raglan Estuary (mentioned in the text) are identified.

example, than about water clarity. A possible reason for the apparent neglect of absorption in estuaries is that mineral suspensoids tend to dominate light attenuation in estuarine waters (e.g. Colijn 1982; Vant 1990). For example, in nine northern New Zealand estuaries the PAR waveband ratio of scattering to absorption (b/a) ranged from 7 to 17 (Vant 1990; see Fig. 4.20). This is not to say that absorption in estuaries is unimportant; rather that the consequences of scattering have been more conspicuous, and have therefore received more attention.

In one of the few published studies of absorption in estuaries, Gallegos *et al.* (1990) showed how water constituents contributed to spectral absorption in Chesapeake Bay. Absorption was greatest at blue wavelengths, with yellow substance, phytoplankton and other particulates all contributing. The particulates and water itself also caused appreciable loss of red light. Several observations of the spectral dependence of irradiance attenuation have shown a similar pattern (Kirk 1979; Champ *et al.* 1980; McPherson & Miller 1987; Stross & Sokol 1989; Vincent *et al.* 1989), with green to yellow wavelengths being attenuated least (e.g. Fig. 4.21). Hence estuarine waters are often green to yellow in hue. Furthermore, irradiance is lower at the wavelengths which are generally more useful for photosynthesis—blue to blue–green and red—than in the clear, blue waters of the ocean or unproductive lakes, for example.

A small fraction (3–11%) of the inflowing yellow substance, particularly the humic fraction, may be precipitated as a result of flocculation in the saline environment (Sholkovitz 1976). More importantly, yellow substance is diluted by the seawater so that its concentration generally decreases with increasing salinity (McPherson & Miller 1987). Concentrations therefore tend to be lower than those observed in nearby inland waters. Values of g_{440} were 0.1–0.6 m^{-1} in the northern New Zealand estuaries (Vant 1990), with those in Pelorus Sound being similarly low (<0.2 m^{-1}) and showing a marked gradient down the estuary to the much lower values in the sea nearby (Vincent *et al.* 1989). Such

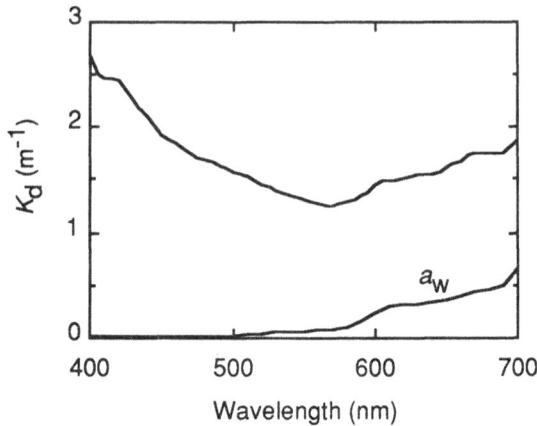

Fig. 4.21. Spectral variation in irradiance attenuation near the entrance of Manukau Harbour, April 1991 (authors' unpublished results). The absorption of pure water contributes to the red light attenuation, and this spectrum is shown for comparison.

values are at the low end of the range of concentrations found in New Zealand inland waters (Vant & Davies-Colley 1984; Davies-Colley & Vant 1987), although they are higher than those found off shore (typically 0.01 to 1 m^{-1} (e.g. Kirk 1983, p. 57; Davies-Colley 1992)). As a result yellow substance is generally only a minor cause of absorption in estuarine waters (Gallegos *et al.* 1990), contributing an average of only 24% of PAR waveband absorption in the northern New Zealand estuaries (Vant 1990).

Absorption was dominated by phytoplankton and other particulates in Chesapeake Bay, with major contributions (up to 4 m^{-1} averaged over the PAR waveband) in the more turbid, eutrophic areas (Gallegos *et al.* 1990). These particulates also dominated in northern New Zealand estuaries, although values of total absorption were rather lower there (<0.6 m^{-1}, Vant 1990). The absorption cross-sections of the phytoplankton in Chesapeake Bay and the northern New Zealand estuaries were similar. Expressed as PAR waveband averages the values were 14 and 16 m^2 (g chlorophyll a)$^{-1}$ respectively, similar to those found for cultures of certain marine phytoplankton (Bricaud *et al.* 1983), and for communities of phytoplankton in lakes (Section 4.2.3).

4.4.3 Clarity and brightness
Table 4.2 summarizes information on visual clarity and euphotic depths from studies on some New Zealand estuaries. Spatial variability within the estuaries is evident, as is the fact that Pelorus Sound waters tend to be clearer than those of the shallow northern North Island estuaries. By way of comparison, euphotic depths from saline waterbodies elsewhere in the world are broadly comparable with those in Table 4.2. Cloern (1987) found that euphotic depths in a group of 25 estuaries ranged from less than 1 m to 20–30 m, with mean values in individual estuaries usually 1–5 m. Sounds and fjords are clearer (10–45 m), while euphotic depths of 120 m have been recorded in clear oceanic waters (e.g. Sargasso Sea), where suspended solids concentrations are typically 2–3 orders of magnitude lower than the estuarine values listed in Table 4.2 (e. g. Simpson 1982).

Table 4.2. Measures of the clarity of some New Zealand estuaries (although the distinction is somewhat arbitrary, separate results for outer and inner estuary sites are shown)

	Secchi depth (m)	Beam attenuation coefficient, c (m^{-1})	Euphotic depth (m)	Turbidity (FTU)	Reflectance (%)	Suspended solids (g m^{-3})
Pelorus Sound[a]						
Outer	7.2	0.9	21	—	7	<1
Inner	2.0	2.8	8.5	—	7	8
Waitemata Harbour						
Outer[b]	1.7	3.6	6.1	3.5	9	10
Inner[c]	0.5	—	—	16	—	47
Manukau Harbour[b,d]						
Outer	1.9	4.2	5.9	4.1	8	12
Inner	0.7	—	2.4	19	—	53
Seven other shallow northern estuaries[b]						
Outer	1.2–4.4	1.3–5.2	4.4–16	1.8–5.6	6–12	4–13

[a] Vincent et al. (1989).
[b] Vant (1990).
[c] van Roon (1983).
[d] Vant (1991).

The high scattering-to-absorption ratios in the northern estuaries (Fig. 4.20) gave rise to high reflectances (6–12%) and a bright appearance. This was probably due to the relatively high concentrations of mineral suspensoids (mean values 3–11 g m^{-3}) which tend to be highly scattering and weakly absorbing (Kirk 1983, p. 30). Similarly high values of b/a (18) and reflectance (7%) were found in Pelorus Sound (Vincent *et al.* 1989), while scattering was also usually greater than absorption in Chesapeake Bay, particularly when mineral suspensoid concentrations were >5 g m^{-3} (Gallegos *et al.* 1990). Reflectances are likely to be high, making estuarine waters appear bright, whenever mineral suspensoid concentrations are high (>5 g m^{-3}), especially if concentrations of absorbing constituents such as phytoplankton are low.

Scattering cross-sections for estuarine phytoplankton assemblages have been found to range from 40 m^2 (g chlorophyll a)$^{-1}$ (from multiple linear regression of the data in Gallegos *et al.* 1990) to 160 m^2 g^{-1} (Vant 1990), at the low end of the range reported by Morel (1987) for marine phytoplankton (60–600 m^2 g^{-1}). Scattering cross-sections for mineral suspensoids in estuaries also vary, from 0.29 (Vant 1990) to 2.4 (Gallegos *et al.* 1990) m^2 g^{-1}, with a value of 0.72 m^2 g^{-1} being obtained for an estuary in Florida (C. L. Gallegos, personal communication). This difference is apparent from Fig. 4.22, with a

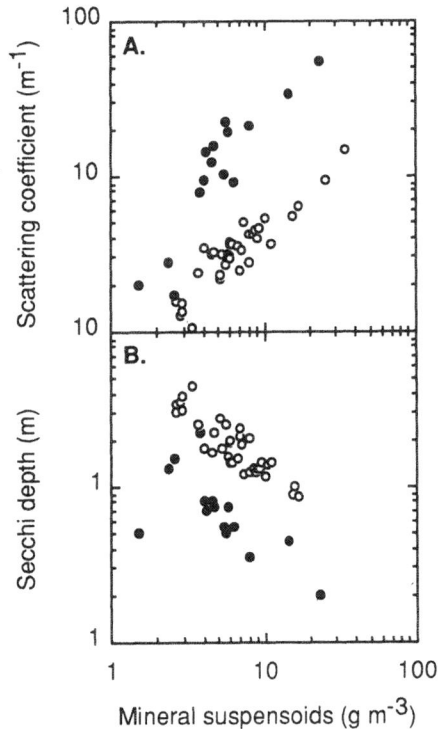

Fig. 4.22. Mineral suspensoids and (A) scattering and (B) Secchi depth in Chesapeake Bay (●, Gallegos *et al.* 1990) and nine northern New Zealand estuaries (○, Vant 1990).

given quantity of suspensoids causing more scattering (and as a result, lower visual clarity: Fig. 4.22B) in Chesapeake Bay than in the northern New Zealand estuaries.

Estuarine water clarity, as beam attenuation, can be estimated from absorption (Section 4.4.2) and scattering. Alternatively attenuation cross-sections of the constituents can be determined directly, as in the study of the northern New Zealand estuaries. The following equation described light attenuation in these estuaries (Vant 1990):

$$c = 180B + 0.30M + 0.94$$

where B is the chlorophyll a and M the mineral suspensoids concentration, and both coefficients are in $m^2 g^{-1}$. The constant ($0.94 m^{-1}$) represents the attenuation by organic detritus and, to a lesser extent, dissolved yellow substance and water itself. In these estuaries, mineral suspensoids caused 56% of beam attenuation on average, four times that of the sparse phytoplankton (chlorophyll a: 0.5–5 mg m^{-3}). Fig. 4.23 shows the partitioning of beam attenuation for one of the estuaries during summer. Phytoplankton (mean chlorophyll a 2.9 mg m^{-3}) contributed 15% of c overall while mineral suspensoids (mean 6.6 g m^{-3}) contributed 58%.

Even in nutrient-enriched estuaries, phytoplankton biomass is often relatively low, with growth being limited by other factors including flushing, grazing, and light limitation (e.g. Malone 1977; Schindler 1981). Blooms of phytoplankton do occur in estuaries (e.g. Paerl 1988), but their effect on colour and clarity is probably often masked by that of the abundant mineral suspensoids. Thus blooms may go undetected unless a programme of monitoring is underway.

As the waters of many estuaries, particularly shallow ones, are naturally turbid (McCave 1979), only major changes to constituent composition will usually have a noticeable effect on estuarine clarity. However, inflows whose optical characteristics are markedly different from background estuarine waters—such as highly coloured effluents—will be exceptions. For example, large waste stabilization ponds treating sewage from the City of Auckland discharge characteristically bright green and turbid effluent

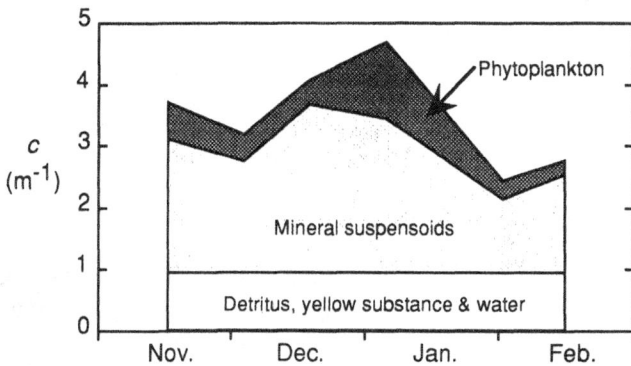

Fig. 4.23. Contributions to beam attenuation in the Raglan estuary, summer 1988–1989 (after Vant 1990, with permission). Only an average beam attenuation ($0.94 m^{-1}$) can be given for detritus, yellow substance and water.

into the waters of the inner Manukau Harbour near Auckland, New Zealand; this effluent can be observed, particularly from the air, as a distinct plume dispersing near the discharge point.

A number of studies have ignored the distinction between phytoplankton, detritus and mineral suspensoids and simply described the effect of total suspended solids (SS) on beam attenuation. The ratio $c/[SS]$—approximately equal to the attenuation cross-section for total particulates (ignoring the minor attenuation due to yellow substance and water)—has been found to be broadly constant at 0.3–0.5 m^2g^{-1} in a range of coastal waters (Jones & Wills 1956; Di Toro 1978; Krause & Ohm 1984; Vant 1990). However, the attenuation cross-section can be higher in some estuaries. For example, in Chesapeake Bay, $c/[SS] \sim 1.4$ m^2 g^{-1} from data in Gallegos et al. (1990) (Fig. 4.22)—suggesting dispersed clay mineral particles.

While the contributions to beam attenuation may be readily related to the constituent concentrations, the same is not true for the attenuation of diffuse light, particularly when scattering is relatively high (Kirk 1983, p. 236). Therefore PAR attenuation in highly scattering (high b/a) estuarine waters is not linearly dependent on constituent concentrations, although this has often been assumed to be so (e.g. McPherson & Miller 1987; Stross & Sokol 1989). The difficulty can be avoided by expressing the irradiance attenuation coefficient as a function of inherent optical properties (e.g. Kirk 1983, equation 6.7), which can be rigorously partitioned according to constituent concentrations. Values of the absorption and scattering coefficients in the presence and absence of particular constituents, and thus the corresponding values of K_d, can then be calculated. For example, Vant (1991) used the absorption and scattering cross-sections determined for the northern North Island estuaries (Vant 1990) to calculate the contributions of phytoplankton and mineral suspensoids to K_d in northern Manukau Harbour. Phytoplankton were minor contributors to irradiance attenuation (15–18%), while mineral suspensoids were rather more important (47–67%), as they were for beam attenuation.

4.4.4 Light climate for plant growth

Plants growing in turbid estuarine waters occupy a highly shaded environment, particularly those located at the turbidity maximum (e.g. Jackson et al. 1987). As in lakes, plants attached to the bottom in estuaries are unable to grow at depths beyond which average energy levels are less than those required for biological maintenance (the 'compensation point irradiance' at which photosynthetic oxygen production is balanced by respiratory requirements so that no net growth can occur). Phytoplankton are also unable to grow if they cannot resist being circulated through depths substantially greater than those in which photosynthesis can occur (i.e. the euphotic zone). Light limitation of plant growth has been described for a number of estuaries, and some illustrative results are summarized below.

Seagrass (e.g. Zostera) and other submerged, rooted macroscopic plants such as macroalgae (e.g. Ulva, Gracilaria, Enteromorpha) are common plants of the intertidal and subtidal regions of estuaries. Compensation point irradiances for these plants are such that they are rarely found below the 1% subsurface PAR level (e.g. Drew 1979; Josselyn & West 1985; Sand-Jensen 1988; Lumb 1990). Reduced light availability has been linked to observed declines in submerged estuarine vegetation (e.g. Peres & Picard 1975; Orth & Moore 1983; Cambridge & McComb 1984; McPherson & Miller 1987)

The resulting absence of a protective cover of vegetation in shallow areas exposed to wind wave and tidal currents can lead to resuspension of bottom sediments, thereby further increasing the turbidity of, and light attenuation in, the overlying waters. This increased turbidity can then be expected to further exacerbate vegetation decline (see Section 6.3). A short-term (2–5 year timescale) decline in macroalgal biomass was observed in the Peel Inlet, with a moderate recovery occurring in subsequent years (McComb & Lukatelich 1990). Throughout this period of change the biomass of macroalgae and of phytoplankton varied inversely, suggesting that shading by the phytoplankton was probably restricting the growth of the macroalgae.

The growth of estuarine phytoplankton can also be light limited (e.g. Colijn 1982; Pennock 1985). In situations where the mixed depth exceeds the euphotic depth, nonmotile cells will spend some of the time at depths where irradiances are lower than the

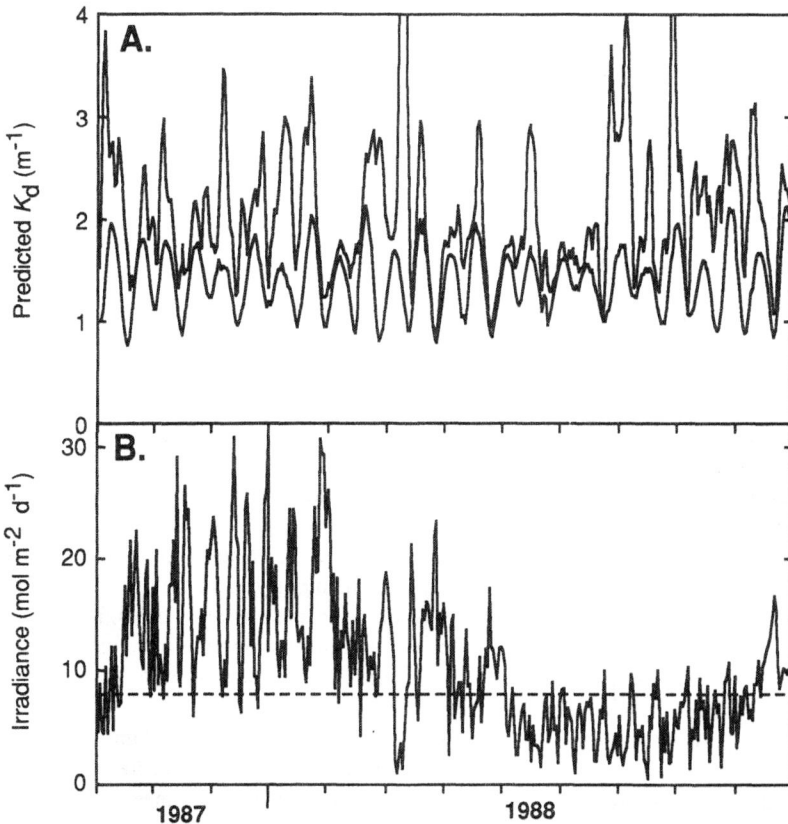

Fig. 4.24. Daily attenuation and underwater irradiance in the inner Manukau Harbour, October 1987–September 1988. (A) Predicted PAR attenuation (upper line). The lower line distinguishes the contributions to attenuation from tidal currents (below the line) and wind stress (above it). (B) Area-weighted, depth-averaged underwater irradiance (PAR). — — —, irradiance saturating to phytoplankton growth in San Francisco Bay in summer (Alpine & Cloern 1988). (Redrawn from Vant (1991), with permission.)

compensation value. The ratio of euphotic depth to mixed depth thus provides an index of the light climate for phytoplankton growth (cf. Fig. 4.8). Some studies have indicated that the critical value of this ratio is about 0.2 for estuarine waters (Cloern 1987; Alpine & Cloern 1988), as it is for lakes (Section 4.2.5). If the ratio is below this value, growth is precluded, while in the range 0.2–1 growth is light limited.

A more rigorous procedure is to determine actual irradiance levels and to compare these with the known requirements of the plants. Vant (1991) calculated daily irradiance attenuation caused by tidal current and wind wave resuspension for northern Manukau Harbour (Fig. 4.24A), and these results were combined with the solar radiation record to calculate average levels of irradiance throughout the year (Fig. 4.24B). Calculated irradiances mostly (>96% of the time) exceeded compensation point values determined elsewhere for marine phytoplankton (1.4 mol m^{-2} d^{-1} for San Francisco Bay (Alpine & Cloern 1988)). Irradiances known to saturate the growth of certain estuarine phytoplankton were also exceeded for much of the time, especially during the summer (Fig. 4.24B). It was therefore concluded that phytoplankton growth in northern Manukau Harbour is not absolutely light limited, and some other factor, such as grazing, may explain the comparatively low phytoplankton biomasses in this nutrient-enriched water body.

4.5 COMPARISONS

This section summarizes the material presented in this chapter by comparing the optical characteristics of three types of surface water bodies: lakes, rivers and estuaries. The morphometry and hydrodynamics of a given water body, together with climate and catchment geology and land use, are probably the main features which ultimately determine its characteristic water colour and clarity. Light-attenuating particulate constituents are less likely to settle out of the water column in flowing waters (rivers, estuaries) than in deep and slowly flushed lakes (but note that particle settling can also be prevented by wind-induced water movement in shallow lakes). Flowing waters generally tend to be more turbid than lakes, although rivers with a stable channel draining a catchment of hard, erosion-resistant rocks and pristine vegetation cover can be relatively clear. Furthermore, phytoplankton usually cannot form dense populations in flowing waterbodies as their cells are rapidly lost by flushing. They are therefore likely to be more prevalent, and thus more important, optically, in lakes than in rivers or well-flushed estuaries. Dissolved salts do not attenuate light, and saline waters are barely more light attenuating than freshwaters (Fig. 2.6). However, in these high ionic strength environments, primary particles tend to aggregate into flocs, which promotes their settlement or capture by vegetation, and reduces their light attenuation cross-sections.

The comparisons in this section are based on results from New Zealand water bodies. The diversity of lakes, rivers and estuaries present in New Zealand, together with the availability of comparable and reasonably comprehensive datasets, means that the comparisons are likely to have some generality, although they are not claimed to be universally representative. However, it must be acknowledged that some of the studies have been selective: rivers have been examined mainly at baseflow; and rather more is known about characteristics of outer estuary sites than of the more turbid inner estuary waters. Furthermore, temporal and spatial variability in rivers and estuaries is likely to be marked. By contrast, conditions in individual lakes are generally more uniform, so their optical properties can be characterized reasonably well from a smaller number of observations.

Most of the comparisons are based on the results from 61 lakes (Howard-Williams & Vincent 1984, 1985; Vant & Davies-Colley 1984), 96 rivers studied at baseflow (Davies-Colley & Close 1990) and 10 estuaries (Vincent *et al.* 1989; Vant 1990), although some information from a larger group of New Zealand lakes is also used (Livingston *et al.* (1986), $n = 138$), as is that for 77 rivers whose clarity has been measured approximately

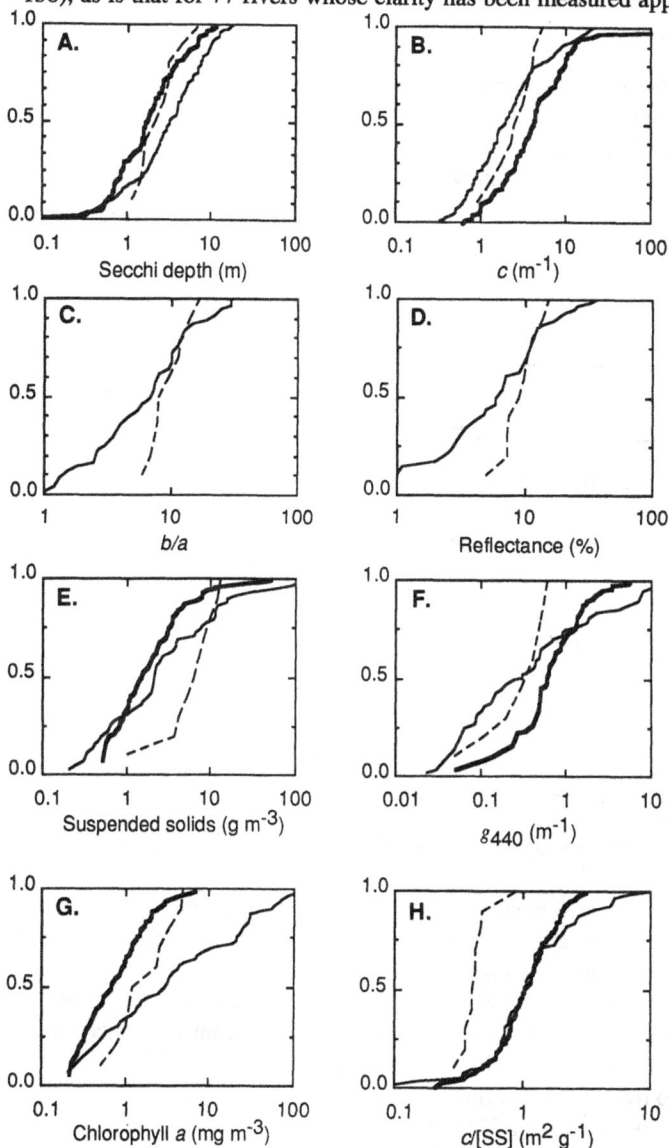

Fig. 4.25. Cumulative frequency distribution curves for optical properties and water constituents in 61 lakes (———), 96 rivers (———) and 10 estuaries (– – –). The Secchi depths in the rivers were estimated from the beam attenuation coefficients (see Table 4.3).

monthly (~25 times up to December 1990) over a wide range of flows in New Zealand's National Water Quality Network. Fig. 4.25 and Table 4.3 compare the distributions of certain of the optical properties and light-attenuating water constituents in these water bodies. The three groups had similar values of median beam attenuation coefficient (2.0–3.9 m^{-1}) and Secchi depth (1.6–3.3 m), although the lakes and rivers were more diverse than the estuaries (higher index of variability). In all the estuaries, the scattering-to-absorption ratio was high (b/a >6), while b/a was generally lower and was much more variable in the lakes. As a result reflectance was generally higher in the estuaries (median 9.5%) than the lakes (median 6.0%, but with a high variability), causing the estuarine waters to appear generally brighter than the lakewaters (except for a small number of particularly bright lakewaters: Fig. 4.25D). This is probably due to the prevalence of highly scattering and weakly absorbing mineral suspensoids in the estuarine waters (Vant 1990).

Table 4.3. Optical properties and constituent composition of New Zealand lakes, rivers and estuaries

	Medians			Index of variability		
	Lakes	Rivers[a]	Estuaries	Lakes	Rivers	Estuaries
Secchi depth (m)	3.3[b]	1.6[c]	2.4	1.5	1.5	0.8
c (m^{-1})	2.0[b,c]	3.9	2.8	1.4	1.5	0.9
R (%)	6.0	—	9.5	1.3	—	0.5
b/a	7.2	—	8.8	1.2	—	0.6
SS (g m^{-3})	2.2	1.4	6.8	3.5	1.6	0.9
Chlorophyll a (mg m^{-3})	3.0	0.7	1.8	7.2	1.5	1.6
g_{440} (m^{-1})	0.3	0.6	0.3	3.6	1.5	0.9
c/SS (m^2 g^{-1})	1.1	1.0	0.4	1.1	0.7	0.25

Values are medians and an index of variability (interquartile range/median). See text for data sources.
[a] Clarity measured ~25 times at a wide range of flows in 77 rivers (median used to characterize conditions in each), otherwise data are for 96 rivers at baseflow.
[b] n = 138 (from Livingston *et al.* 1986).
[c] Calculated assuming that the product of Secchi depth and beam attenuation coefficient is 6.4 (Kirk 1988).

Despite the broad similarity in values of beam attenuation coefficient, the constituent compositions of the lakes, rivers and estuaries differed appreciably. Median suspended solids concentrations were higher in the estuaries (6.8 g m^{-3}) than the lakes (2.2 g m^{-3}) or the rivers at baseflow (1.4 g m^{-3}), although once again the fresh waterbodies were more diverse, with concentrations differing by orders of magnitude between some of the lakes (Fig 4.25E). Yellow substance (as g_{440}) also varied widely from lake to lake and from river to river, while concentrations in the estuaries were more uniform (Fig. 4.25F). As expected, phytoplankton biomass (as chlorophyll *a)* was much higher in many of the lakes than the estuaries (Fig. 4.25G), where it was only a minor cause of beam attenuation on average (Vant 1990). Still lower chlorophyll a was found in the rivers (where much of what was measured is probably sloughed periphyton).

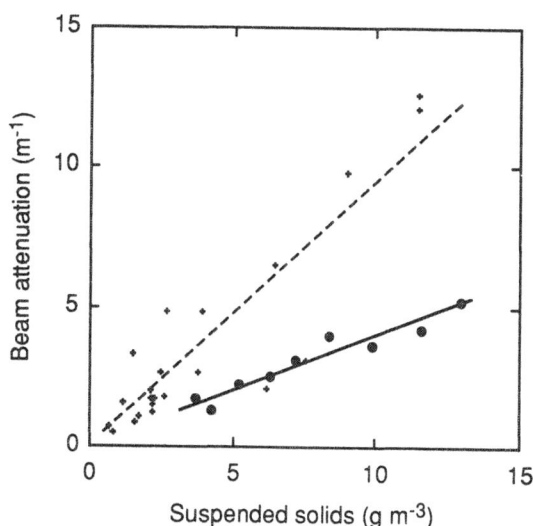

Fig. 4.26. Mean values of beam attenuation coefficient (c) and suspended solids (SS) in nine estuaries (●) and 22 lakes (+). The equations of the lines are $c = 0.40$ SS (———) and $c = 0.95$ SS (— — —). (Redrawn from Vant (1990), with permission.)

The ratio of beam attenuation to suspended solids, a rough measure of the beam attenuation cross-section for total particulates, was low (median 0.4 $m^2\,g^{-1}$) and reasonably constant in the estuaries (Fig. 4.25H). Median values were appreciably higher in both rivers and lakes (1.0–1.1 $m^2\,g^{-1}$). The generally larger suspended particles present in estuarine waters attenuate light less efficiently than those in lakes (Vant 1990; see Fig. 4.26). Particle aggregates and larger silt grains in the estuaries are maintained in suspension in the generally turbulent conditions there.

The way in which colour and clarity vary within a particular water body is also dependent on water body type: characteristic patterns of both temporal and spatial variability are found within lakes, rivers and estuaries. Spatial variability in a river is usually caused by the increasing loads of light-attenuating constituents which are added as the river proceeds from its often pristine headwaters to the more-developed lower reaches (Davies-Colley 1990a). Incomplete mixing of optically distinct coastal and riverine water masses within estuaries causes spatial variability in colour and clarity in these waterbodies. Resuspension of bottom sediments and the hydrodynamic trapping of particulates within the turbidity maximum add to this variability. By contrast, lakes are usually more optically homogeneous.

Temporal variability in light attenuation is illustrated for a North Island lake, river and estuary in Fig. 4.27. Although observations were only made at monthly intervals in the lake, it is likely that shorter-term variability is only minor in this particular water body. Beam attenuation here was dominated by phytoplankton biomass (Vant & Davies-Colley 1986) and major changes in this tended to reflect seasonal (i.e. longer-term) cycles. Not surprisingly then, seasonality was apparent in the record for the lake (Fig. 4.27A). Seasonality was also apparent in the river record (Fig. 4.27B) where it reflects the broad seasonality in river flow, but was not obvious in the estuary (Fig. 4.27C).

The other difference between the three water bodies is in the short-term (i.e. daily) fluctuations apparent in the records for the river and the estuary. Attenuation changes from day to day as river flow responds to rainfall, and as wind velocities vary over the estuary. Such short-term variability tends to be uncommon in lakes unless they too are

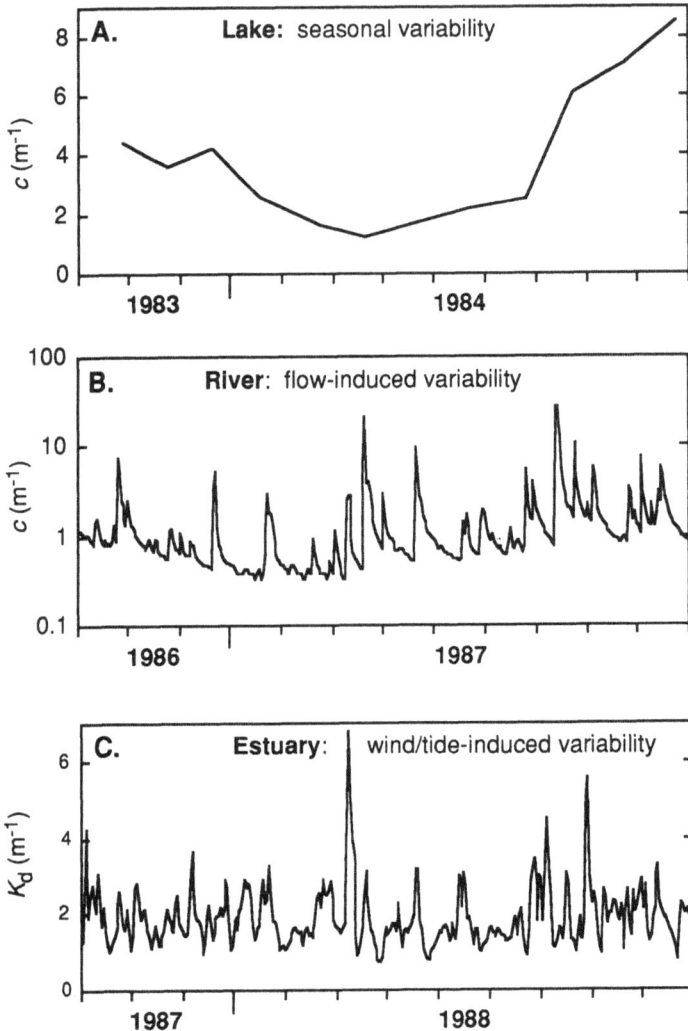

Fig. 4.27. Attenuation timeseries in three water bodies. (A) Measured beam attenuation in Lake Okaro. (Redrawn from Vant & Davies-Colley (1986), with permission.) More information on this lake is given in Section 6.4. (B) Predicted beam attenuation in the Kauaeranga River (calculated from river flow and the relationship in Davies-Colley (1990a)). Note the log scale of c. (C) Predicted irradiance attenuation in the inner Manuaku Harbour. (Redrawn from Vant (1991), with permission.)

very rapidly flushed or contain large areas of shallows over which wind-induced resus-pension can occur. Conversely, attenuation would probably fluctuate less rapidly in deeper estuaries or in rivers where flow variability is slight.

The potential for light limitation of plant growth also varies between the three types of water bodies. New Zealand river waters at baseflow are typically relatively shallow and clear so light attenuation within the water column is generally not great (Section 4.3.3), although shading by mining wastewaters has been shown to reduce periphyton growth in some small streams (Davies-Colley *et al.* 1992). Otherwise, shading by ripari-an vegetation is probably more generally important in rivers. Information from many diverse lakes indicates that the degree of light limitation varies markedly from lake to lake, being lowest in shallow, clear-water lakes (Fig. 4.8). High spatial and temporal variability in estuaries has made the underwater light climate difficult to characterize, and rather little is known about the light-limitation in New Zealand estuaries in general. Although turbid, these water bodies are typically also shallow, so sufficient irradiance for plant growth may be common (Vant 1991).

To summarize, mineral particles tend to be maintained in suspension by turbulence in flowing waters, and phytoplankton growth is limited by hydraulic flushing, so the light-attenuating constituents present in lakes, rivers and estuaries usually differ appreciably. Phytoplankton abundance is usually lower in rivers and estuaries than in lakes. Even though light scattering from the larger particles suspended in turbulent estuarine waters tends to be inefficient, the high concentrations of mineral suspensoids found there mean that scattering is relatively high, so that estuarine waters are often turbid and are general-ly bright in colour because of relatively high b/a ratios. Concentrations of light-attenuat-ing particulates are lower in many lakes and rivers at baseflow. Finally, a high degree of spatial and temporal variability in optical properties is typically observed in individual rivers and estuaries, while lakes are usually more optically homogeneous.

5

Managing colour and clarity

5.1 INTRODUCTION

In this chapter we consider the management of optical quality of natural waters.

The two categories of water use most sensitive to colour and clarity are recreation (including both passive aesthetic use and contact recreation) and habitat for aquatic life. However, water supply (including industrial supply) is also sensitive to optical quality of water.

The next section (5.2) critically reviews the criteria, guidelines, and recommended standards for colour and clarity to protect uses of natural water which have been produced by several agencies around the world. The review discusses the overseas colour and clarity recommendations under three main headings—aesthetics and recreational water use (including bathing), aquatic life protection, and, for completeness, other uses (i.e. potable supply, and industrial uses). Colour and clarity are probably not an issue for agricultural uses. Because almost any raw water can be treated for potable supply and industrial use (at a cost, of course), the first two use categories are by far the most important here. The recommendations developed by the major review agencies (e.g. the EEC, USEPA) are discussed, in part as a historical note, but also to set the scene for development of improved guidelines for colour and clarity.

In Section 5.3 we outline our approach to the development of improved guidelines for the protection of the optical quality of waters. Most of the guidelines are given as maximum relative changes, following the recommendations of Kirk (1988). Management tools for calculating the changes in visual clarity, light penetration, and colour of waters induced by effluent discharge or diversion of waters are then considered. Given the optical characteristics of the inflow water (effluent or tributary) and those of the receiving water, together with the dilution ratio, it is possible to calculate the change in optical character of the receiving water to be expected following a management change.

Although this book is about natural waters we need to consider the optical properties of effluents when discharged to natural waters. In Section 5.4 we consider the optical character of various effluents (point sources) which can affect the colour and clarity of natural waters. Many effluents are strongly light attenuating so it is somewhat surprising that waste treatment engineers or, for that matter, the stewards of receiving water quality have not paid more attention to the optical impact of wastewaters. For example, in their otherwise valuable text 'Wastewater engineering', Metcalfe & Eddy (1979) almost ignore this aspect of wastewaters and give most attention to oxygen demand and solids content and, to a lesser extent, infectious micro-organisms, nutrients and toxic materials. In their Table 3.2 listing constituents of concern in wastewaters, suspended solids are regarded as of concern because of their propensity to form sludges and to create anaerobic conditions, but their optical water quality degrading potential is apparently not recognized. Two exceptions to the general neglect of the optical properties of wastewaters are, firstly, in the pulp and paper industry with respect to highly coloured bleached Kraft effluent (Gellman 1982), and, secondly, in the paper and textiles industries where dye wastes can produce highly 'unnatural' colours in receiving waters.

The chapter then goes on to consider the control of those constituents of waters and wastewaters which are primarily responsible for degrading optical quality: suspended solids and phytoplankton. Section 5.5 discusses suspended solids generated from soil erosion—that is diffuse or non-point sources of suspended solids. Of most concern is the suspended matter near the attenuation cross-section maxima at around 1 μm diameter for mineral solids and about 3 μm for organics. Light scattering is usually dominated by the inorganic components, but the light absorption by organic coatings adsorbed on the mineral surfaces, or that of organic detrital material eroded along with the mineral particles, may also be important. Preventing degradation of optical quality of a given water body then becomes a problem of controlling soil erosion in its catchment, particularly those contributing areas such as bare soil in construction areas, roads, mine overburden, and active mass erosion scars. Cure of optical quality degradation by removal of fine-grained solid particles from water is much more difficult than prevention of their mobilization in the first place.

The other main source of optical water quality problems usually derives from the entry into standing waters (especially lakes) of the nutrient elements nitrogen and phosphorus and the consequent growth of phytoplankton to 'nuisance' levels. The nuisances posed by excessive phytoplankton biomasses can include filter blockage, toxicity of cyanophytes (blue–green algae) and surface scum formation. However, paramount in lakes used for recreation is the optical quality degradation related to light attenuation by phytoplankton. Preventing degradation of optical quality of a given water body then becomes a problem of controlling both point and diffuse nutrient sources. Control of phytoplankton growth, mainly by controlling nutrients, is discussed in Section 5.6.

5.2 REVIEW OF INTERNATIONAL CRITERIA AND GUIDELINES

The terms: criteria, guidelines and standards, much used in the field of water quality management, require definition at this point. The definitions we will use henceforth are as follows. *Criteria* (singular, criterion) are scientific findings used to derive guidelines or standards for water use. For example, in the field of aquatic toxicity the finding that half of a sample of test organisms dies after a given time of exposure to a certain concentration of a toxicant is a criterion. Empirically derived curves relating suitability for use to an appropriate measurement of some attribute, e.g. water clarity, would constitute a set of criteria. Until recently there were few examples of such criteria for colour and clarity. *Guidelines* are numerical or narrative statements whose application should maintain a certain water use. Numerical guidelines should be based on criteria. In effect a guideline translates criteria into a form which recommends how a standard could, or should, be written. This may call for a value judgement on acceptable risks or requirements. *Standards* are guidelines which have statutory force to maintain certain water uses. Standards should be supported by criteria.

The terms criteria, guidelines and standards are sometimes, confusingly, used interchangeably and one author's 'guidelines' may be another author's 'criteria'. Also documents developing 'recommendations' (i.e. guidelines) to support certain water uses have sometimes used the term 'criteria' in their titles (e.g. NTAC 1968; USEPA 1973a). The above definitions are broadly consistent with those of the Task Force on Water Quality Guidelines of the Canadian Council of Resource and Environment Ministers, entitled 'Canadian water quality guidelines' (Anon 1987).

5.2.1 Aesthetics and recreational water use
In 1968 the National Technical Advisory Committee to the US Secretary of the Interior produced a review document entitled 'Water quality criteria' (NTAC (1968), commonly referred to as the 'Green book'). In their summary of recommendations on general aesthetic requirements they state that 'surface waters should be free of substances attributable to discharges or wastes as follows: ...substances producing objectionable color... or turbidity'. This document was far sighted in providing for explicit protection of the optical quality of water. However, there are two problems with this recommendation. Firstly it is simply not analytically meaningful to state that a water should be 'free' of anything; secondly there is no definition of 'objectionable'.

In their summary for primary contact recreation waters (under the heading 'criteria for desirable factors'), NTAC (1968) states '...clarity should be such that a Secchi disc is visible at a minimum depth of 4 feet. In "learn to swim" areas, the clarity should be such that a Secchi disc on the bottom is visible'. There is an additional non-numeric stipulation for diving areas which states that 'the clarity shall equal the minimum required by safety standards, depending on the height of the diving platform or board'. No examples of such safety requirements were presented, however. NTAC (1968) provide no support for their guidelines ('criteria') other than to state that 'clarity in recreational waters is highly desirable from the standpoint of visual appeal, recreational enjoyment, and safety'. These recommendations seem to be concerned mainly with safety considerations rather than aesthetics, but this is not explicitly stated. We will return to this point later in this section.

The US EPA 'Blue book' (USEPA 1973a) amplified the above recommendations and stated that, for bathing and swimming, water should 'be clear enough for users to estimate depth, to see subsurface hazards easily and clearly, and to detect submerged bodies of swimmers or divers who may be in difficulty'. Also, 'the clearer the water, the more desirable the swimming area'. Both of these statements are sensible as far as they go, but for water management they are of little help in predicting the effects of discharges on use of such waters. As with NTAC (1968), the separate effects of colour and clarity are combined, probably because the optical basis of these concepts was not understood.

The Europeans separated colour and clarity in their Directive on the quality of designated bathing waters as follows (EEC 1976).

Colour: Guideline requirement—no stipulation.
> Mandatory (a minimum) requirement—no abnormal change except in the event of exceptional geographical or meteorological conditions (to be carried out by visual inspection or photometry with standards on the Pt–Co scale).

The mandatory requirement is interesting in referring to change in colour. However, the baseline from which a 'change' can occur is not defined and nor is the term 'abnormal'. These requirements do not address the possibility that some 'abnormal change' may result in a water just as acceptable to users. Presumably the colour after filtration is to be measured ('true colour', Section 3.5.1), but this is not specified.

Transparency: Guideline requirement—2 m Secchi disc depth.
> Mandatory requirement—1 m Secchi disc depth except in the event of exceptional geographical or meteorological conditions.

Here we have absolute numerical requirements (and a fortnightly sampling frequency is specified), but there is no apparent justification for these values. There is also an unsubstantiated sample percentile requirement for each numerical value (95% conformity for the mandatory value, 90% for the guideline), an approach apparently only employed in Europe.

The most recent US EPA commentaries on the topic (USEPA (1976), the 'Red book', and USEPA (1986), the 'Gold book') make no further advance. Indeed, the 1986 publication merely repeats, verbatim, the sections in the 1976 document on 'color' and 'solids' and 'turbidity', and omits entirely the section on 'aesthetic qualities'. In their Aesthetic Qualities section USEPA (1976) state 'All waters free from substances attributable to wastewater or other discharges that: ...produce objectionable color or turbidity...'. Again, as with NTAC (1968), it is not defensible analytically to refer to 'free from' without, for instance, definition of a detection limit or at least a defined analytical method from which a detection limit could be estimated or derived. However, at least the concepts of colour and turbidity (= lack of clarity) have been separated. In their specialized section on colour, and when considering aesthetics, the statement changes to 'waters shall be virtually free from substances producing objectionable color for aesthetic purposes'. 'Objectionable' here is defined as a significant increase over natural background levels. However, no comment is made as to what constitutes 'significant' (a change could be statistically significant, but quantitatively and qualitatively of minor consequence), and, further, we now have the phrase 'virtually free from' which is also not defined and is possibly of less value than the earlier 'free from' which at least could have some analytical meaning if combined with a defined detection limit. Additionally, in the Rationale section the following statement appears: 'non-natural colors such as dyes should not be perceptible by the human eye', but the term 'perceptible' occurs without amplification.

USEPA (1976, 1986) also stated that 'because of the extreme variations in the natural background amount of color, it is meaningless to attempt numerical limits' and that 'the aesthetic attributes of water depend on one's appreciation of the water setting'. However, the possibility that scientific research into these topics might be of value in deriving numerical limits was not recognized. USEPA (1976, 1986) acknowledged that turbid water interferes with the recreational use and aesthetic enjoyment of water, and that there are potential dangers for swimming, but they stated only that 'the less turbid the water the more desirable it becomes for swimming and other water contact sports'. No attempt was made to develop a relationship between suitability of water for use and clarity.

The American Fisheries Society (AFS 1976) reviewed the 'Red book' and felt that the sections referred to above dealt too briefly with the issues of colour and clarity. However, no improvement to the guidelines was suggested except to insert the phrase 'in amounts' after 'discharges' in the aesthetic quality 'criterion' quoted above. This obviates the analytical nonsense referred to but begs the issue of defining the 'amounts'.

The Victorian EPA (Australia) has produced a compilation of 'Recommended water quality criteria' (EPAV 1983). They stated that 'clear water is desirable for "learn to swim" areas from a safety viewpoint for primary contact recreation waters' (their class 1 waters). For such waters the criteria are: 'a Secchi disc should be visible at a maximum [*sic*] depth of 1.2 m or approximately 50 Jackson Turbidity Units. In "learn to swim" areas a Secchi disc should be visible on the bottom'. The use of the term 'maximum' rather than minimum is surely an error, and no evidence is presented to support the 50 JTU stipulation which is inconsistent with the 1.2 m Secchi disc depth. Typically turbidity as measured on a Hach 2100A nephelometer in NTU (NTU ~ JTU (McCluney 1975)) is approximately $4/z_{SD}$, so that a 1.2 m Secchi depth corresponds to about 4 NTU. (Note that, as discussed in Section 3.3.3, turbidity is a highly instrument-specific measurement.)

In terms of colour, the EPAV (1983) 'criterion' is 'colour should not exceed 100 Pt–Co units'. This value seems extremely high (corresponding to a very dark water of orange hue—the visually very degraded Tarawera River downstream of the discharge from a Kraft pulp mill (Plate 11) has a colour of 80 Pt–Co units (Wilcock & Davies-Colley 1986)). EPAV (1983) state that 'in many cases there is insufficient data to quantify population preferences with respect to aesthetic properties of water'—a more cautious approach than that of the USEPA (1976, 1986) who asserted, without basis, that it is 'meaningless to attempt numerical limits'.

For 'secondary contact' recreation waters (i.e. for activities including wading, boating and fishing in which some direct contact with water may occur, but in which the probability of bodily immersion or the intake of significant amounts of water is minimal) and also for aesthetic appreciation waters, there are no clarity criteria, but the colour 'criteria' are as for primary contact.

Canada has published the 'Canadian water quality guidelines' (Anon 1987). This major document is frequently updated as new evidence comes to light. In the General Requirements section it is stated that 'visual impact of the whole area is as important as the quality of the water'. Anon (1987) do not substantiate this assertion by citing any panel studies. Their aesthetics guideline states that 'all waters should be free from substances attributable to wastewater or other discharges in amounts that would interfere with the existence of life forms of aesthetic value: these include ...substances producing objectionable colour... or turbidity' (unchanged in the April 1992 update). The emphasis here is not on human aesthetic perception of the water *per se*, but on protection of habitat for aquatic life forms which are considered desirable.

The clarity guideline merely repeats the NTAC (1968) recommendation that a Secchi disc should be visible at a minimum depth of 1.2 m. Anon (1987) point out that 'it is important that swimming areas are clear enough for users to estimate depth and to see subsurface hazards easily'. However, while this justifies a visual clarity guideline, no evidence is presented to support the particular value stated. Neither is there evidence presented supporting the statement referring to a Secchi disc on the bottom being visible in 'learn to swim' areas. (In fact, if the Secchi disc is on the bottom, the optical basis of the measurement, which requires that this target be viewed against the water background, is invalidated—refer to Appendix 2.5.)

For turbidity, the Anon (1987) guideline is 'the turbidity of water should not be increased more than 5.0 NTU over natural when turbidity is low (< 50 NTU)'. This guideline represents an important advance over previous bathing guidelines in that, rather than defining a maximum value, a maximum change in clarity is stipulated. However, a large percentage increase could be permissible in a clear water (e.g. >100% change in waters < 5 NTU). This is unsatisfactory because very clear waters are likely to have great aesthetic value as a consequence of their high clarity and therefore require a high level of protection. Again the numerical values recommended are not substantiated by any research findings. A suggestion was made that both turbidity and clarity determinations should be made when assessing visibility and safety of swimmers, but without further explanation.

In their section on colour, Anon (1987) stated that 'a guideline for the colour of recreational water largely depends on the preference of users, and it is impossible to put an absolute value on it'. Again, this statement (echoing USEPA (1976, 1986)) is made without substantiation. However, Anon (1987) did state that colour should not be so intense as to impede visibility in areas used for swimming. Their use of the word 'intense' is vague. They cite a 1972 limit of 100 Pt–Co units proposed by Environment Canada and state that no rationale for this was given, but themselves make no advance.

Most recently, water quality guidelines for use in Australia have been published by the Australian and New Zealand Environment and Conservation Council (ANZECC 1992). The section on recreational water quality and aesthetics seeks to protect both visual clarity and colour and gives guidelines to protect visual clarity and colour for aesthetic purposes, and visual clarity for swimming, based on our research findings. These guidelines will be discussed in Section 5.3.

5.2.2 Protection of aquatic life

The US EPA 'Blue book' (USEPA 1973a) recommends that for freshwater aquatic life and wildlife 'the combined effect of color and turbidity should not change the compensation point more than 10 percent from its seasonally established norm, nor should such a change place more than 10 percent of the biomass of photosynthetic organisms below the compensation point'. However, no research findings are cited as evidence to support these values which appear rather arbitrary. (Presumably compensation 'point' is the compensation depth, i.e. that depth below which the time-averaged production of oxygen is exceeded by respiratory consumption.) In addition, the compensation depth may not be defined in shallow, clear rivers and streams, which nevertheless still require some protection of their light climate.

USEPA (1976, 1986) stated in the 'criterion' section that 'increased color (in combination with turbidity) should not reduce the depth of the compensation point for photosynthetic activity by more than 10 percent from the seasonally established norm for aquatic life'. This is a rephrasing of the USEPA (1973a) recommendation. However, there is also a separate, similar, statement for turbidity which commences 'settleable and suspended solids should not reduce the depth...', and then continues as above. This correctly recognizes that both turbidity (= light scattering) and colour (= light absorption) contribute to irradiance attenuation in water bodies.

In their commentary on USEPA (1976) the American Fisheries Society (AFS 1976) felt that the colour guideline ('criterion') did not adequately address the effects of colour on aquatic life and that only a weak case was presented. In particular they noted that the use of a 'compensation point' to indicate a zone of 'effective photosynthesis' was 'only vaguely substantiated' as was the maximum value of 10% change. Also, the difficulty of measuring the 'compensation point' was not considered. AFS (1976) felt that USEPA (1976) had oversimplified the USEPA (1973a) recommendation in omitting the suspended solids requirement for fishing protection. They also criticized USEPA's (1976) synonymous use of 'suspended solids' and 'turbidity' and suggested a separate consideration of these concepts. For turbidity they suggested use of a nephelometric analytical method but recommended more research before stating any numeric guideline.

The EEC Directive to support freshwater fish life (EEC 1978) does not provide any stipulations for colour and clarity, but the Directive to protect marine shellfish waters (EEC 1979) surprisingly does. There is no guideline value but there is a mandatory colour stipulation which states 'a discharge affecting shellfish waters must not cause the colour of the waters after filtration to deviate by more than 10 mg Pt/l (i.e. 10 Pt–Co colour units) from the colour of the waters not so affected'. The minimum sampling frequency is quarterly and 75% of samples must conform to the numerical value given. No justification is presented in the Directive. There is no turbidity stipulation, but there is a mandatory suspended solids requirement of no more than a 30% increase, again without substantiation. The relative change structure of this guideline is noteworthy.

The Victorian EPA (EPAV 1983) specify several levels of protection for the maintenance of aquatic ecosystems, and for each present recommendations for turbidity and colour. For Level I (maximum protection corresponding to a natural or pristine state) and Level II (high level of protection) the recommendation follows USEPA (1973a): 'the combined effect of turbidity and colour should not reduce the depth of the compensation point for photosynthetic activity by more than 10% from the seasonal background value'. For Level III protection waters ('medium level of protection—water quality will undergo some degradation due to waste discharge but not sufficient to cause harmful effects to biota') a 25% change is permitted. For Level IV waters ('minimum level of protection consistent with unimpaired functioning of the ecosystem—water degraded to the threshold level of harmful effects') a 50% change is permitted. All the above guidelines are presented without scientific justification.

The Canadian Water Quality Guidelines (Anon 1987) present no colour specifications for aquatic life, but there is discussion on suspended solids and their effect on clarity. Of interest here is the statement that 'the Ontario Ministry of the Environment (1984) recommended that suspended matter should not be added to surface water in concentrations that will change the natural Secchi disc reading by more than 10%'. This is an interesting recommendation which was apparently ignored in the final Anon (1987) guideline document which makes no mention of Secchi depth or indeed any measure of visual water clarity and refers only to limiting suspended solids (SS) increase, ignoring the argument that a major reason for limiting SS is concern with visual clarity. Possibly a clarity guideline as above was omitted because defining a 'natural' Secchi disc clarity would require considerable monitoring effort.

The most recent Australasian guidelines for aquatic life protection (ANZECC 1992) present recommendations in terms of the euphotic depth based on our research. These guidelines will be discussed in Section 5.3.

5.2.3 Other uses

The USEPA (1973a) recommended a maximum colour of 75 Pt–Co units for raw water (presumably after filtration, i.e. 'true colour', but this is not stated). This recommendation was based on the limitations of the defined treatment process which includes coagulation, sedimentation, filtration, and disinfection. No recommendation was made for turbidity because 'customary methods for measuring and reporting turbidity do not adequately measure those characteristics harmful to public water supply and water treatment processing' and that turbidity in water should be readily removable by treatment. USEPA (1973a) states that turbidity 'should not be present to an extent that will overload the water treatment plant facilities' nor cause 'unreasonable treatment costs'.

The latest US EPA commentaries on the subject (USEPA 1976, 1986) repeated the above USEPA (1973a) recommendation and noted that water can be treated using standard coagulation, sedimentation and sand filtration to remove colour to substantially less than an acceptable 15 Pt–Co units when the source water does not exceed 75 Pt–Co units.

USEPA (1976, 1986) also maintained that, because of the ability of common water treatment processes (as above) to remove suspended matter and to achieve acceptable final turbidities, and because turbidity is a function of the composition of suspended material as well as its concentration, it is not possible to state a raw water turbidity 'criterion' of general application to municipal (and industrial) water supply.

The AFS (1976) review noted that the 75 Pt–Co units guideline was either acceptable to, or not commented on, by their reviewers.

In Europe, EEC (1975) derived characteristics of surface water intended for the abstraction of drinking water. These range from 10 Pt–Co units (after simple filtration) as a guideline value for water receiving 'A1' simple physical treatment (i.e. rapid filtration and disinfection), to 200 Pt–Co units as a mandatory maximum value for water receiving intensive physical and chemical treatment. The higher value given here (as with other mandatory values) was for exceptional climatic or geographical conditions.

EEC (1975) give a suspended solids guideline value (of 25 mg l^{-1}) for 'A1' waters only but specify a turbidity stipulation (in mg SiO_2 l^{-1} and in Jackson units) for treated water (EEC 1980). Surprisingly, there is also a Secchi disc 'Guide Level' of 6 m and a 'maximum admissible concentration' [sic] of 2 m (EEC 1980); in addition, there is a colour 'guide level' of '1 mg/l Pt/Co scale' and a 'maximum admissible concentration' of 20.

The Victorian EPA (EPAV 1983) raw water colour 'criteria' for water requiring no treatment, disinfection only, or special treatment are a 'current acceptable' value of 50 Pt–Co units and a 'long term desirable' value of 5 Pt–Co units. The corresponding recommendations for turbidity are 25 NTU and 5 NTU respectively. For raw water requiring 'complete treatment' the sole colour criterion is 50 Pt–Co units and the sole turbidity 'criterion' is 10 000 NTU.

The 'Canadian water quality guidelines' (Anon 1987) make no recommendation for raw water.

ANZECC (1992) stipulate a guideline value of 15 Pt–Co units for raw waters for drinking purposes that are subjected to coarse screening. No guidelines were suggested for turbidity because of site-specific requirements for chlorine disinfection (chlorine concentration and contact time).

The available compilations suggest that industrial water users are not very sensitive to optical quality of raw water. Consistency of supply and other quality aspects are usually more important (e.g. EPAV 1983). For example, Anon (1987) presented no recommendations for colour and clarity of industrial supply waters presumably because they expected that most raw waters would receive appropriate treatment prior to use.

5.3 DEVELOPMENT OF IMPROVED GUIDELINES FOR OPTICAL WATER QUALITY

5.3.1 Philosophy
We have seen in the previous section that, although many criteria and guidelines have been published over the past two decades, there have been few advances in that time. The main problems are as follows.

(1) The narrative statements are often vague while others ignore any possibility that research on user preferences might assist in formulating guidelines.
(2) There is often no clear distinction of the concepts of colour and clarity, and frequently there is a lack of recognition of their particular characteristics (e.g. visual clarity as opposed to light penetration).
(3) The desirability of relative change guidelines (so as to provide equal protection for pristine and optically degraded water bodies) has seldom been recognized.
(4) There is a lack of substantiation of the suggested numeric guidelines, so they often appear to be the product of guesswork rather than being based on empirical evidence.

Kirk (1982) appears to have first introduced the concept of 'optical water quality' with which we are concerned here. Later Kirk (1986) suggested a minimum set of measurements which are needed to characterize inland waters optically. He recognized the need to consider *inherent* as well as *apparent* optical properties, because, while the latter measures quantify the optical attributes of water requiring protection, the inherent properties are additive (Section 2.2) and thus predictable on mixing of different waters or wastes in waters. More recently Kirk (1988) considered management of the optical quality of waters and suggested a conceptual framework for the setting of standards for wastewaters based on the three optical attributes of receiving waters: visual clarity (measured by Secchi depth, z_{SD}), light penetration (measured by the vertical attenuation coefficient for solar irradiance, K_d(PAR)), and brightness of the water colour (measured by the irradiance reflectance just below the surface, $R(0, PAR)$). He outlined procedures to calculate the effects of discharges on these apparent optical properties following laboratory measurements of the inherent properties a and b of the effluent.

Kirk (1988) further suggested that allowable change in the apparent optical properties of a receiving water should be expressed as *maximum relative change*. This provides for comparable protection of the optical character of all waters, regardless of their baseline clarity or colour. Kirk did not suggest values for maximum relative changes but stated 'these should certainly be low'.

Kirk (1988) did not suggest what baseline a permissible change should be measured from, but it seems implicit in his paper that this should be the 'natural' or pristine situation rather than the existing condition. Otherwise, when there are multiple discharges to a river, for example, the baseline light attenuation would increase below each individual discharge so permitting cumulative 'optical pollution' to an unacceptable extent. Defining the natural condition poses something of a problem when the optical quality has been degraded by changes in catchment land use, rather than by point sources. However, there are often means by which the original optical character of a water body can be modelled, or estimated by comparison with neighbouring unmodified systems.

We have extended Kirk's concepts and, where there is some supporting evidence, have gone on to formulate guidelines to protect optical water quality. We agree that relative change guidelines for protection of optical water quality are generally appropriate, but for some kinds of water use an absolute guideline is required, notably a minimum visual clarity for safety in bathing waters.

Here we address three categories of water use, i.e. bathing, aesthetics, and aquatic habitat, and propose numeric guidelines for management where there is some scientific basis for doing so. Table 5.1 gives a list of water uses versus the relevant optical attributes for which guidelines appear necessary. It is not necessary, or appropriate, to have guidelines for all possible combinations. For example, the colour attributes of water do not seem likely to affect bathing *per se*, although they may well affect the aesthetic appreciation of water involved in bathing.

5.3.2 Visual clarity
Contact recreation
Earlier in this chapter, water clarity criteria relating to bather safety were briefly discussed. The requirement, from safety considerations, is that the bottom should be visible in areas which are of wadeable depth, and also in areas set aside for diving, or at least in the latter case the water clarity should be such that people diving into the water should have an adequate depth of vision to enable them to dive safely. As we have seen, the Secchi disc has been recommended for assessing visual clarity in this context. However, it is inadequate protection that a Secchi disc be visible on the bottom because this is a high contrast target and most substrates (and hazards!) are much less contrasting. (In fact, as stated earlier, the optical basis of the Secchi measurement is invalidated if the disc is observed on the bottom.) A just-visible Secchi disc on the bottom means that potential hazards of lower inherent contrast (for instance, broken bottles, holes, snags) will not be visible.

A recent study in which the turbidity of a small stream with a cobble–gravel bed was

Table 5.1. Colour and clarity guideline requirements for water uses.

	Bathing	Aesthetics	Aquatic habitat
Clarity			
Visual	*For safety purposes and human perception requirements	*Absolute value, plus noticeable change value	*For fishing purposes and vision of aquatic fauna
Light penetration			*Maximum change requirement to protect light climate for photosynthesis
Colour			
Hue		*Absolute value, plus noticeable change value	*For light quality effects on biota and photosynthesis
Saturation		?	
Brightness		*	

Some uses may overlap, for instance a light penetration requirement may be required for aquatic habitat protection but in addition an aesthetic requirement may also be needed.
*, Guideline proposed in the text.

artificially manipulated while being viewed by small panels ($n = 20$) of students (Davies-Colley & Smith 1990) showed that the black disc sighting range (y_{BD}) when the bottom disappeared was approximately 35% greater than the stream depth. The Secchi depth (z_{SD}) is approximately 25% greater than y_{BD} (Davies-Colley & Close 1990), so the depth at which the bottom disappeared was about 60% of z_{SD}. These findings may be generally applicable. Therefore if the bottom is to be just visible at 1.2 m (i.e. adult chest) depth, the corresponding value for z_{SD} is about 2 m (and the optically better-defined y_{BD} is about 1.6 m). This can be more generally expressed as follows: if, for safety reasons, the water bottom is to be just visible at depth z, the black disc visibility (y_{BD}) should exceed $1.35z$, and the Secchi disc depth (z_{SD}) should exceed $1.7z$.

Bather preferences with respect to water clarity presumably combine aesthetic and safety considerations in some way. Recently, user preferences have been the subject of a small-scale study (20 respondents at each of eight sites) in New Zealand (Smith *et al.* 1991) which investigated these preferences at the same time that water clarity measurements were made. This study showed that a value for y_{BD} of 1.6 m (equivalent to substrate visibility at about 1.2 m depth, and z_{SD} of about 2 m) was regarded by about 85% of respondents as suitable, or better, for bathing (Fig. 5.1). The clarity range 0.8–1.4 m appears to be very critical, encompassing a shift from about 30% to 80% of respondents regarding water as suitable or better for bathing. This is broadly consistent with the finding by Vant & Davies-Colley (1988) that bathing was uncommon in otherwise regularly used lakes of mean Secchi depth <0.9 m in central North Island, New Zealand.

Additionally, a Delphi-style exercise involving the 15 offices of New Zealand's Water Resources Survey indicated that a black disc range of 1.6 m corresponds to water

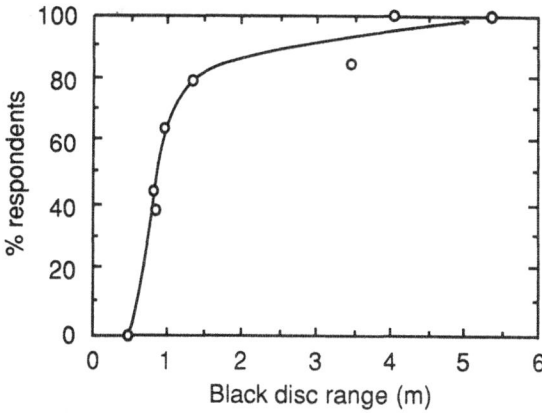

Fig. 5.1. Percentage of respondents rating a water's clearness as 'suitable' (or better) for bathing, versus black disc sighting range (from Smith *et al.* (1991)). ○, data from a selection of New Zealand rivers and lakes used to establish the relationship; ——, hand-drawn fit.

being described, on the average, as just suitable for bathing (Fig. 5.2; Smith & Davies-Colley 1992). At a value of 1.1 m, waters are regarded as just marginally suitable for bathing. Because the respondents (hydrologists) were very familiar with the black disc method, considerable weight can be given to their opinions. Their response curve for aesthetic appreciation of waters is similar to that for bathing (Fig. 5.2) suggesting that the guideline recommended above for bathing (i.e. $y_{BD} > 1.6$ m) will also satisfy aesthetic requirements.

These clarity requirements would normally be applicable for wadeable water depths. For diving areas, generally applicable guidelines are difficult to give but the water clarity would generally need to be greater than that required for wading, to an extent dependent

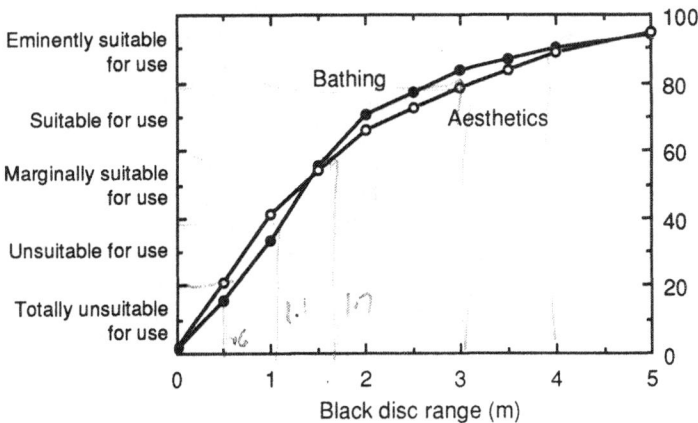

Fig. 5.2. Clarity response curves for bathing (●) and aesthetics (○) water use. (From Smith & Davies-Colley (1992), with permission.) The numbers on the right-hand axis indicate 'percentage suitability'.

on the physical nature of the site. For underwater swimming (sometimes also called 'diving'), we suggest that visual clarities greater than about 3 m black disc visibility may be generally desirable for safety reasons, although the aesthetic value of water for this sport probably continues to increase with clarity up to the visual ranges encountered in the very clearest waters.

Aesthetics
In addition to the minimum requirement stated in the previous section, a maximum relative change guideline for visual clarity is required to protect the natural aesthetic value of waters. Such a guideline value must necessarily be greater than the threshold of detectability of clarity change. This threshold was examined in an on-site panel study (Davies-Colley & Smith 1990) in which the turbidity of a small New Zealand stream was artificially manipulated. The median threshold for detection of a reduction in visual range (y_{BD}) was found to be 10–15%, corresponding to an 11–18% increase in the beam attenuation coefficient, c. This finding is consistent with the experience of the authors, and we expect it will have reasonable transferability, that is, a threshold of 10–15% should apply elsewhere.

Based on this limited evidence, we suggest that adequate protection of visual clarity for aesthetic uses may be provided in the range 15–25% reduction in clarity, corresponding to a 18–33% increase in light beam attenuation. A 20% reduction in visual clarity (corresponding to a 25% increase in c) is considered to be a reasonable guideline value.

Visual clarity can be measured with either a Secchi disc or a black disc to administer this guideline, although the latter is generally preferable. Alternatively the beam attenuation coefficient can be measured with a transmissometer.

Aquatic life
The visual range of aquatic animals is limited, ultimately, by the optics of the water, just as it is for humans (Lythgoe 1979). We are unaware of clarity criteria to protect visual requirements of aquatic fauna but it seems evident that such protection is required. For example, sight-feeding salmonid fish are known to avoid turbid waters (Lloyd 1987, Table 1). (Visual clarity of waters is also likely to affect the management of fisheries, rather than the fish *per se*, because of the difficulty encountered in assessing fish stocks in turbid waters.) Predatory aquatic birds, such as kingfishers and herons, are also expected to be sensitive to visual clarity of waters, but little seems to be known about specific requirements. Part of the problem is that many aquatic animals can use senses other than vision to locate prey for example, so that their functioning may not be very noticeably impaired when the visual clarity of water is reduced.

In the absence of any firm scientific criteria we speculate that protection of visual clarity from the aesthetic perspective (by the requirement that visual range not be decreased more than 20%) may confer adequate protection on the 'visual habitat' of aquatic fauna. However, research on the visual clarity requirements of 'keynote' species is urgently required.

5.3.3 Light penetration
A reduction of irradiance penetration can affect the growth of fixed and suspended plants thereby affecting the basis of the whole aquatic ecosystem. Therefore protection

of light penetration into water bodies is very important. As we have seen, some major criteria-developing agencies have suggested the use of compensation depth (e.g. USEPA 1973a, 1976; EPAV 1983) as a management tool to protect aquatic ecosystems from changes in primary production due to changes in the light climate. However, a simpler mechanism is to protect the euphotic depth, z_{eu}, the depth at which photosynthetically active radiation (PAR) is reduced to 1% of the value found immediately below the water surface. As we have seen, z_{eu} is related to the diffuse attenuation coefficient for downwelling light, K_d ($z_{eu} \approx 4.6/K_d$), which is weakly dependent on the ambient light field, as well as being strongly dependent on the optical properties of the water. In practice, euphotic depth and compensation depth are often very similar, but the former is much more readily measured by simple light profiling as opposed to detailed productivity measurements.

Strictly speaking K_d and hence z_{eu} are apparent optical properties of water (Sections 2.3 and 3.2). However, K_d is largely determined by the inherent absorption coefficient of water a, while also being affected by the scattering coefficient, b. This stronger dependence on a than b can be seen in an equation given by Kirk (1984a) linking K_d to these two inherent optical properties:

$$K_d = \frac{a}{0.847}\left(1 + \frac{0.170b}{a}\right)^{1/2}$$

(for a clear sun at 45° altitude). In practice, z_{eu} is not very sensitive to changes in the incident solar flux caused by changes in cloud cover and solar altitude (Kirk 1984a). Because z_{eu} can be related to inherent water properties, changes in a natural water resulting from, say, discharge of an industrial wastewater can be predicted prior to its discharge provided that we can obtain values for a and b for the waste.

A standard for visual clarity alone would protect light penetrability of water in many, but not all, situations. In general, the proportional reduction in visual clarity induced by an effluent discharge, for example, will not be the same as the proportional reduction in light penetration. Addition of a highly light-absorbing, but weakly scattering, material such as a pulp mill effluent will affect light penetration more severely than visual clarity, whereas addition of a highly scattering, but weakly absorbing, material such as suspended clay mineral particles will affect visual clarity relatively more than light penetration. For example, in Lake Pukaki (Plate 2) with a high content of scattering, but non-absorbing, glacial flour, the Secchi depth is around 0.5 m but the euphotic depth is about 4 m (Vant & Davies-Colley 1984). This underlines the importance of distinguishing the two main aspects of water clarity and separately providing for protection of these two aspects in water legislation and management.

The problem of how to specify a numerical guideline for the euphotic depth remains. Kirk (1988) recommended specifying a maximum relative change in the vertical attenuation coefficient of PAR, K_d(PAR). USEPA (1973a, 1976) suggested a maximum relative change in compensation depth of 10%, but did not furnish any supporting evidence for selection of this particular value. The same maximum relative change is recommended here, but applied instead to the euphotic depth, following Kirk's recommendation. However, we cannot, at present, justify this particular numerical value. A 10% reduction in euphotic depth sounds fairly restrictive but it should be recognized that this corresponds to a 40% reduction in irradiance (PAR) at the (natural) euphotic depth (lighting drops from 1% to 0.6% of the surface irradiance). A 10% reduction in euphotic depth will result in a 10% contraction in the depth range of benthic plants whose maximum depth of colonization is controlled, as is common, by light penetration (e.g. Vant *et al.* 1986). As with visual clarity, we are faced with the problem of establishing, or estimating, a baseline from which to measure change, perhaps in the presence of appreciable natural variability.

In shallow water bodies such as rivers, the *virtual* euphotic depth (calculated as $4.6/K_d$) may be deeper than the maximum water depth. However, it is still important to protect the primary production in such water bodies, so protection for the light penetration is needed. For example, Davies-Colley *et al.* (1992) found that reduction in streambed lighting of between 12% and 73% (mean 44%), induced by discharge of clays from alluvial gold mines on small streams of the West Coast of New Zealand's South Island, reduced productivity of benthic algae almost proportionately, and also had a marked effect on benthic invertebrate densities. To protect the light climate of benthic plants in shallow waters we propose a maximum permissible reduction in PAR at the bed of 20%.

For waters shallower than half their (virtual) euphotic depth (i.e. $z_{max} < 0.5z_{eu}$) this '20% maximum change in PAR' guideline is less restrictive than the '10% maximum change in euphotic depth' guideline. Where $z_{max} = 0.5z_{eu}$ the two guidelines are approximately equivalent (a 40% reduction in lighting at the euphotic depth corresponding to about a 20% change in lighting at the midpoint of the euphotic zone). For waters deeper than the midpoint of the euphotic zone ($z_{max} > 0.5z_{eu}$) the '20% change in PAR' guideline is the more restrictive, and the '10% maximum change in euphotic depth' guideline would normally be applied instead. To summarize, we propose that for waters shallower than $0.5z_{eu}$ the PAR should not be changed by more than 20%; for waters deeper than $0.5z_{eu}$ the euphotic depth should not be changed by more than 10%.

Of course, the underwater light field is important, not just for photosynthesis but for the illumination of submerged features for the vision of both aquatic fauna (including predatory birds as well as fish) and people (Plates 9A, 9B). An impact which reduces the euphotic depth of a water body will always reduce the penetration of illuminating light, usually by a similar factor. Any difference in the degree of change arises from the different spectral response implied by 'visible' light (technically illuminance—light as measured by a sensor with the same spectral response as the human eye) and PAR. Since penetration of PAR into water bodies is very closely linked to that of visible light, protection of euphotic depth may also serve to protect the penetration of illuminating radiation into water bodies. Thus the 'maximum 10% change' guideline for euphotic depth may also serve to protect the light field for vision of aquatic fauna—and also the light field for visual detection of submerged features in water by human recreational users.

5.3.4 Colour
Hue
There is some evidence that an absolute guideline for hue may be desirable, for example for the purpose of considering waters for designation as recreational water bodies. A survey of the 15 offices of the Water Resources Survey in New Zealand (Smith &

Davies-Colley 1992) suggested that blue to green hues are preferred over yellow to red hues in waters from an aesthetic perspective. A hue of 40–45 Munsell units (i.e. a green) appears to be the threshold (Fig. 5.3). However, there are reasons for expecting that aesthetic response to hue is affected by perceptual interactions with other features, including water clarity and brightness. Smith & Davies-Colley (1992) speculated that yellow-hued waters might be more aesthetically acceptable if visually clear and dark than if bright and turbid ('muddy' coloured). Unpublished studies by the authors seem to confirm this hypothesis, but further research is required before firm recommendations can be made regarding guidelines for absolute hue of recreational waters.

The need for a relative change guideline to protect natural hue of waters seems evident, however. Human observers can distinguish spectral colours separated by as little as 2 nm in wavelength, but undoubtedly people are much less sensitive to changes in hue of typically low-saturation water colours. No panel studies appropriate to this problem appear to have been published, but in our experience water colours about 5 Munsell hue units apart can usually be distinguished. Davies-Colley & Close (1990) reported that the root mean square deviation of river water colour descriptions from a colour standard match was 4 Munsell units, suggesting reasonable precision.

In practice a change in natural Munsell hue of more than about 5–10 units is likely to be associated with an unacceptable change in natural water clarity. That is, we would expect clarity to be generally more sensitive to change than hue so that a clarity standard would also, in most instances, serve to protect colour. Although few situations seem likely to occur where hue would be significantly affected but not clarity (an obvious exception would be highly light-absorbing, but weakly scattering, Kraft pulp mill effluent), it is still prudent to protect water hue explicitly. We think that a maximum Munsell colour change of 5–10 units may be appropriate. This corresponds, for example, to the change from a green (Munsell number 45) to green–yellow (Munsell number 35).

A more restrictive guideline of 5 Munsell units maximum change (e.g. from blue–green (Munsell number 55) to bluish green (Munsell number 50)) might be appropriate for

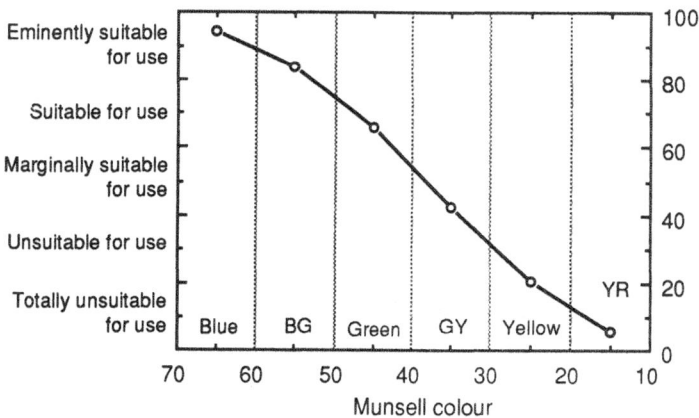

Fig. 5.3. Colour response curve for aesthetic water use. (From Smith & Davies-Colley (1992), with permission.) The numbers on the right-hand axis indicate 'percentage suitability'.

waters of great scenic value. In practice, special optical instrumentation, ideally a spectroradiometer, would need to be deployed to administer such a guideline (e.g. Davies-Colley *et al.* 1988). For long-term monitoring of trends, absorption measurements for total absorption at 340 or 440 nm will be more reliable than visual comparison with, say, Munsell colour patches.

We expect that sensitivity to hue change will increase with saturation. The above guidelines are expected to be appropriate in typically low saturation waters (say, Munsell chroma <4). However, more restrictive guidelines could be appropriate in waters of high saturation.

The hue of water colour, as we saw in Chapter 2, is a manifestation of the spectral quality of light in water, and this optical characteristic is expected to affect aquatic organisms as well as human users of water. Therefore, there is a need to protect the spectral quality of the natural light field in waters in order to protect aquatic habitat. In particular, it has long been known that the spectral sensitivity of the eyes of some fishes is well matched to the colour of their typical aquatic habitat (e.g. Lythgoe 1979). This match maximizes the fishes' visual potential in the light-restricted underwater environment. Presumably, therefore, fish are sensitive to change in the colour of their underwater illumination. Thus we may expect that change in water colour will have subtle, but potentially serious, impacts on aquatic habitat. Unfortunately, no criteria seem to be available. Until the appropriate research is conducted, we suggest that the proposed guideline of a maximum 5–10 Munsell unit change in hue to protect natural aesthetic quality may also serve to protect the aquatic habitat, inasmuch as it is affected by the spectral quality of the underwater light field.

Saturation
Water colour saturation cannot be easily matched by eye so it is difficult to recommend a guideline for this attribute of colour. This is not a serious deficiency because, while hue and brightness are unlikely to be appreciably affected without significant reduction in water clarity, it is even more unlikely that saturation of a water colour could be changed without affecting one or more of the other optical aspects: brightness, hue or clarity. For this reason no guideline for saturation is given.

Brightness
Typical water reflectances are about 0.03 to 0.05 (3% to 5%) but reflectance can range widely from <0.01 in very dark waters to >0.3 in extremely bright waters such as Lake Pukaki, New Zealand (Plate 2) (Vant & Davies-Colley 1984). There is no hard information, but our experience suggests that people are not particularly sensitive to changes in brightness of waters. The detection threshold may be in the region of a 50% change in reflectance for relatively dark waters (say, $R(0-) < 0.05$). For example, a change in reflectance from 0.05 to 0.075 or to 0.025 might be detectable. Thus, where brightness requires protection, we propose, at this stage, that the natural reflectance should not be changed by more than 50%. There is some anecdotal evidence that a more restrictive guideline (e.g. 30% maximum change) may be appropriate for relatively bright waters.

In practice, accurate enumeration of a guideline for reflectance may not be very critical because situations where a change in reflectance, and thus brightness, of a water could be induced without appreciably changing the clarity (either euphotic depth or visual

Plate 1. Crater lake, Cascade Range, Oregon, USA. This lake is one of the clearest in the world (Secchi depth about 40 m). The striking blue–violet colour of the water viewed from above (Munsell 5.5 PB, dominant wavelength 472 nm — Smith *et al.* 1973) is essentially the hue of water itself. (Photo — RJD-C, January 1981.)

Plate 2. Lake Pukaki, Canterbury. This large glacial lake has very turbid water (Secchi depth about 0.45 m (Vant & Davies-Colley 1984)) and a bright blue–green colour (PAR reflectance >30%) owing to intense scattering by fine silt-sized glacial flour derived from glaciers in the Southern Alps (background). New Zealand's highest mountain, Mt Cook (Maori — *Aorangi*, 'the cloud piercer'), is the prominent peak just left of centre in the photograph. (Photo — RJD-C, 2 March 1983.)

Plate 3. Water of Lake Hakanoa, northern Waikato. The algal soup character of the water is clearly evident in the foreground, closer than the rushes. This hypertrophic lake has a very high phytoplankton biomass (mean annual chlorophyll a = 173 mg m^{-3} (Vant & Davies-Colley 1987)) and is very turbid (mean Secchi disc visibility = 0.2 m). The yellow–green intrinsic water colour (Munsell 5 GY) can be seen where the water surface is viewed at an acute angle, but where the water is viewed at a larger angle of incidence, the sky reflection predominates. (Photo — RJD-C, 11 January 1983.)

Plate 4. Underwater photograph in Lake Rotomanuka, Waikato. This small peatland lake is visually fairly clear (Secchi depth about 3.6 m, authors' unpublished data), but the high content of yellow substance (g_{440} = 1.3 m^{-1}, (Davies-Colley & Vant 1987)) produces a yellow hue which is strikingly evident underwater. (Photo — John Nagels, Water Quality Centre, 11 November 1983.)

Plate 5. Lake Waahi, northern Waikato. This shallow lake is very turbid (Secchi depth about 0.30 m—RJD-C unpublished data) owing to frequent resuspension of clay-rich bottom sediments by wind waves. The muddy yellow colour (Munsell 5Y) results from the high content of yellow substance and detritus. Before 1978 this lake was heavily infested by aquatic macrophytes (mainly *Egeria densa*). Decline of the plants has exposed the bottom sediments to wave action. (Photo — RJD-C, 16 January 1985.)

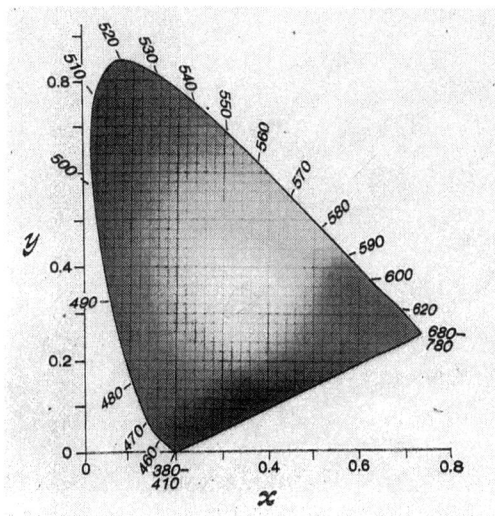

Plate 6. The CIE chromaticity diagram. The colours shown are merely indicative spectral colours that do not represent the true colours within the horseshoe-shaped locus bounding all real colours.

Plate 7. Use of a black disc to measure visual water clarity in Kapitea Creek, West Coast, South Island. The water in this stream is coloured orange by a high yellow substance content (g_{440} = 3.3 m^{-1}, RJD-C unpublished data), but the water is visually clear during baseflows (black disc visibility about 2.0 m). (Photo — RJD-C, 16 February 1989.)

Plate 8. Rafting in the lower gorge, Motu River, North Island. The bluish-green hue of this water is typical of baseflows in New Zealand rivers draining pristine, forested catchments. Aesthetically attractive water colour and clarity increase the recreational appeal of such waters. (Photo — RJD-C, 14 October 1983.)

Plate 9A. Snorkel divers in the Waihou River, Waikato Region. This spring-fed river is very clear (black disc visibility about 7.5 m, RJD-C unpublished data) at this site. An impression of the intrinsic water colour (a blue–green hue) is evident in the pool beyond the snorkel divers. (Photo — Chris Hickey, Water Quality Centre, 23 October 1985.)

Plate 9B. Underwater photograph of the Waihou River (same site as the previous photograph, but on a different occasion). Macrophyte stands and the sand and gravel substrate can be clearly seen. The fish in the middle distance are rainbow trout. (Photo — Chris Hickey, Water Quality Centre, 4 December 1986.)

Plate 10A. Alluvial gold mine, Red Jacks Creek, Grey River Valley, West Coast, South Island. This typical small operation comprises a tracked excavator feeding alluvium to a trommel (rotating screen) which sorts material at 12.5 mm size. Coarse alluvium is dumped by the conveyor as a tailings heap (foreground) whilst the finer material is sluiced over riffle tables which concentrate gold. The excavation of alluvium exposes the unconfined groundwater as a pond which provides the water supply for sluicing. The inevitable disturbance of clay and silt by heavy machinery generates very high turbidities in the pond water (of order 1000 NTU) resulting in seepage of fines into the nearby stream. (Photo — RJD-C, 3 April 1987.)

Plate 10B. Seepage from the alluvial gold mine shown in Plate 10A into Red Jacks Creek, Grey River Valley, West Coast, South Island. The stream water is a dark orange colour owing to a high yellow substance content ($g_{440} = 4.1$ m^{-1}, RJD-C unpublished data). The sluice pond has no surface connection with the stream channel; however, the pond water moves with the local piezometric gradient carrying clay particles into the stream where the groundwater streamlines intersect the channel. The visual effect is rather like adding milk to tea. (Photo — RJD-C, 3 April 1987.)

Plate 11. Swimming in the Waikato River, Hamilton. The water clarity at this site (black disc range about 1.5 m, unpublished data from NZ's National Water Quality Network) is marginal for swimming safety. The intrinsic water colour is a dark greenish yellow (mean Munsell hue = 2.5 GY (Davies-Colley 1987)). An impression of this colour can be seen where the water is shadowed by trees. (Photo — DGS, February 1988.)

Plate 12. Aerial photograph of the mouth of the Tarawera River, Bay of Plenty Coast, North Island. The dark-coloured river plume, stained with effluent from a large Kraft pulp mill, is clearly seen contrasting with the blue seawater of the Pacific Ocean. The hue of the river water is actually a yellow–orange (Munsell 10 YR) but this is not easily seen because of the low brightness of the water colour. (Photo — W. A. Taylor, Bay of Plenty Catchment Commission — now Regional Council, Whakatane, New Zealand, 1976.)

Plate 13. Cattle damage to the riparian zone, Mangakino River, central North Island. (The culprits are seeking shade under trees on the left in the middle distance.) During rainstorms, or with any rise in water level, suspended matter can be easily eroded from the trampled and compacted bare soil areas adjacent to the stream channels. (Photo — RJD-C, 2 February 1987.)

Plate 14. Lake Coleridge, inland Canterbury, South Island. This beautiful blue-coloured alpine lake has been subjected to optical quality changes resulting from diversion to the lake of turbid river waters laden with glacial flour (see Section 6.2). (Photo — RJD-C, 23 September 1986.)

range) are likely to be rare. By way of example, consider the impact of adding two extreme types of waste material to a water body. At one extreme we have highly light-absorbing (but weakly scattering) Kraft pulp mill effluent (Section 5.4.5). This effluent will certainly lower the reflectance of the receiving water, but will also lower the euphotic depth. Since $R \propto b/a$, the absorption coefficient, a, would have to be doubled on addition of this effluent to lower the reflectance by 50%; but this would also (nearly) halve the euphotic depth—which would be unacceptable. At the other extreme, a highly light-scattering (but weakly absorbing) mine waste water will increase the reflectance of the receiving water, but will also decrease the visual clarity. If this effluent increased the reflectance by 50%, the scattering coefficient, b, would be increased by approximately 50% also. Typically the b/a ratio of waters is greater than unity so that the beam attenuation coefficient ($c = a + b$) would be increased by between 25% and 50%. This would result in the black disc range (y_{BD}) falling by between 20% and 33%, which again would be unacceptable.

Perhaps the only area for concern here is with diversion of a (natural) inflow which results in improved clarity but adversely changes natural reflectance. For example, consider the bright ($R \sim 30\%$) greenish-blue colour of New Zealand's Lake Pukaki (Plate 2), the brightness of which is undoubtedly part of its scenic appeal. Suppose diversion of the glacial Tasman River (the main inflow to Lake Pukaki and the source of almost all its light-scattering particles) were contemplated, say for hydro-electric power generation. A concern might be that the clarity of the lake would increase, but at the cost of reduced brightness, which some people might regard as an undesirable, or even unacceptable, change. Therefore, in some rare instances brightness of water colour may need explicit protection.

5.3.5 Summary of the guidelines
The guidelines developed above are summarized in Table 5.2, and amplifying comments follow.

Clarity
For bathing waters, the guideline for visual clarity is that the black disc visibility should exceed 1.6 m for safety reasons (based on visibility of the bottom at a wadeable depth of 1.2 m) (Plate 10). Equivalently, the Secchi depth should exceed 2.0 m. A black disc visibility of 1.6 m corresponds to 85% of a field panel regarding the water clarity as suitable for swimming. If, for example, a different suitability rating is required, Figs 5.1 or 5.2 could be used to derive the appropriate guideline. For safety in high diving areas the visual water clarity may need to be appreciably greater, as indeed it should be for underwater swimming.

For aesthetics, we suggest that the absolute requirement for visual clarity be the same as for bathing (i.e. $y_{BD} > 1.6$ m). This should be considered a minimal requirement because aesthetic quality increases with clarity. To protect natural aesthetic quality, it is recommended that the natural visual clarity should not be changed more than 20%. In practice, this is likely to be the most restrictive of all the guidelines, but fortunately it is perhaps the easiest to administer because visual clarity is easily measured and is linked immediately to the inherent optical property, c.

For aquatic habitat we speculate that the same maximum reduction in natural visual clarity as for aesthetics (i.e. 20%) may provide adequate protection of the visual range of

Table 5.2. Summary of the colour and clarity guidelines for water uses (as developed in this chapter)

	Bathing	Aesthetics	Aquatic habitat
Clarity			
Visual	$y_{BD} > 1.6 \text{ m}^{a,b}$	$y_{BD} > 1.6$ m	
	$z_{SD} > 2.0 \text{ m}^a$		
		$\Delta y_{BD} < 20\%$	$\Delta y_{BD} < 20\%^c$
Light penetration			$\Delta z_{eu} < 10\%$
Colour			
Hue	$—^e$	Hue > 40–45 Munsell units (green)	
		Δ(hue) < 10 Munsell units (Δ(hue) < 5 Munsell units in waters of high scenic value)	Δ(hue) < 10 Munsell units^d
Saturation		$—^e$	
Brightness		Δ(reflectance) < 50%	

The guidelines suggested should not be construed as limits to which the water can be degraded. If colour requirements are deemed necessary for bathing waters, those for aesthetics should be used.
[a] For bather safety (i.e. for bottom visibility from the surface) for a water depth of 1.2 m (chest depth for adults).
[b] For user preference: corresponds to ≈85% of people stating water suitable.
[c] To protect visual requirements of aquatic fauna.
[d] To protect light quality for vision and photosynthesis.
[e] Guideline considered, but deemed unnecessary.

animals, but acknowledge that we cannot justify this particular value. A maximum change in natural euphotic depth of 10% is suggested to protect the natural light climate, but again this cannot be adequately justified given the present state of knowledge. We speculate that this protection for the photosynthetic light climate in water will also serve to protect natural levels of underwater illumination.

Colour

For bathing waters, no colour guidelines are regarded as necessary.

For aesthetics, a minimum Munsell hue number of around 40–45 (i.e. green) seems to be desirable. To protect natural aesthetic quality of waters a maximum hue change of 5–10 Munsell units is recommended. No recommendation can be made regarding saturation. To protect natural brightness of waters we propose a maximum reflectance change of 50%, recognizing that future research may result in this guideline being revised, probably downwards.

To protect aquatic habitat as regards the spectral quality of the light field we have made the same recommendation as for aesthetics, namely a maximum 10 Munsell unit change in natural hue. Again we are not aware of criteria to justify particular numerical limits.

5.3.6 Implementation of the guidelines

The optical quality of natural waters can be affected by a wide variety of actions, including particularly discharge of wastewaters, change in land use or actions on land (e.g. forest harvesting), and diversion or impoundment of river waters as discussed in Sections 5.4–5.6. Where consents to these actions are being sought from a water management agency, that agency will generally need to go through the following sequence of actions:

(1) estimate the change in light attenuation (light absorption and scattering) in the receiving water that will result from the proposed action;
(2) estimate the changes in various optical attributes (visual clarity, light penetration hue, brightness) of the receiving water that will arise owing to change in the light-attenuating properties;
(3) compare the estimated changes in optical characteristics with the relevant guidelines.

If, in step (3) above, the relevant guidelines will be met in the receiving water, the activity would normally be consented to (unless of course there are other water quality or ecological concerns). If the guidelines will not be met, it may be necessary to do inverse calculations (or modelling) in order to work back from the guideline values (which essentially define the maximum *assimilative capacity* of the receiving water) to define what lesser degree of activity than that originally proposed would be permissible. For example, a smaller effluent discharge or the same discharge of an effluent of improved quality might be possible without breaching the guidelines.

Consider firstly step (1) above: estimation of the change in light absorption and scattering in the receiving water that will result from the proposed action.

In the case of impoundment of river water or activities on land, step (1) is generally rather difficult to carry out. This is because estimating the change in light absorption and scattering may involve consideration of all the complex, interacting factors affecting the delivery of light-attenuating substances (especially suspended solids) from the catchment to the water body, and the behaviour of these materials once in the receiving water. Sometimes the generation of light-attenuating substances within the water body may need to be estimated (e.g. the erosion of shallow lake sediments by wind waves or the growth of phytoplankton to higher biomasses with increased residence time of nutrient-enriched river water in an impoundment). Estimation of the optical quality impacts of land use changes may also involve predicting the delivery of algal-growth-promoting nutrient elements to the (standing) water body. The control of phytoplankton growth and suspended solids generation from the perspective of control of optical water quality degradation will be considered in Sections 5.5 and 5.6.

In the case of diversion of natural water (e.g. diversion of part or all of a river flow from one site to another site—which may be in another catchment), or discharge of a wastewater, the calculations involved in step (1) above are comparatively straightforward. All that is needed, in principle, is to know the light-absorbing and scattering properties of both the receiving water (before the change) and the inflow (diverted water or wastewater), together with the flows of the inflow and the receiving water (or indeed, simply, the dilution ratio). The calculations rely on the additivity principle (Section 2.2) applied to inherent optical properties of water (Section 2.2), which in turn assumes implicitly that the light-attenuating materials do not undergo chemical or physical changes such as flocculation and are not removed from the water column by processes such as sedimentation (as with the coarser mineral solids in lakes) or filtration (as with suspended material in waters passing through macrophyte beds).

A balance on total light attenuation cross-section in a mixture (subscript, mix) of two waters (or a water and a wastewater, subscripts 1 and 2) can be written as follows:

$$c_{mix}V_{mix} = c_1V_1 + c_2V_2 \tag{5.1}$$

where V is volume. Each term in equation (5.1) has the units of optical cross-section ($m^{-1} \times m^3 = m^2$). If the beam attenuation coefficients c_1 and c_2 and volumes V_1 and V_2 are known, the beam attenuation coefficient of the mixture, c_{mix}, is easily calculated. Equations of identical form can be written for the absorption and scattering coefficients considered separately.

A similar 'mass balance' type equation applies to the mixing of an effluent discharge or diverted water into a receiving stream, except that in this case volume flow rate appears in the equation rather than volume. The balance on attenuation cross-section flow rate (units: $m^2 s^{-1}$) in the river can be stated:

$$c_d\left(Q_u + Q_i\right) = c_u Q_u + c_i Q_i \tag{5.2}$$

where Q is discharge, and the subscripts d, u, and i denote downstream, upstream and the inflow, respectively. Again, equations of identical form can be used to calculate absorption and scattering coefficients separately.

Note that these equations apply at a given wavelength in the spectrum. For the purposes of visual clarity calculations it will usually be permissible to do calculations for the wavelength of peak sensitivity of the human eye (555 nm), or for some other wavelength in the case of vision of some aquatic animal. However, for calculations of colour changes in waters, in particular, the attenuation balance calculations will need to be performed across the spectrum, in practice at 10 nm or smaller wavelength intervals. Appendix 3 gives worked examples of both single-wavelength and spectral calculations of the changes in inherent optical properties of a river due to different effluent discharges.

Once the inherent optical properties in the receiving water after the proposed management change have been calculated from equations (5.1) or (5.2) (or their equivalents in terms of a and b), we can turn our attention to step (2) above: estimating the changes in various optical attributes (visual clarity, light penetration, hue, brightness) of the receiving water that will arise due to change in the light attenuating properties.

Calculation of the visual clarity impact is simplest because this attribute is inversely related to beam attenuation, c. Thus, once the beam attenuation coefficient (at 555 nm) in the receiving water has been estimated, the black disc visibility is simply calculated: $y_{BD} = \Psi/c$. For calculation of visual clarity change in rivers it is often convenient to restate equation (5.2) directly in terms of black disc visibility:

$$\frac{Q_u + Q_i}{y_{BDd}} = \frac{Q_u}{y_{BDu}} + \frac{Q_i}{y_{BDi}} \qquad (5.3)$$

These equations represent a very powerful tool in practical water management. They permit direct calculation of the clarity-degrading effect of a wastewater or other inflow on the receiving water. They also permit the reverse calculation. That is, given a permissible visual clarity change, it is possible to calculate effluent standards. This is a common problem in processing of applications for consents to discharge wastewaters. Appendix 3 gives worked example calculations illustrating the use of these equations. Note that calculations for rivers should usually be done for a low flow condition when the available dilution in the river flow is relatively low, and the water is clear. That is, the calculations should be performed for times when the assimilative capacity of the river (proportional to upstream flow of attenuation cross-section, $c_u Q_u$) is low.

Calculation of the other optical attributes of natural waters from the inherent properties is not quite so straightforward. Fortunately, a number of workers have given empirical equations relating the apparent optical properties of water to the absorption and scattering coefficients based on the results of Monte Carlo simulations of light photon penetration into waters, including Gordon et al. (1975) and Kirk (1981a,b, 1984a). Kirk (1988) indicated how his expressions could be used in practical problems of water management and it is his contributions that we discuss below.

Light penetration of water bodies, as we have seen, is characterized by the irradiance attenuation coefficient, K_d, which is inversely related to the euphotic depth. Kirk (1984a) gave the following expression for the average value of the irradiance attenuation coefficient over the euphotic zone (the 45° version of this equation has already been given):

$$K_d = \frac{1}{\mu_0}\left[a^2 + (0.425\mu_0 - 0.190)ab\right]^{1/2} \qquad (5.4)$$

where μ_0 is the cosine of the angle of the direct solar beam to vertical in water (i.e. after refraction at the water surface), denoted by θ. μ_0 ($= \cos \theta$) is easily related to the zenith angle of the sun θ', by Snell's law: $\sin \theta' = n \sin \theta$, where n is the relative refractive index of water ($= 1.33$). Kirk's (1984a) analysis showed that light penetration falls somewhat as solar zenith angle increases; that is, as solar altitude above the horizon decreases. The irradiance attenuation coefficient is most sensitive to solar zenith angle at low values of the scattering to absorption ratio, b/a. That is, sun position is expected to have its greatest effect on light penetration in visually clear, but coloured, waters. Under an overcast sky the average value of μ_0 is 0.856 which is equivalent to a direct solar beam at a zenith angle of 43.6°. For the purposes of calculation of a 'standard' euphotic depth in water management applications it may often be convenient to assume a uniformly overcast sky and to set $\mu_0 = 0.856$ in equation (5.4). Almost equivalently, a 'reference' solar altitude of 45° could be used for such calculations, at least in temperate latitudes (as in Section 5.3.3: $\mu_0 = 0.847$).

Kirk (1984b) has shown how the penetration into water of broad-band photosynthetically available light (PAR, 400–700 nm) can be calculated by summing the spectral irradiances at each of a number of wavebands (e.g. at 10 nm intervals) across the spectrum at each depth interval. An accurate estimate of the impact on euphotic depth of changes in the optical character of water can then be obtained. Appendix 3 gives a worked example calculation.

Dubinsky & Berman (1979) and Jewson *et al.* (1984) have shown that the irradiance attenuation coefficient for PAR is only slightly greater than the irradiance attenuation coefficient for the most penetrating wavelength in waters, λ_{max}. Thus, so long as the colour of the water (and thus the value of λ_{max}) is not likely to be greatly shifted by a given management action, it is possible to estimate the change in euphotic depth (but not the absolute magnitude) from the change in the irradiance attenuation coefficient at λ_{max}. Again Appendix 3 gives worked examples.

Colour of water bodies, as we saw in Chapter 2, is related to the irradiance reflectance coefficient at the water surface, $R(0)$. Brightness is related to the magnitude of illuminance reflectance at the water surface, which will generally be very similar to the reflectance at 555 nm. Hue is related to the spectral variation of the reflectance and is more difficult to model.

A very simple relationship between the reflectance coefficient at the water surface and the ratio of the backscattering coefficient, symbol b_b, to the absorption coefficient, a, was first suggested by Morel & Prieur (1977) for the purpose of water colour modelling (Section 2.6):

$$R(0-) = 0.33 \frac{b_b}{a} \tag{5.5}$$

Kirk (1991) showed from Monte Carlo modelling that this relationship is accurate, irrespective of optical character of the scattering particles, except in very highly scattering waters. The backscattering coefficient is simply the total scattering coefficient multiplied by the proportion of photons that scatter at angles greater than 90°. This proportion is dependent on the volume scattering function, $\beta(\theta)$, which quantifies the angular dependence of scattering, but b_b/b is typically about 1.9% in natural waters (Kirk 1991). Thus, as we saw in Section 3.5, the reflectance coefficient is roughly proportional to the ratio of the scattering and absorption coefficients, and we can write the *relative* spectral reflectance:

$$R_r(0-, \lambda) = \frac{b(\lambda)}{a(\lambda)} \tag{5.6}$$

The hue of the water can then be estimated following the application of chromaticity analysis to this relative reflectance spectrum (refer to Section 2.6.4). The brightness can be quantified by the value of the spectral reflectance at 555 nm: $R(0, 555) \approx 0.0063b(555)/a(555)$.

The reader is referred to Appendix 3 which illustrates many of the principles discussed above for implementation of the proposed guidelines for optical water quality. The appendix does not give any examples of detailed calculations of water colour changes, but it is expected that such calculations would be required only in the (comparatively rare) situation where there is a problem with hue when light penetration and visual clarity of the water body are not changed beyond guideline values.

We next turn our attention to the sources of light-attenuating materials of management concern: discharges of wastewater, activities on land that cause soil erosion—and generate suspensoids in water, and the growth of phytoplankton.

5.4 POINT DISCHARGES OF LIGHT-ATTENUATING MATERIALS

A point discharge is one by which wastewater enters the aquatic environment through an identifiable pipe, diffuser, channel, or drainage ditch. The discharged wastewater may be of municipal or industrial origin, the latter including waste products originally from the agricultural or forest industry.

Table 5.3 lists some major types of point discharge of wastewater that are significantly light attenuating and have caused degradation of receiving water colour and clarity in New Zealand. Absorption, scattering and beam attenuation coefficients are given at reference wavelengths of 440 and 550 nm, together with 'traditional' measurements of nephelometric turbidity and, in one instance, platinum–cobalt 'colour' (median values are given where there is significant variability and sufficient data are available to define this variability).

5.4.1 Urban stormwater

Established modern practice in large cities and towns is to collect most non-toxic effluents, including all household sewage, into the municipal sewers separately from stormwater. Formerly, stormwaters were often combined with sewage in the one sewerage system. We include urban stormwater here under point discharges because it is usually collected and discharged at identifiable points despite arising from diffuse sources.

Municipal stormwaters are recognized as being significantly polluting since they have high suspended solids levels, and contain toxic materials and nutrients, and faecal indicator microorganisms (e.g. Cordery 1977; Pitt & Field 1977). However, the degraded optical character (highly turbid, Table 5.3) of these discharges, especially the 'first flush' after a dry spell, appears not to have been addressed, although it has been recognized by a few authors (e.g. USEPA 1983). Urban stormwaters are often discharged to water bodies which are potentially very valuable as a recreational resource precisely because of their proximity to population centres. Visual quality degradation of certain of these urban waters (especially lakes) is likely to be severe and in the past may have been largely ignored, or passively accepted as an unfortunate but inevitable consequence of urban land use.

Much information is available in the literature on the suspended solids concentrations and yields from urban catchments of different characteristics (e.g. USEPA 1983). Unfortunately it is difficult to estimate precisely light attenuation from suspended solids concentrations because the grain size distribution and composition of the solids varies widely with an associated wide range in optical cross-sections (Section 2.4). Published data on solids mass yields are even less useful. It is to be hoped that data on the light attenuation or light-scattering coefficients of urban stormwaters will be collected in future studies.

Table 5.3. Optical characteristics

Effluent type	$a(440)$ (m^{-1})	$a(550)$ (m^{-1})	$c(550)$ (m^{-1})
Urban stormwater	Variable?	—	High
Domestic sewage Screened only	10.5 (9.8)	6.9 (5.9)	109 (92)
Primary treated	9.7	7.3	149
Secondary (Activated sludge)	3.9	2.4	42
Wastewater stabilization ponds (municipal sewage)	26	9	40
Meat processing wastes (untreated)	38	21	284
Pulp and paper Bleached Kraft[a]	58	19	55
Other (chemomechanical)	Low	—	—
Mining Alluvial (incl. aggregate)	—	—	Variable (can be high)
Hard rock	Low	Low	31
	—	—	—
Coal fines[b]	56 (0.41)	41 (0.30)	240 (1.7)
Other point discharges Dairy	Low	Low	Variable (can be high)
Dye	High (variable)	High (variable)	Variable

[a] Pt–Co colour = 625.
[b] Optical cross-sections obtained by dividing optical coefficients by suspended solids concentration (= 137 g m^{-3}).

of selected wastewaters

$b(550)$ (m^{-1})	Turbidity (NTU)	Site location and reference
High	214 (range 31–340)	Wairau Creek, Auckland, NZ (Hutchinson 1980)
102 (86)	78 (53)	
141	90	Sewage effluents from Tauranga City, North Island, NZ (Davies-Colley unpublished data)
39	15	(the second set of screened sewage results are for effluent from Bromley near Christchurch, South Island, NZ)
31	26	11 domestic sewage oxidation ponds in North Island, NZ (Davies-Colley, unpublished data)
263	125	Canterbury Frozen Meat Co. abattoir, South Island, NZ (Davies-Colley, unpublished data)
36	27	Mean of four Kraft effluent samples from Tasman Mill, Kawerau (Davies-Colley 1984; 1990b; Wilcock & Davies-Colley 1986)
—	~10	Caxton Mill effluent, Kawerau, NZ (Davies-Colley 1984, 1990b)
Variable (can be high)	Variable (can be high)	Small placer gold operations in Westland (Davies-Colley et al. 1992)
—	39	Waihi Gold Mine Co. dewatering effluent (Davies-Colley & Smith, unpublished data)
—	1700	Mean of seven samples of dewatering effluent from State Coal Co. open-cast mines near Huntly (Davies-Colley, unpublished data)
200 (1.4)	140	Huntly West Mine (State Coal Co.), Waikato (Davies-Colley unpublished data)
Variable (can be high)	Variable (can be high)	(Authors' observations)
Low	Low	(Authors' observations)

The high turbidity values given for urban stormwater in Table 5.3 must be considered only roughly indicative of scattering properties in urban storm waters generally, since great variability in the turbidity of stormwaters occurs, both between catchments and within the one catchment over hydrological events (R. B. Williamson, Water Quality Centre, personal communication). The variation between catchments may be expected to reflect variables such as soil types, drainage works, vegetation and land use—particularly the impervious proportion of total catchment area. As could be expected, construction which exposes bare soil areas is particularly significant in generating high suspended sediment concentrations. The control of suspended solids from construction sites and other sources will be discussed in Section 5.5.

The colour of urban stormwaters can also vary appreciably. Rapidly urbanizing catchments such as Wairau Creek in Auckland, New Zealand (studied by Williamson (1986); turbidity data are given in Table 5.3) are often heavily loaded with suspended solids derived from subsoils exposed during construction. These suspended solids are typically rather lightly pigmented (sometimes yellow owing to iron pigments), but are highly scattering, and give rise to bright cream or yellow-coloured stormwaters. In contrast, 'mature' urban catchments with relatively little active construction activity such as Hillcrest in Hamilton, New Zealand (Williamson 1985, personal communication) contain suspended solids rich in organic detritus eroded from surface soils or litter that are highly light absorbing, but rather less strongly scattering, and give rise to relatively dark orange ('brown') water colours. Relatively large (silt-sized) organic-rich particles may absorb essentially all impinging light and scatter an equal quantity of light by diffraction (van de Hulst 1957). Unfortunately, no data on the absorbing properties of urban stormwaters are available with which to assess further the colouring impacts of these discharges, although we speculate that absorption coefficients, like scattering coefficients, may be highly variable, both within and between catchments.

5.4.2 Municipal sewage
The traditional concerns with domestic sewage effluents have been oxygen demanding organics, pathogenic microorganisms and nutrients, and most developed countries now treat sewage to remove these materials. This treatment, inasmuch as it involves reduction in suspended solids content, also reduces the light attenuation by sewage effluents while somewhat changing their colour.

Table 5.3 summarizes optical measurements made by the authors on some municipal (primarily domestic) sewages at various stages of treatment. (Samples were mainly from the activated sludge plant at Tauranga, North Island, New Zealand.) Raw or preliminarily treated (screened, de-gritted) domestic sewage is typically grey in colour and is sometimes referred to as 'grey water'. This wastewater is intensely scattering of light as shown by the high scattering coefficient but is rather weakly light absorbing.

Primary treatment of sewage involves removal of coarse and comparatively rapidly settling solids (typically about 50–70% of total suspended solids (Metcalfe & Eddy 1979, p. 336)) by sedimentation. Floatable solids are also typically removed by surface skimming. These relatively coarse wastewater solids, including floatables, in raw sewage, are generally of more immediate concern than the fine suspended solids which remain in effluent from primary sedimentation tanks. The optical properties of primary treated

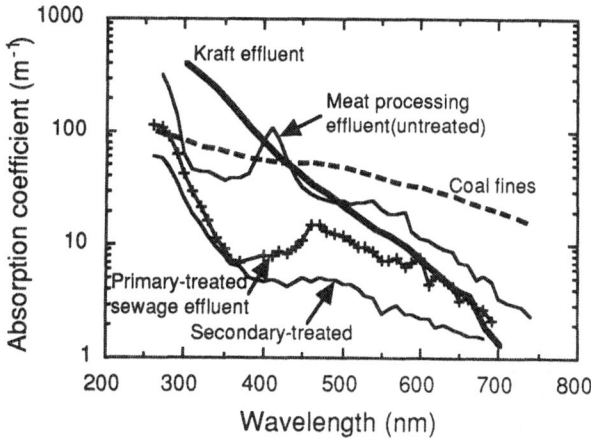

Fig. 5.4. Absorption spectra for some selected effluents. The suspended solids content of the mine dewatering effluent loaded with coal fines was 137 g m^{-3}. (Authors' unpublished data.)

sewage are rather similar to those of raw sewage, probably because the content of fine (and efficiently light-attenuating) solids is unchanged by passage through the sedimentation tanks. The absorption spectrum given in Fig. 5.4 is for primary treated sewage, rather than for raw sewage. The grey colour of this effluent, as with the raw sewage, arises because the absorption of visible light is only weakly spectrally selective (Fig. 5.4) while there is intense scattering (Table 5.3). Absorption in the ultraviolet, as could be expected, is far more significant than in the visible (Fig. 5.4) and the absorption peak of genetic materials (DNA and RNA) (and also proteins (e.g. Harm 1980)) can be seen at around 270 nm. An interesting feature in the absorption spectrum of primary treated sewage is the small broad peak at about 470 nm. We do not know which chromophore is responsible for this feature.

The biological processes occurring in conventional secondary treatment (activated sludge and trickling filter plants) result in conversion of much of the biochemically oxidizable organic matter into bacterial biomass and refractory humic compounds (Rebhun & Manka 1971). The secondary effluent is slightly yellow coloured, in contrast to the grey raw wastewater, and this colour (attributable to yellow substance type material) is consistent with the 'earthy' humus smell that is usually evident (Metcalfe & Eddy 1979). The scattering and absorption of light by secondary treated sewage is appreciably lower than that of raw or primary treated sewage (Table 5.3). The absorption spectrum of the Tauranga secondarily treated sewage effluent in Fig. 5.4 is somewhat different from that of primary effluent. In particular, the spectrum has lost some of its detail and is starting to resemble the exponentially shaped, but otherwise featureless, absorption spectrum of humified detritus. This is consistent with the more 'humified' status of secondary treated sewage, by comparison with raw or primarily treated effluents.

Large sewage outfalls can cause conspicuous visual (aesthetic) impacts where untreated wastes are discharged close to or on the shore of receiving water bodies. Discolouration problems (mainly hue contrasts with the receiving waters) have been reported for all of the large (daily mean flow >0.1 m^3 s^{-1}) shoreline discharges in New Zealand, and surface slick formation for most (Smith *et al.* 1987).

Where solely domestic wastes are discharged well off-shore, reported discolouration problems are rare. This is consistent with Newton's (1975) conclusion that visible discolouration of receiving waters by discharges of raw domestic sewage is unlikely where initial dilution exceeds 50-fold. Recent practice is to design outfalls to achieve high initial dilution (e.g. 100-fold) in the receiving water which, together with increasing treatment prior to discharge, should largely solve colour problems in future.

High intial dilution factors will also largely solve any problem with impact on visual clarity. Dilution by 100 times will result in a change in beam attenuation within the effluent plume of about 1.5 m^{-1} for a primary treated effluent, and about 0.4 m^{-1} for a secondary treated effluent, based on data in Table 5.3. Increments of light attenuation of this magnitude will cause only local visual clarity degradation in relatively clear receiving waters (say waters with $c < 1$ m^{-1}, corresponding to a horizontal black disc visibility >5 m) before subsequent dilution reduces the impact below detectable thresholds (see Section 5.2).

5.4.3 Waste stabilization pond effluents

Waste stabilization ponds, also known as 'waste lagoons' or 'oxidation ponds' (e.g. Metcalfe & Eddy 1979), are a very common method of treatment for organic wastewaters in warm, temperate, and tropical parts of the world (Wrigley & Toerien 1990). They are often used for domestic sewage treatment, sometimes without primary treatment (sedimentation) or even preliminary treatment (screening), and are also much used for treatment of animal wastes where these are concentrated as in farm feedlots or piggeries. Waste stabilization ponds, even where they are used to treat municipal sewage, are considered separately from municipal sewage treated by other means (refer to Section 5.4.2) because their optical and other characteristics are quite distinct.

Waste stabilization ponds can be likened to shallow (generally well-mixed) hypertrophic lakes (Wrigley & Toerien 1990). Within these ponds putrescible organic wastes are mineralized by saprophytes (bacteria and fungi) and the released nutrients are in part incorporated into phytoplankton biomass (Metcalfe & Eddy 1979). Production of oxygen by the pond phytoplankton during daylight hours augments the oxygen supplied by atmospheric aeration. The waste stabilization pond effluent is essentially pond supernatant, a complex mixture of particulate and dissolved materials including some highly light-attenuating constituents. The high suspended solids concentrations in waste stabilization pond effluents (often exceeding 100 g m^{-3}) have been noted by several authors (e.g. Wrigley & Toerien 1990) and discounted by some (e.g. Gloyna & Tischler 1980) as being relatively innocuous by comparison with sewage solids for example. However, these high solids concentrations, mainly of algal biomass, should be of concern, not least because of their potential optical impact.

Most domestic waste stabilization ponds are facultative (following the classification of Metcalfe & Eddy (1979, Table 10.10, p. 551)); that is, both aerobic and anaerobic processes of waste stabilization occur, the latter mainly in the settled sludge on the pond bottom. Animal manures are much 'stronger' in terms of oxygen demand than domestic sewages, so waste stabilization ponds used to treat these wastes usually follow an anaerobic pond in which much particulate organic removal occurs by sedimentation and digestion.

Hickey *et al.* (1989a,b) have summarized the wastewater characteristics of a number of waste stabilization ponds (oxidation ponds) in New Zealand used to treat dairy wastes (mainly cattle faecal matter) and municipal sewage. They showed that effluent quality, including optical quality, is rather poor and is highly variable because of the stochastic nature of the major forcing variables to which ponds are subjected, particularly wind and insolation. The data that were summarized by these authors included suspended solids measurements but, unfortunately, did not include optical characteristics with the exception of some turbidity and Secchi depth observations. In municipal ponds in the Auckland (North Island, New Zealand) area, median turbidity was 28 NTU (5 and 95 percentiles: 11 and 152 NTU), and median Secchi depth was 200 mm (50 and 450 mm).

Hickey *et al.* (1989b) suggested that water clarity impacts on receiving waters from pond effluents may be 'substantial', although obviously the clarity impact will depend on the background clarity of the receiving water as well as available dilution.

Table 5.3 gives statistics for optical coefficients of domestic waste stabilization ponds in New Zealand. The median beam attenuation coefficient at 550 nm was 40 m^{-1} and the 95 percentile was 108 m^{-1} showing that municipal ponds are fairly highly light attenuating. Most (94% on average) of the suspended solids in these municipal pond effluents was algal biomass and associated detritus (Davies-Colley, unpublished data). This explains the fairly bright, green to yellow colour typical of the water in these ponds, which appear very similar to productive lakes. The absorption spectrum of waste stabilization pond waters is very similar to that of eutrophic lake waters (e.g. Fig. 3.9) and shows the well-known chlorophyll *a* absorption peaks at 440 and 676 nm. The colouring impact of waste stabilization pond effluents seems likely to be generally less severe than the clarity impact, if only because the hue is mostly in the 'acceptable' green range (see Section 5.2). However, at low dilutions waste stabilization pond effluents can produce conspicuous hue shifts (to yellow–green) and much increased brightness, as well as visual clarity impacts in clear receiving streams.

The largest waste stabilization pond system in New Zealand at Mangere, Auckland (300 000 m^3 d^{-1} from 650 000 people—being the larger part of the population of Auckland City) produces a plume that is noticeable from the air when discharged to the waters of Manukau Harbour. The plume is visible because the colour of the effluent differs appreciably from that of the estuarine receiving waters. However, the effluent has little effect on the beam attenuation coefficient (and, therefore, the visual clarity) of the already turbid receiving waters near the point of discharge (Vant 1990).

5.4.4 Meat processing wastewaters

Table 5.3 gives optical coefficients for a typical meatworks effluent, prior to treatment. The light attenuation is considerably higher than in domestic sewage, consistent with the general picture of meatworks wastes as 'stronger' than domestic sewage in terms of both solids and oxygen demand. However, more important than the clarity impact of such a wastewater is the red-colouring impact of the blood content. Fig. 5.4 shows the absorption spectrum of an untreated meatworks wastewater. The strong absorption peak at 415 nm is attributable to blood haemoglobin. Selective absorption of blue through yellow wavelengths (<580 nm) is responsible for the characteristic red hue.

Meat-processing plants (abattoirs) are a major category of industrial discharge in New Zealand. In the past aesthetic problems, particularly the visual ingress of wastewaters, have been a major concern with meat processors in this country. One of the most infamous is the discharge (still occurring at the time of writing) of effluent from a large plant at Gisborne, in the North Island. This discharge is readily visible as a reddish plume from the air or from Kaiti Hill, a prominent and popular tourist viewpoint above Gisborne City. In the last decade the meat industry has moved significantly to improve effluent treatment, and the more obvious visual insults to receiving waters should soon be a thing of the past.

5.4.5 Pulp and paper wastewaters

Bleached Kraft pulp mill effluents are notoriously highly light absorbing and their discharge to the aquatic environment is significantly visually impacting (e.g. Gellman 1982). Table 5.3 gives optical coefficients for the effluent from a large Kraft mill at Kawerau, New Zealand. This effluent is discharged into the Tarawera River (Plate 12) and is discussed in Section 6.6. Similar values would probably apply to Kraft mills generally. The scattering (and nephelometric turbidity) of this wastewater is about the same order, numerically, as the light absorption at 550 nm in the middle of the visible spectrum. Thus Kraft effluent is typically more significantly colouring than clarity degrading.

The absorption spectrum of Kraft effluent is very similar to that of natural yellow substance (Davies-Colley & Vant 1987), being essentially featureless except for an exponentially shaped rise with decreasing wavelength (Fig. 5.4). Absorption is high throughout the visible spectrum but is highest at the blue extreme. Thus Kraft effluent, like natural yellow substance, imparts a yellow hue to waters. However, the absorption of light throughout the spectrum is so high that this hue is not readily apparent in receiving waters which give an impression of 'inky' blackness owing to very low reflectances (very dark colour). The strong light absorption by Kraft effluent markedly reduces light penetration into receiving waters (refer to the worked example in Appendix 3). This light reduction may have far-reaching ecological effects as well as a significant visual impact.

The impact of Kraft effluent on visual clarity of waters is perhaps of less concern than that on colour and light penetration, but visual clarity impacts should also be considered if new Kraft mills are being planned. Kraft effluent reduces visual clarity by a lesser factor than light penetration, because the latter quantity is less dependent on light scattering. However, visual clarity reduction can still be significant. For example, the visual clarity of the Tarawera River is reduced to one-third of that upstream, while reducing light penetration to about one-fifth of the upstream value (Davies-Colley, unpublished data) (see Section 6.6).

Other discharges from the pulp and paper industry are generally insignificant in comparison with Kraft effluent. Paper fibres may cause turbidity problems in the immediate vicinity of discharges from non-Kraft mills but they are otherwise comparatively innocuous. In particular, effluents from other pulp processes are weakly coloured, with the exception of dye wastes from the production of dyed paper products. Fig. 5.5 compares the absorption spectrum of the Kraft effluent discharged to the Tarawera River (Section 6.6) with that of a chemomechanical effluent discharged to the same river. The Kraft sample, being very strongly absorbing, was diluted ten times before the absorption scan was run in the UV. As could be expected from its lack of colour, the chemomechanical effluent is negligibly absorbing in the visible spectrum although there is significant

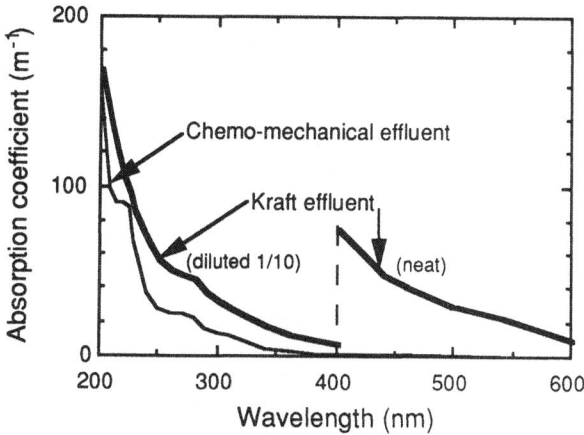

Fig. 5.5. Absorption spectra for pulp effluents. The Kraft effluent spectrum was measured on the neat effluent at visible wavelengths (>400 nm) and diluted 1/10 at ultraviolet wavelengths (<400 nm). The effluent from the chemomechanical mill was run neat. (Authors' unpublished data.)

absorption in the ultraviolet. The two spectra have some coincident peaks in the UV, probably related to common chromophoric groups in wood waste material.

5.4.6 Mining wastewaters
The mining industry is notorious in many parts of the world for discharge of inorganic suspended solids (Lyle 1987). Some optical measurements on New Zealand mine wastewaters are given in Table 5.3. We need to make a distinction between alluvial mining and hard-rock mining, and the latter category further subdivides into open-cast and underground mining, hence the categorization in Table 5.3.

Alluvial mining
Alluvial gold mining (placer gold mining) is an important industry in New Zealand, as it is elsewhere, notably in Alaska (Pain 1987; Reynolds *et al.* 1989). Typical small (e.g. two- or three-person) operations in New Zealand comprise a tracked excavator delivering gold-bearing alluvium to a rotating screen (trommel) that separates coarse gravels and cobbles from finer (gold-bearing) particles, usually at 12 mm diameter (e.g. USEPA 1988) (see Plate 10A). Very often the trommel is placed on a pontoon floating on a pond (a surface expression of unconfined groundwater) created by the excavation. The pond water is used as a supply of process water for operating the sluices which concentrate the gold in the fine fraction along with dense mineral grains (USEPA 1988).

Depending largely on local geology and day-to-day operating variables, loads of suspended solids derived from alluvial gold mining operations can have severe impacts on the ecology and visual water quality of small receiving streams (e.g. Lloyd *et al.* 1987; USEPA 1988; Reynolds *et al.* 1989). Even where there are no direct surface discharges of wastewater and the active stream channel is not disturbed directly by the mine machinery, the highly turbid pond waters can seep through the alluvium (Plate 10B). Filtration by the alluvium usually removes only the coarser silts, leaving the clays to seep into the adjacent stream. Clays can only be removed efficiently if physicochemical conditions are such as to promote flocculation of particles into coarser aggregates which are filterable by porous media (McDowell-Boyer et al. 1986).

Davies-Colley et al. (1992) investigated the ecological impacts, including those resulting from changed optical water quality, of six small alluvial gold mine operations on the west coast of New Zealand's South Island near Greymouth. The pond waters were typically extremely turbid (up to about 4000 NTU), thus providing a source of turbidity which often seeps through the alluvium into the nearby stream channel. Since the 'discharges' from these mines were almost always subsurface it was difficult to characterize their turbidities or attenuation coefficients. The optical impact of the mines varied greatly, between operations, and downstream of the same operation at different times. However, the impact was often very severe and visually gross (Plate 10B), with turbidities in the receiving streams up to 460 NTU. The median increase in beam attenuation of six small streams was about 30-fold, resulting in median sighting ranges of the order of 0.2 m over six weeks of monitoring.

Another form of alluvial mining, albeit for a far less valuable product than gold, is aggregate mining. Sand or gravel for concrete manufacture and other purposes is often mined from alluvial deposits, sometimes directly from active river channels. Again depending on the local geology, this type of mining activity, and effluents from associated aggregate grading and washing facilities, has the potential to impact severely optical quality of streams (e.g. turbidities exceeding 400 NTU were measured downstream of an operation in Southland, New Zealand (SCB 1986)). Aggregate extraction should be recognized for its potentially severe impact on optical quality of waters. Where possible, extraction should be carried out above water level and away from the active river channel.

Hard-rock mining
Open-cast mining has long been recognized as having the potential to cause severe environmental damage, notably to optical water quality, because these operations inevitably involve exposure of large areas of bare soil. The overburden heaps are exposed to water erosion, and high suspended solids yields may be generated by runoff of storm water (e.g. Lyle 1987; Addis et al. 1984). Those solids in the silt size range can be fairly easily removed by simple settling in sedimentation ponds (Lyle 1987), but clay-sized particles cannot be removed in such ponds without physical–chemical treatment with coagulants or flocculants. The impact on optical quality of receiving waters therefore depends on local geology (and hence nature of the overburden material), and efficiency of treatment (if any), as well as receiving water dilution.

Usually water will enter the mine workings of open-cast mines by seepage from intercepted groundwater aquifers (and perhaps old mine drives) as well as from direct rainfall on the mine 'catchment'. To prevent flooding of the mine workings this water is removed periodically by pumping (e.g. Addis et al. 1984). Open-cast mine dewatering effluents are typically very turbid, although this again depends markedly on local soils.

A single grab sample of the (untreated) dewatering effluent from an open-cast gold mine at Waihi, North Island, New Zealand (Table 5.3) had a moderate turbidity, but was rather strongly coloured (orange), probably by iron pigments. In contrast, the dewatering effluent from an open-cast coal mine, near Huntly, North Island, New Zealand, where coal seams are associated with kaolinite clays, was extremely turbid (mean turbidity of seven samples = 1700 NTU (Davies-Colley, unpublished data)). This latter mine, and other open-cast coal mines in the vicinity of Huntly, had very severe impacts on optical quality of receiving waters until moves were made to institute treatment of dewatering effluents and overburden runoff. The treatment typically consists of coagulation with alum or poly-aluminium chloride and sedimentation of the resulting flocs, sometimes aided by addition of polyelectrolytes. Turbidities lower than about 30 NTU can be achieved routinely in the treated effluents.

Underground mining, by its nature, generally poses less of a potential impact on the optical quality of water than does open-cast or alluvial mining. Much smaller quantities of rock and soil are disturbed so there is less potential for generating high loadings of suspended solids. The environmental concerns are more likely to focus on other issues, particularly acid drainage and toxic metal mobilization. However, dewatering effluents from underground mines, together with runoff water from associated surface facilities and spoil heaps, can still produce impacts on optical quality of receiving waters, depending on the nature of these wastewaters (in turn dependent on geology) and the available dilution in the receiving waters.

One unusual type of optically impacting material that is sometimes discharged from coal mines and associated coal-handling facilities is coal fines (coal dust). Coal fines are sometimes discharged in significant quantities in dewatering effluents, particularly from underground mines. Table 5.3 gives some data for a grab sample of dewatering effluent heavily laden with coal fines from the underground Huntly West Mine. The sample was intensely light absorbing and also strongly light scattering. The coal particles, as could be expected, behaved like optical black bodies and had a very severe optical impact on a small receiving stream that appeared dark grey in colour and very turbid. The absorption spectrum of the dewatering effluent (Fig. 5.4) was rather featureless and almost non-selective through the visible region of the spectrum. A significant degree of 'packaging' (Section 2.4.3) of pigments in the coal particles may have been responsible for the spectral shape rather than the intrinsic absorption spectrum of the coal constituents.

5.4.7 Other point discharges

A great variety of industries are likely to produce optically degrading effluents on occasions and in particular situations. The textile industry, and, as has already been mentioned, the paper industry, can sometimes produce highly 'unnatural' colours in receiving waters following dye runs. A large number of industries (e.g. brick, tile, pipe, pottery, paper, paint, pharmaceutical manufacturers) use clay minerals in various processes. Flushing of waste clays or clay slurries can cause severe, albeit usually temporary, optical quality problems in receiving waters. There may well be other processes and waste products that can be optically impacting, but are outside the authors' experience.

In New Zealand, the dairy industry is a major wastewater discharger to natural waters, although in the last two decades there has been a strong trend towards spray irrigation on land (Smith 1986). Milky wastes, being intensely scattering, although insignificantly absorbing, of visible light, have caused severe optical degradation on occasion. Indeed, it is of interest to note that milk, being an emulsion of particles with scattering properties rather similar to those of natural water particulates, has been used as a convenient model material for experiments on light transmission (radiative transfer) in waters (Timofeeva 1974).

Milky wastewaters from dairy factories are variable in character because many dairy operations are batch mode. Concern on the part of management authorities mainly for oxygen depletion in receiving waters has driven a substantial clean-up of dairy wastewaters in New Zealand. However, in some situations and on occasions, dairy factory wastewaters may still cause undesirable turbidity and associated clarity impacts in receiving streams.

The dredging of harbours and other navigable waterways, and the subsequent disposal of the dredge spoil, can cause increases in turbidity in the vicinity of the operations. The impact on water clarity depends on the type of dredge, the character of the spoil and the dredging and disposal method adopted (Herbich & Brahme 1991). However, turbidity increases appear to be of short duration and natural conditions are quickly restored; open water disposal causes more turbidity than dredging (Herbich & Brahme 1991). Apparently, measures to reduce sediment resuspension at the dredge site have been only partially successful. We are unaware of any detailed studies examining the optical effects of dredging and spoil dumping. In the 44 dredge operations that were examined by Herbich & Brahme (1991), only five specifically mention measurement of turbidity which was increased by up to 55 JTU above ambient in one case.

5.5 CONTROL OF SUSPENDED SOLIDS FROM SOIL EROSION

Suspended solid matter is often regarded as the most significant water pollutant, not just volumetrically but in terms of environmental damage and economic significance (e.g. ASCE 1975; Clark *et al.* 1985). Suspended solids have a multitude of environmental impacts on water bodies, including siltation of both man-made and natural channels, infilling of reservoirs, sludge formation, decreased quality for consumptive uses (leading to increased treatment costs), smothering of benthic habitat, and transport of other pollutants, notably nutrients, heavy metals and toxic organics (Clark *et al.* 1985). However, possibly the most ecologically significant, and certainly the most visually obvious, impact of suspended solids in water is on optical quality.

The source of light-attenuating suspended solids in waters, with the exception of wastewater solids (Section 5.4) and phytoplankton (Section 5.6), is erosion by water. Of course a low rate of erosion (termed *geological* erosion) occurs naturally, so we are concerned here with accelerated erosion, usually referred to as *soil* erosion. Erosion of soils by water is greatly accelerated over background rates in four main types of activities (Goldman *et al.* 1986): agriculture, forestry (logging), mining and construction. USEPA (1973b) gives a table indicating intensity of erosion on lands subjected to these different types of activities (Table 5.4). The impact on quality of water draining from a catchment will obviously depend on the areal extent of activity in the catchment, as well as intensity of erosion. Thus the values in Table 5.4 indicate that grazing of a whole catchment may degrade optical water quality as much as building or road construction in only 0.5% of the catchment. Moreover, the grazing impact may continue indefinitely whereas the construction impact is of limited duration. Location of soil-disturbing activity in relation to the stream channel network also influences the severity of impact on water quality.

There is a vast literature on the subject of suspended solids in waters, and on soil erosion and the control of suspended loads from various activities on land. Space precludes

Table 5.4. Representative rates of erosion from various land uses

Land use	Erosion rates	
	t km^{-2} yr^{-1}	Relative to forest
Forest	8.5	1
Grassland	85	10
Abandoned surface mines	850	100
Cropland	1 700	200
Harvested forest	4 250	500
Active surface mines	17 000	2 000
Construction (buildings, roads)	17 000	2 000

Source: USEPA (1973b).

a comprehensive review of this literature here, although reference is made to some useful texts. Our concern, of course, is with the impact that this largest of all pollutant categories has on optical water quality, and this perspective leads to some important philosophical differences from the typical soil conservation approach to erosion. Here we discuss the application of erosion control principles, keeping in mind the ramifications for optical water quality, which necessarily emphasizes the fine-grained 'wash load' far more than the readily settleable coarse silt and sand.

5.5.1 Philosophy

Much of the concern in the hydrological, geomorphological and soil conservation literature has been with particulate material yield from catchments (in t km^{-2} yr^{-1}) (e.g. Bordas & Walling 1988). Such yield figures correspond to a denudation rate (in mm yr^{-1}) averaged over the whole catchment. Suspended solids yield is an important geomorphological consideration, and a major agronomic concern inasmuch as it reflects soil losses, but usually yield of solid matter is not the relevant quantity to consider in relation to the downstream (sometimes referred to as 'offsite' (e.g. Clark et al. 1985)) environmental impacts of the eroded material, including particularly, optical water quality impacts. Typically the concentration of suspended solids is a more relevant quantity than the yield, as can be illustrated by considering the optical impact of turbid flood waters.

Most of the suspended solids yield from a typical catchment is discharged during floods occurring for only a small fraction of the total time in a year (e.g. Leopold et al. 1964; ASCE 1975, pp. 485–487). Thus the exposure time of recreational users or aquatic life in rivers to the high suspended loads of floods is usually quite limited. During high turbidity flood events, recreational users are unlikely to wish to use river waters for contact sports, partly because of the poor aesthetic quality, although safety concerns related to strong currents and poor visibility will usually be overriding.

Impoundment of turbid flood waters may reduce downstream yield of suspended solids owing to sedimentation, but the fine-grained silt and clay particles responsible for most of the light-attenuation (i.e. turbidity) do not settle readily, so that the yield of light-attenuating material may be little changed. However, impoundment may greatly increase the exposure time of both aquatic organisms and recreational users to turbid waters. Arguably, from an optical water quality perspective, a long exposure time at moderate turbidities is more damaging than a short exposure to extreme turbidities. These concepts are illustrated in Fig. 5.6.

We see, therefore, that the optical impact of flood waters is likely to be less severe in uncontrolled river waters than in lakes and river impoundments, or the rivers downstream of such standing water bodies, or, for that matter, rivers downstream of flood control structures. Perhaps a re-evaluation of environmental priorities for the handling of

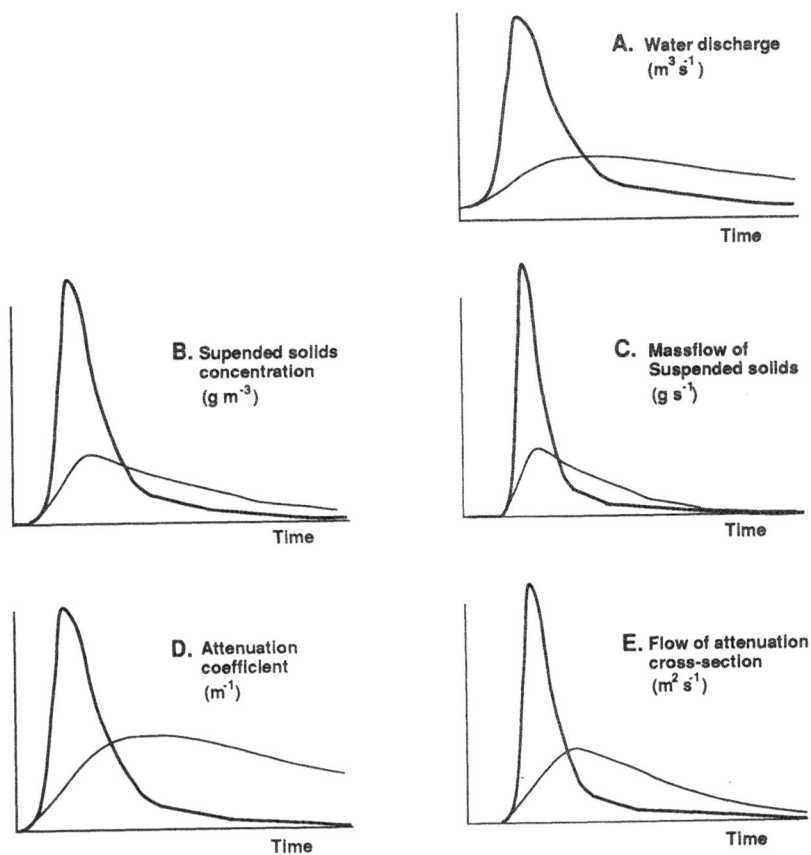

Fig. 5.6. Schematic diagram illustrating the effects of impoundment of flood waters on suspended solids yield and turbidity of waters downstream. (A) Flood hydrographs (——, original hydrograph; ——, hydrograph downstream of the impoundment). (B) Suspended solids concentration versus time. (C) Massflow of suspended solids versus time. Some suspended material is assumed to be removed by settling such that the suspended solids yield (the time integral of suspended solids massflow) is reduced. (D) Attenuation coefficient versus time. The exposure time to relatively high turbidity (high attenuation coefficient) is greatly increased by impoundment even though the peak is lowered. (E) Flow of attenuation cross-section versus time. Since most of the attenuation is by non-settleable solids the yield of attenuation cross-section (the time integral of attenuation cross-section flow) is unchanged by impoundment.

flood waters is required—at least for modest floods. If exposure of downstream water bodies, particularly lakes, to optical water quality damage is likely to be increased by impoundment or diversion of flood waters, there may be a case for risking some over-bank flooding so as to permit rapid routing of the flood water turbidity out of the system.

The above discussion illustrates that SS concentrations are more relevant than yield. However, SS concentrations alone are not sufficient to assess environmental impact generally, and optical quality impact in particular. The impact of SS is strongly dependent on the nature of the suspended particles as well as on their quantity. Most important among characteristics of suspended particles, as regards their environmental behaviour and optical impact, are the particle size distribution, which depends on supply factors (i.e. the range of particle sizes available in the source materials) and surface interactions—as well as hydraulic forces, and the composition of the particles as it affects density relative to the water, surface forces at the sediment–water interface and optical properties such as refractive index and absorption of light. The surface chemical character of suspended particles in water, which largely determines whether or not the particles will form a stable suspension or will agglomerate into larger particles, depends greatly on the water composition and on adsorbed materials on the particle surfaces.

The optical impact of suspended solids is often not recognized, in spite of the fact that the turbidity of waters with high suspended solids concentrations is probably their most obvious feature (Clark *et al.* 1985, Chapter 1). Very often, measurements of suspended solids are made in situations where the optical impact is of paramount importance, and it would be more appropriate (and usually much simpler) to measure water clarity or turbidity. As we saw in Section 2.4, the optical cross-section of suspended particles depends strongly on particle size, as does their settling velocity. Fig. 2.9 shows that the maximum attenuation cross-section of suspended particles in waters occurs at about 1.2 μm diameter for inorganic particles and about 5 μm for organic particles. Particles around the size range of, say, 0.2 to 5 μm for mineral particles, and 1 to 10 μm for organic particles, being efficiently light attenuating, contribute most turbidity in waters. These particles (clay to fine silt size range) have low fall velocities and so settle slowly, if at all, in natural bodies of surface water which always have some degree of turbulence.

5.5.2 Factors affecting soil erosion

Since the particles which attenuate light most efficiently in waters have little if any tendency to settle once suspended, the emphasis must be to prevent their entrainment by water in the first place so as to avoid optical water quality impacts. That is, measures to protect optical water quality from the impact of suspended solids must focus on prevention of erosion by water. This is particularly important on clay-rich soils which, while usually cohesive and difficult to erode (requiring much greater entraining shear stresses than silts or fine sands which are comparatively easily eroded), yield large quantities of highly light-attenuating particles once they are eroded. Where erosion of clay soils is unavoidable, as in some mining operations, turbid waters may need to be treated with chemicals (coagulants and flocculants) to promote agglomeration of fine particles into larger settleable flocs.

Before discussing control measures we consider the phenomena involved in soil erosion and the factors that affect erosion rates. Detailed discussion is given in numerous texts (e.g. Dunne & Leopold 1978; Selby 1982); here we give only a brief outline.

On well-vegetated slopes, especially steep slopes, the main form of erosion is typically mass movement (Fig. 5.7) such as landsliding. Mass movements occur mainly during rainstorms, when the extra weight of water in soil, together with the reduction in slope-normal stress by pore water pressure, destabilizes weathered material. Mass movement does not of itself supply suspended solids to waters, except where it delivers eroded material directly to stream channels. However, mass movements expose areas of bare soil to the direct impact of raindrop or to running water entrainment. Thus SS concentrations are typically high in water draining basins that suffer severe mass movement.

When vegetative cover of soils is reduced or removed, the soil surface is exposed directly to the action of rainwater or running water (Fig. 5.8). The partial or total removal of vegetation may arise for a variety of reasons, including mass movement, trampling and grazing by animals (Plate 13), cultivation of arable land, tracking by vehicles, or deliberate removal as in mining and construction activities (Toy & Hadley 1987). Once a rainstorm commences the first type of erosion on bare soil is *splash erosion* by raindrops. Splashing of raindrops can erode directly by moving soil particles downslope, and also contributes indirectly to erosion by damaging the soil structure at the surface, thereby inhibiting infiltration and promoting overland flow.

Overland flow of water may commence as a shallow 'sheet'. Sheet flow is not usually very erosive in itself, but it transports the products of rainsplash erosion. Obviously this *sheet erosion* is confined to soils bare of vegetation. Once the 'sheet' has moved a metre

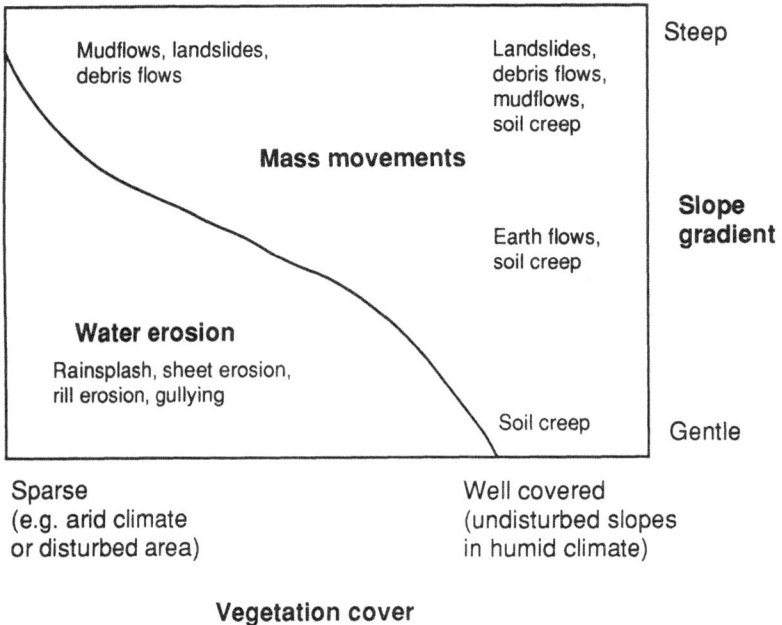

Fig. 5.7. Conditions of vegetation cover and slope under which various erosion processes dominate on hillslopes. (Adapted after Dunne & Leopold (1978).)

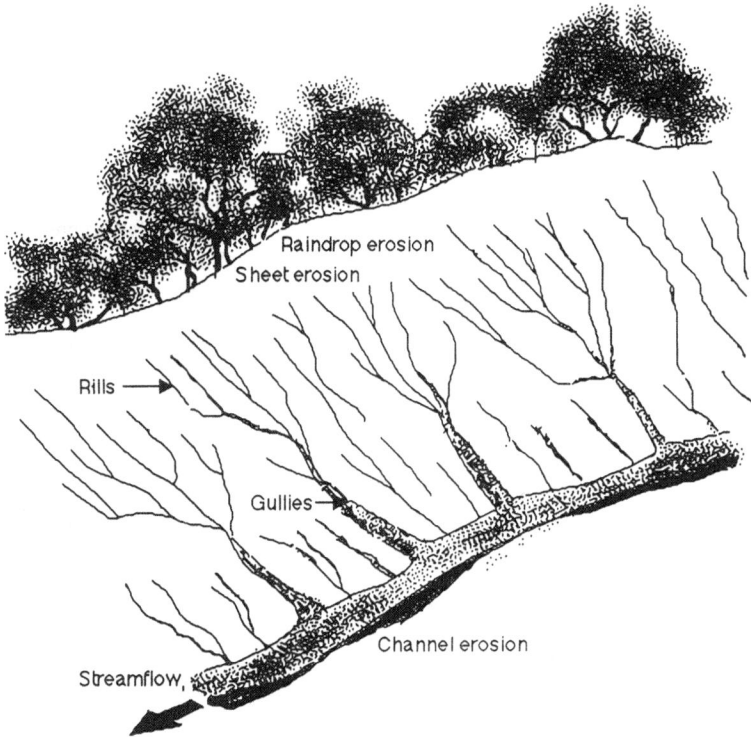

Fig. 5.8. Types of soil erosion by water on slopes denuded of vegetation.

or so, the water tends to become channelized on the soil surface and its velocity increases. When the shear stress that this water flow exerts on the soil exceeds a critical value, depending mainly on particle size, but also on a wide range of other factors including degree of cohesion and soil aggregation, the flow begins entraining soil particles. *Rill erosion* is characterized by the cutting of small channels known as rills (Fig. 5.8).

Gully erosion involves movement of water at rapid velocity in defined channels (Fig. 5.8). These high water velocities are very erosive and entrain soil particles of a wide range of sizes, thereby rapidly enlarging the gully. Gullies may form by enlargement of rills and also by slumping of oversteepened banks.

Channel erosion processes, for example the slumping of undercut streambanks, occur naturally in large floods, but can be greatly accelerated if the magnitude of flood flows is increased by catchment land use changes, notably urbanization (Fig. 5.9).

The four main factors affecting soil erosion are climate, soil characteristics, topography and ground cover. The first two are largely 'givens' for a particular site, so erosion control must focus on manipulation of topography and ground cover. These factors are perhaps best discussed within the framework of the equation most used for calculation of soil erosion: the universal soil loss equation (USLE) developed by the US Soil Conservation Service (Wischmeier 1976). The USLE can be stated as

$$A = RKL\text{-}SCP$$

where A is soil loss rate (dimensions: mass per unit area per unit time). (Note that some of the symbols appearing in this equation are used elsewhere in this book for other quantities.) The various factors are as follows.

R is the *rainfall erosivity factor* (the climate effect), the product of two characteristics for all rainstorms in an average year (the maximum 30 min rainfall intensity and the kinetic energy of the raindrops—itself a function of the rainfall intensity) summed for the year. Mitchell & Bubenzer (1980) discuss calculation of R values. Obviously R is an unmanageable 'given' for a particular site.

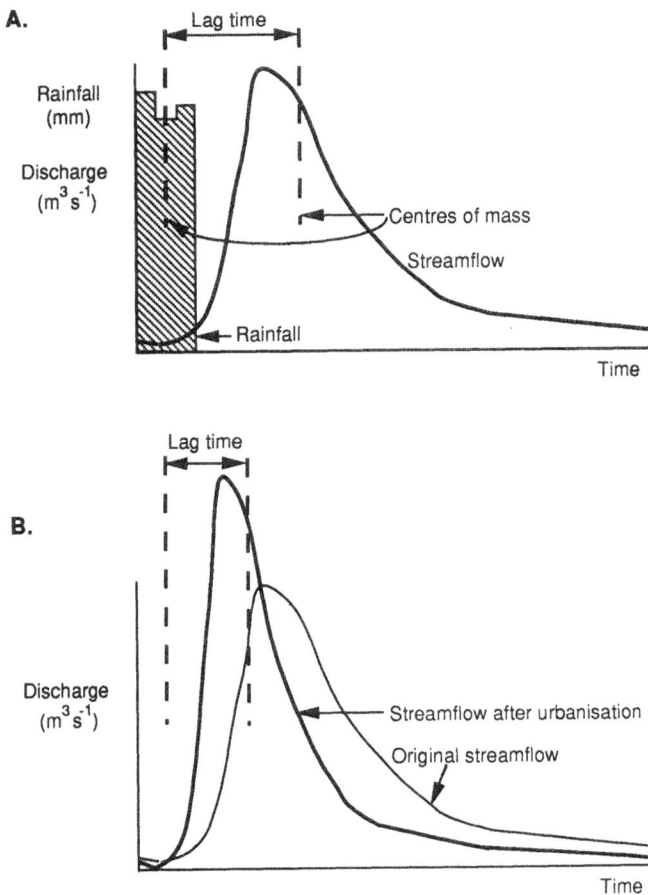

Fig. 5.9. Schematic illustration of the change in the pattern of streamflow after urbanization of a catchment. (A) Pattern of streamflow in relation to rainfall in the natural system. (B) Change in streamflow pattern after urbanization. (Adapted after Leopold (1968), with permission.)

K is the *soil erosivity factor* (the soil characteristics effect). This factor is usually taken as a function of soil texture, silty soils being the most erodible (yielding about 6 times the soil loss of sands and clays). However, soil aggregate size is physically more relevant to erodibility since it is aggregates of primary particles that are initially entrained (e.g. Thornes 1980). The preferred means for estimating K is by plotting the soil textural analysis (per cent sand, silt and clay) in a ternary (triangular, 3-axis) diagram on which iso-erodibility lines are superimposed (e.g. Goldman *et al.* 1986, p. 515). Soil erodibility generally decreases with increase in soil organic content (e.g. Mitchell & Bubenzer 1980), and a correction to estimated K can be made if organic content is known. K, like R, is largely a 'given' although some local manipulation may be possible. From an optical water quality perspective, the texture of soil is especially relevant since soils high in clay or fine silt are liable to yield low clarity runoff water (see Fig. 2.9).

L-S, the *length–slope factor* (topographic effect), combines the effect of slope length and gradient. Empirically, soil loss increases as a quadratic function of gradient and as a power function of slope length (the exponent itself being dependent on gradient). The L-S factor is actually the ratio of soil loss from a site to that from the same site standardized to a 21 m slope at 9% gradient. Tables and nomograms are given in standard texts to estimate this factor, which can often be manipulated by slope grading or by construction of mid-slope diversion ditches or terraces.

C is the *cover factor* (termed the *cropping management factor* in the context of cropland soil conservation). This factor is the ratio of soil loss under the specified cover to that from bare soil and is often the factor most amenable to control. Native vegetation gives a high level of soil protection (e.g. $C = 0.01$, i.e. 99% reduction in soil loss (Goldman *et al.* 1986)). Grass cover is rated at $C = 0.1$ while mulching of bare soil is generally less effective ($C = 0.1$ to 0.5). Extensive tables are given for croplands (e.g. Mitchell & Bubenzer 1980).

Finally, P is the *erosion control practice factor*. This is the ratio of soil loss from a site to that with up and downslope tillage. P is generally the least variable factor (and therefore least important). Compacted, smooth soils are assigned $P = 1.3$ (Goldman *et al.* 1986) while a slope disced parallel to the contours is assigned $P = 0.8$. Contouring, and especially terracing, are cropland soil conservation measures that reduce P more significantly (e.g. Beasley 1972).

The USLE is an empirically based tool for the prediction of annual average soil loss. The actual erosion process is not represented and, indeed, the equation is strictly applicable only to sheet and rill erosion, not gully erosion. The equation was developed originally for use on cropland, but can also be applied to exposure of soils in construction, mining and unpaved road surfaces. Of the factors appearing in the USLE the L-S and C factors are most amenable to manipulation so as to minimize erosion, as is illustrated below.

There are a number of good texts on erosion and sediment control although, as we have noted, they emphasize quantity of solids eroded or catchment yield rather than downstream water quality, and thus fail to give due recognition to the disproportionate water quality significance of the clay and fine silt fraction, compared with settleable solids. UNESCO (1982) reviewed the problems of sedimentation from a comprehensive global perspective. Beasley's (1972) 'Erosion and sediment pollution control' deals very comprehensively with control of solids mainly from agricultural land use. Chapters 4 'Sources and yields' and 5 'Sediment control methods' in ASCE (1975) are also useful although emphasizing engineering rather than soil conservation solutions. 'Applied hydrology and sedimentology for disturbed areas' by Barfield *et al.* (1981) has been written mainly from the perspective of surface mining problems but, again, this text has more general applications, as does Lyle's (1987) 'Surface mine reclamation manual'. 'Geomorphology and reclamation of disturbed lands' by Toy & Hadley (1987) is written from the perspective of applied geomorphology and appears to be one of the few texts that discusses erosion of lands scarred by recreational activities (hiking trails and offroad vehicles: motorcycles and dune buggies). An excellent text is that of Goldman *et al.* (1986), 'Erosion and sediment control handbook', which is written mainly from the perspective of urban construction problems but, again, has general application to surface mining and roading and, to a lesser extent, agriculture and forestry. Much of the following discussion has been distilled from Goldman *et al.* (1986).

5.5.3 Construction and surface mining

The inevitable denudation of soils during building or road construction can increase solids yields up to 40 000 times that from undisturbed land, and is typically 10 to 20 times that from agricultural land (Table 5.4). Similarly, surface mining involves exposure of bare soil areas and has the potential for very intense erosion (e.g. Addis *et al.* 1984). Goldman *et al.* (1986) give ten 'common-sense' principles for the control of erosion:

(1)   fit development to the terrain;
(2)   time grading and construction works to minimize soil exposure;
(3)   retain existing vegetation whenever feasible;
(4)   vegetate and mulch denuded areas;
(5)   divert runoff away from denuded areas;
(6)   minimize length and steepness of slopes;
(7)   keep runoff velocities low;
(8)   prepare drainageways and outlets to handle concentrated or increased runoff;
(9)   trap sediment (suspended solids) on site;
(10) inspect and maintain control measures.

These general principles are discussed in turn below.

Principles (1) and (2) apply mainly to construction including road construction, but also to mine planning, and are intended to reduce erosion potential on a disturbed area. Urban land planning and also highway planning should take into account the topography, existing vegetation and erodibility of soils, and minimize soil disturbance by minimizing grading. As far as possible, activities that inevitably result in bare soil exposure should be scheduled for dry seasons of the year.

Principle (3) is possibly the most important from an optical water quality perspective since vegetation is by far the most effective (and cost-effective) form of soil erosion control. Vegetation reduces erosion rather than merely dealing with eroded material and the water conveying it (Goldman *et al.* 1986). Vegetation protects soils from erosion by absorbing raindrop impact, reducing runoff volume (by interception, and promotion of infiltration), reducing the velocity of runoff water, and binding the soil with roots and cohesive organic matter. Re-establishing vegetation, once removed, can be a long-term and costly process, even where the soil fertility is not degraded by erosion or removal of topsoil. Therefore, existing vegetation, however scrubby and aesthetically unappealing, should be retained as much as possible. The removal of existing vegetation for construction should, ideally, be restricted to the foundation site and access area (Goldman *et al.* 1986).

Principle (4) applies equally to surface mining as to construction, and possibly also to agricultural land scarred by bare soil areas. A properly re-vegetated soil is protected from erosion. Bare soils can be covered with temporary vegetation, particularly hardy drought-resistant species and nitrogen-fixing legumes for nutrient-impoverished soils, before permanent landscaping is carried out. Goldman *et al.* (1986) list plant species recommended for erosion control. Mulching with materials such as straw, wood chips or fabric mats is usually best regarded as temporary erosion control, and a means for improving the micro-environment (e.g. by moisture conservation) until erosion-controlling plants are established.

Principles (5) to (7) have to do with the use of simple structures (a) to minimize contact of storm water with erodible soils and (b) to prevent runoff water from reaching erosive velocities and causing severe rilling and gully formation. As far as possible storm flows should be prevented from contact with bare soil areas by interception in diversion ditches (principle (5)). Because slope gradient and length are among the most critical factors affecting erosion (as can be seen in the USLE), minimizing slope length and steepness, e.g. with mid-slope diversion ditches or terracing, prevents runoff water from reaching erosive velocities (principle (6)). Runoff water velocities can also be kept low by deliberately roughening surfaces, for example by 'trackwalking' slopes with tracked vehicles and lining drains with vegetation or crushed rock (riprap) (principle (7)).

Principle (8) recognizes that the nature of storm runoff from disturbed areas is often very different from that under the original vegetation cover. In particular, water yields are increased and the 'lag time' (time from the centre of mass of rainfall to that of storm-flow in the stream) is decreased, so that much larger peak stormflows occur (Fig. 5.9). Drainage structures such as culverts must be designed to handle these greatly increased peak stormflows. Often energy dissipation structures and protection works such as riprap aprons are desirable. Goldman *et al.* (1986) discuss the use of the rational formula relating water runoff to rainfall for the calculation of design flows: $Q = C_R i A_C$, where i is rainfall intensity for the 'design' rainstorm, A_C is catchment area and C_R is a 'runoff coefficient'. There are extensive tabulations of C_R so as to account for the effects on runoff of topography, soil type, and vegetation.

Principle (9) recognizes that some erosion during soil-exposing activities such as construction may be inevitable. Therefore every attempt must be made to prevent solid particles from leaving the site and causing damage to receiving waters. Goldman *et al.* (1986) discuss in some detail the design and use of sedimentation basins and traps, straw bale dikes and silt fences. However, these structures are effective mainly on coarser settleable solids (sands and coarse silt) (Goldman *et al.* 1986, p. 2.13), rather than the clays and fine silts that are efficiently light attenuating. Therefore, from an optical water quality perspective, applying principle (9) is 'shutting the stable door after the horse has bolted'. Where clay or fine silts are inevitably entrained as in some mining activities, treatment of turbid storm waters or dewatering effluents by coagulation/flocculation may be necessary. This is discussed briefly below.

Principle (10) recognizes that prevention of erosion problems is much better than cure. Regular inspection of disturbed soil areas and of water conveyance structures such as ditches and culverts should be carried out, and maintenance (e.g. repeat planting or mulching, or repair of structures) performed immediately if required.

Goldman *et al.* (1986) give recommendations for ordinances to be applied by municipal or other local authorities to control erosion on construction sites. They maintain that all areas of significant soil disturbance or exposure should be subject to permits.

Where entrainment of very fine solids in the clay to fine silt size range is inevitable, as is the case in some mining operations, it is usually not possible to trap the solid particles onsite by simple gravity settling. Settling of the fine solids in sedimentation basins must then be promoted with the use of chemicals. Fine particles such as clay minerals in waters typically carry a negative surface charge which causes mutual electrostatic repulsion (Stumm & Morgan 1981), and this repulsion must be overcome before the particles will agglomerate into larger and settleable flocs. Chemicals which promote agglomeration of particles are termed coagulants or flocculants. The use of these terms is not consistent throughout the chemical, geological and engineering literature (e.g. Weber 1972), but coagulation usually refers to the reduction of (usually negative) surface charge by chemical adsorption of ions (usually cations), while flocculation refers to the production of flocs by interparticle bridging. Removal of solids in waters by coagulation/flocculation is standard practice in water and wastewater treatment.

The most commonly used inorganic coagulants are alum (hydrated aluminium sulphate) and PAC (poly-aluminium chloride), but iron salts are also sometimes used. These chemicals promote agglomeration of particles both by adsorption and charge neutralization, and by adsorption simultaneously to two or more particles. Increasingly, the use of inorganic coagulants is being augmented or even replaced by the use of organic polymers known as 'polyelectrolytes' (referring to their multiple charges) which flocculate particles by adsorption and interparticle bridging, sometimes at much lower dosages than are required for inorganic chemicals.

The SS (and light attenuation) in effluents from sedimentation basins can usually be very efficiently removed by coagulation/flocculation. However, generally applicable procedures for predicting SS removal efficiency and effluent concentrations are not available, so selection of a coagulant/flocculant and optimization of dose and other important parameters, notably pH, usually requires laboratory experimentation (e.g. Weber 1972). So-called 'jar tests' are typically carried out with a stirring platform using 1000 ml beakers containing the turbid water and various doses of coagulant/flocculant at different pH values. These 'jars' are stirred vigorously for a few minutes for mixing and then more slowly for about 10 min to induce particle–particle contact and flocculation. The turbidity of the water above the settling flocs can then be measured to indicate efficiency of clarification of the water.

5.5.4 Agriculture and forestry

The actual process of clearing native vegetation cover for conversion of land use to agriculture or production forestry can severely impact water quality (e.g. UNESCO 1982; McColl & Hughes 1981), including optical water quality. Burning of the existing vegetation can be particularly damaging because it destroys the reserves of organic matter in the soil, thus damaging soil structure and fertility, and reducing the speed with which vegetation cover can re-establish. Tracking by heavy equipment used to clear or crush existing vegetation can lead to severe soil erosion as a result of soil compaction. The following principles for land clearance are adapted after McColl & Hughes (1981):

(1) areas for clearance should be selected having regard to land use capability (i.e. suitability for particular uses);
(2) burning should be controlled to preserve riparian and steepland vegetation, and should be scheduled (with regard to seasonal rainfall and temperature patterns) so that re-vegetation will be rapid;
(3) animals and vehicles should be excluded from likely contributing areas;
(4) vegetation cover should be re-established as rapidly as possible.

The need for erosion control on cropland agriculture in the US prompted the development of soil conservation principles and the USLE (Mitchell & Bubenzer 1980). There is an extensive literature giving recommendations for soil conservation and prevention of erosion on cropping lands (e.g. Beasley 1972). Here we give greater attention to pastoral agriculture (grazing) and production forestry, which are perhaps less well documented as sources of suspended solids and optical water quality degradation.

Pastoral farming and forestry rarely leave extensive bare areas, and pastoral surface soils are stabilized by the presence of shallow rooting systems; forestry soils are protected by litter and residual roots. Therefore entrainment of soil particles in waters draining a catchment occurs mainly from relatively limited areas, where disturbance of soil or vegetation cover coincides with overland flow during rainstorms. The partial contributing area concept, first introduced by Hewlett (1961), describes storm runoff as arising from certain contributing areas adjacent to the perennial stream channels where local waterlogging occurs. The concept, illustrated in Fig. 5.10, is now well established. There are divergent views on process details, but clearly the nature of vegetation and soils in these contributing areas greatly affects suspended solids delivery to streams. As far as possible, therefore, disturbance of soils or vegetation in contributing areas should be avoided to minimize mobilization of solid particles.

Yield of suspended solids from extensive land uses (diffuse sources) such as grazing depends on the condition of vegetation and soils in riparian zones (e.g. Karr & Schlosser 1978; Petersen et al. 1987), which include most of the contributing areas of catchments. Undisturbed vegetation in riparian zones promotes infiltration of runoff water that would otherwise reach water bodies as overland flow, provides an opportunity for suspended material to be filtered by contact with vegetation, and protects streambanks from channel erosion. Wetland areas may be particularly significant in water retention and in clarification of waters (e.g. Novitski 1978). Thus riparian vegetation functions to attenuate flood peaks and total water yields, and also tends to trap or 'filter' suspended matter delivered in overland flow.

Grazing
Soil erosion on grazed land is not usually as severe as on cropland of the same slope, because the soil is more completely covered (UNESCO 1982). However, grazing is often carried out on slopes far too steep for tillage. Soil erosion on grazed pasture is usually higher than on ungrazed pasture because the trampling of grazing animals damages soil

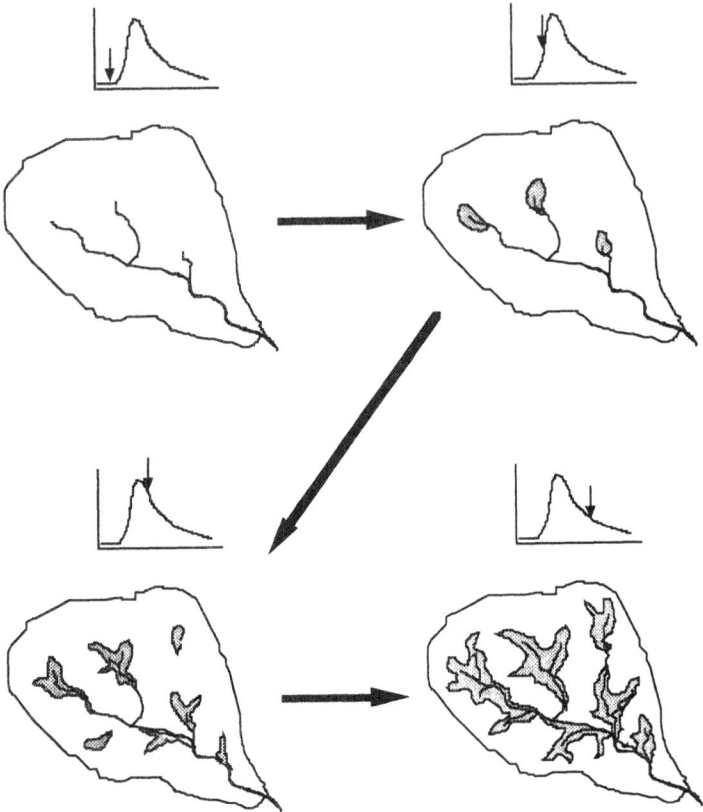

Fig. 5.10. Schematic time lapse view of a basin showing expansion of the areas contributing to storm-flow in relation to the storm hydrograph. (After Hewlett & Nutter (1970).)

structure, and the grazing reduces vegetation cover (USEPA 1979). Thus trampling and compaction of soils by livestock on grazed steeplands may result in high suspended solids concentrations in overland runoff.

There have been few definitive studies on the influence of grazing animals on water quality, including suspended solids concentrations and yields, possibly because of the difficulty in controlling the many factors affecting behaviour of animals and the damage to vegetation, soils and channels that they cause. A review by Skovlin (1984) and an overview article by Elmore & Beschta (1987) focus attention on riparian zones where damage by grazing animals is most significant because these include contributing areas.

Access by livestock into riparian areas apparently reduces water and solids-attenuating functions of these areas (e.g. Elmore & Beschta 1987). Indeed, livestock may force access in order to reach the water supply that they require, and this can lead to severely trampled areas in the riparian zone where there is a high potential for loss of entrained solids directly to the water body during rainstorms (Plate 13). Trampling by livestock, particularly cattle which exert greater hoof pressure than other domestic ungulates

(UNESCO 1982), may destabilize streambanks, leading to accelerated channel erosion.

Skovlin (1984) cites studies in the USA which demonstrate that surface condition and soil erosion in grazed catchments are strongly related to grazing intensity. At 'moderate' grazing intensity the animal production is maximized, but a good pasture cover is maintained and forage yield is sustainable. In the semi-arid western USA, cattle choose to graze most intensely on riparian meadowland because of factors such as proximity to water supply, shading riparian trees, and more palatable grasses, but this puts extra pressure on the riparian zone precisely where the water quality impacts are likely to be most severe (Plate 13). Sheep, in contrast, do not so actively seek out the riparian zone, and they may be less damaging anyway because of lower hoof pressures on the soil.

The research findings discussed by Skovlin (1984) suggest fencing of riparian zones to exclude livestock and to protect the vegetation in these areas, and also to protect the values of the aquatic system beyond. This usually means that alternatives to natural water supplies must be provided. Smith (1987, 1989) demonstrated that suspended solids yield in overland runoff from steep New Zealand hill country was appreciably reduced (lowered by 87%) when cattle and sheep were excluded by fencing. The following general guidelines (adapted after McColl & Hughes (1981, p. 28, 50)) can be given for the control of soil erosion and resulting optical quality impacts as well as other downstream impacts in grazed watersheds:

(1) fence off streambanks and other riparian areas to exclude stock;
(2) provide livestock with water supplies other than natural water bodies;
(3) manage grazing intensity with fencing and rotations;
(4) avoid over-grazing so as to maintain good cover and sustainable forage yield.

Production forestry
Forestry is often regarded as an ideal land use for minimizing erosion and suspended solids generation. In New Zealand, considerable areas of steepland that are intrinsically unsuitable for grazing (i.e. erosion prone) have been retired from grazing by the planting of trees for production forestry, with the major secondary objective of controlling erosion (McColl & Hughes 1981). However, production forestry necessarily involves some soil and vegetation disturbance during road construction and harvesting, so creating the potential for erosion and degradation of the quality of runoff water.

Increases in suspended solids yield and associated stream turbidity during forest harvesting operations have been well documented (e.g. Brown & Krygier 1971; Beschta 1978, 1979; Graynoth 1979; Anderson & Potts 1987; Campbell & Doeg 1989). There seems to be general agreement that the main sources (of order 80%) of elevated suspended solids concentrations in production forests are associated with roading (e.g. Megahan 1975; Beschta 1978; Mosley 1980; Reid & Dunne 1984; Bilby et al. 1989), although some sources may be associated with firebreaks and landings (log-handling areas). During harvesting, suspended solids yields temporarily increase owing to compaction of soils under the weight of logs and heavy equipment, depending on the log recovery procedures and local terrain. Forestry roads are seldom engineered to the same standard as highways (e.g. Toy & Hadley 1987). In particular, water conveyance and hillslope stability are not given the same attention, with the result that mass failures can occur and runoff water may reach erosive velocities over bare soil areas, particularly on (unpaved) road surfaces as well as associated ditches and cut and fill/sidecast slopes. Harvesting activities other than road and track construction give much less dramatic increases in suspended solids (e.g. Likens *et al.* 1977; Anderson & Potts 1987).

Mass movements of soils are often important in supplying material to streams in steep forested terrain (e.g. Swanson *et al.* 1981). Unfortunately the scientific understanding and predictability of mass wasting is less well founded than that of erosion by overland water flow, and general principles are more difficult to enunciate, let alone to quantify (e.g. Selby 1982). However, it seems clear that mass wasting scars bare of vegetation can be important sources of suspended solids generation by overland flow, as can the debris deposits from the mass movements.

The solids yield from forest roads has been found to depend strongly on traffic volume (e.g. Reid & Dunne 1984; Bilby *et al.* 1989). Following harvesting activity which involves intense road use, solids generation from road surfaces generally decreases sharply, although in steep terrain the destabilization of hillslopes by abandoned forest roads may continue to promote mass failures—which provides new disturbed areas then subject to water erosion. Bilby *et al.* (1989) discuss management of forest roads to reduce impacts on water quality in forest streams. 'Problem sections' of forest roads can be identified as those with a steep gradient, and that are subject to heavy use and drain directly to stream channels, so abatement efforts should concentrate on these areas. In particular, road ditches can be directed to drain onto areas of forest floor where there is opportunity for water to infiltrate and solids in runoff to be filtered before reaching stream channels.

5.6 CONTROL OF PHYTOPLANKTON

5.6.1 Phytoplankton targets for lakes

Only in certain types of lakes are phytoplankton commonly the dominant cause of manageable problems of colour and clarity. Phytoplankton do not usually dominate the optical properties of rivers and estuaries (see Sections 4.3 and 4.4), and problems of colour and clarity are not manageable in oceanic waters (where phytoplankton are dominant). In addition, while there is a broad correlation between the fertility (i.e. nutrient status) of lakewaters and the phytoplankton biomass found therein, rivers and estuaries are more complex environments for phytoplankton growth, with factors such as flushing, light limitation and grazing frequently placing additional major controls on the production of algal biomass. As a result, little is known about managing phytoplankton biomass in environments other than lakes. This discussion of the management of phytoplankton-dependent colour and clarity problems is therefore confined to the control of phytoplankton biomass in lakes, a topic which has been well studied for many years (e.g. see OECD (1982), Cooke *et al.* (1986), Vant (1987a), and Sas (1989) for further details of the material discussed here).

When phytoplankton cells are present at sufficient concentration to discolour and reduce the clarity of a lakewater noticeably, the excessive biomass is known as a 'bloom'. Close inspection of a bloom, particularly under direct sunlight, can often reveal the presence of colonies or aggregates of cells appearing as specks or particles (individual cells are too small to be observed by the naked eye). A 'scum' forms if these aggregates float to the water surface, as commonly occurs with gas vacuolated cyanophytes (blue–green algae). While it is usually difficult to control the formation of a bloom or scum of a given type of phytoplankton at particular time, control of the average density of the phytoplankton assemblage may be feasible.

Having determined that phytoplankton are causing poor lakewater clarity (Section 4.2.3) it is necessary to establish as a management target a desirable and achievable lower density of phytoplankton. Table 5.5 shows water clarities and particulate concentrations typically associated with the phytoplankton biomass (as chlorophyll a) found in lakes of different trophic classes. This table provides a guide to the approximate clarity which can be achieved as a result of controlling phytoplankton biomass in lakes whose optical character is dominated by phytoplankton. Because control of phytoplankton density often requires control of plant nutrient concentrations, the table also includes corresponding nutrient concentrations for the different lake classes (the latter being those typical of New Zealand lakes, although it is recognized that the nitrogen concentrations characteristic of each class tend to be rather lower than in lakes in developed, northern hemisphere countries where nitrogen is often present in considerable excess of phytoplankton requirements—see later).

Table 5.5. Lake trophic classes showing measures of water clarity, associated phytoplankton biomass, and the approximate nutrient concentrations giving rise to these biomasses in typical phytoplankton-dominated New Zealand lakes

Lake class	c (m^{-1})	Secchi depth (m)	K_d(PAR) (m^{-1})	Suspended solids (g m^{-3})	Chlorophyll a (mg m^{-3})	Total P (mg m^{-3})	Total N (mg m^{-3})
Oligotrophic	<0.5	>10	<0.2	<0.5	<2	<10	<200
Mesotrophic	0.5–1	5–10	0.2–0.4	0.5–1	2–5	10–20	200–300
Eutrophic	1–3	1.5–5	0.4–1	1–5	5–30	20–50	300–500
Hypertrophic	>3	<1.5	>1	>5	>30	>50	>500

After Vant (1987a). See also Figs 4.4 and 5.11.

Before proceeding to implement any phytoplankton control strategy, it is necessary to consider the probable outcomes of control, and to be satisfied that these will be sufficient for the achievement of noticeable improvements in colour or clarity. In some instances it may be that the desired degree of improvement simply cannot be achieved at reasonable cost. In terms of the classification scheme of Table 5.5, the most that may be expected in many cases will be to shift a lake to an adjacent class of lower phytoplankton density.

Fig. 5.11 shows the beam attenuation coefficient measured in a number of New Zealand lakes plotted against the chlorophyll a concentration. Superimposed on the figure are boundary lines encompassing phytoplankton-dominated waters. These were taken from the lines delineating so-called 'Case I' marine waters from Morel (1987). The lines are functions of [chlorophyll a]$^{0.62}$, another manifestation of the non-linear optical response of phytoplankton discussed in Section 4.2 (Fig. 4.4). This non-linearity probably

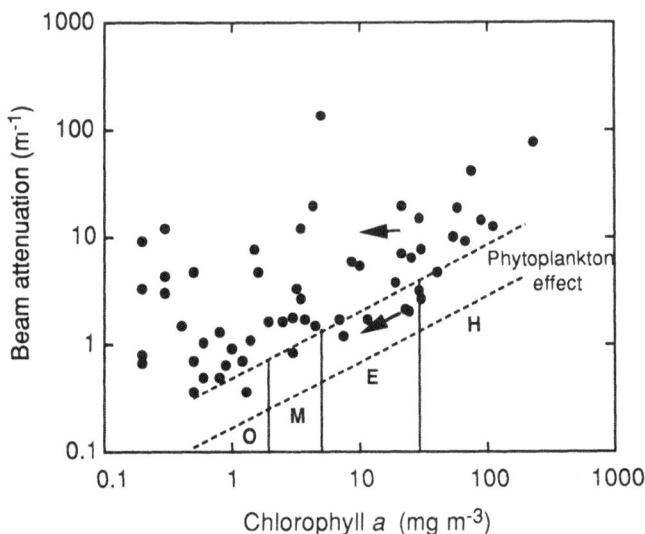

Fig. 5.11. Beam attenuation coefficient and chlorophyll a in New Zealand lakes (data from Howard-Williams & Vincent (1984) and Vant & Davies-Colley (1984)). The dashed lines, which are functions of [chlorophyll $a]^{0.62}$, encompass the band in which phytoplankton dominate attenuation (after Morel (1987)). The letters refer to the trophic classes in Table 5.5, with the vertical lines at the boundaries between the classes. The lower arrow shows how attenuation decreases in a phytoplankton-dominated lake when chlorophyll a falls from 20 to 10 mg m^{-3}, while the upper arrow shows the effect in a lake with appreciable non-algal attenuation.

reflects the trend for larger and less efficiently light-attenuating phytoplankton to domi-nate the flora of productive waters. In oligotrophic waters, detritus and bacteria are rela-tively more important. Fig. 5.11 shows that in many New Zealand lakes non-algal atten-uation is appreciable. Only phytoplankton-dominated lakes are likely to show much response to phytoplankton control, and even here the effect is expected to be non-linear. For example, halving phytoplankton biomass (e.g. from 20 to 10 mg chlorophyll a m^{-3}) in a phytoplankton-dominated lake could be expected to cause attenuation to fall by about a third (from 1.9 to 1.2 m^{-1}: see Fig. 5.11) (Secchi depth change: 3.4 to 5.3 m). If non-algal attenuation is appreciable, then less improvement in water clarity can be expected: if non-algal attenuation were initially five times greater than algal attenuation in the above example (i.e. non algal $c = 9.5$ m^{-1}, $z_{SD} \sim 0.67$ m) then the same decrease in phytoplankton biomass (halving) would only reduce total attenuation by about 6% (to 8.9 m^{-1}, $z_{SD} \sim 0.72$ m) (higher arrow on Fig. 5.11).

Reductions in phytoplankton biomass can be achieved either by dealing with the phy-toplankton community itself (e.g. poisoning, diluting or harvesting), or by controlling the causes of excessive biomass, namely elevated loads of particular plant nutrients from certain identifiable sources (Fig. 5.12). In-lake control may provide short-term relief from the symptoms of excessive phytoplankton biomasses, but nutrient control has the potential to provide lasting improvements.

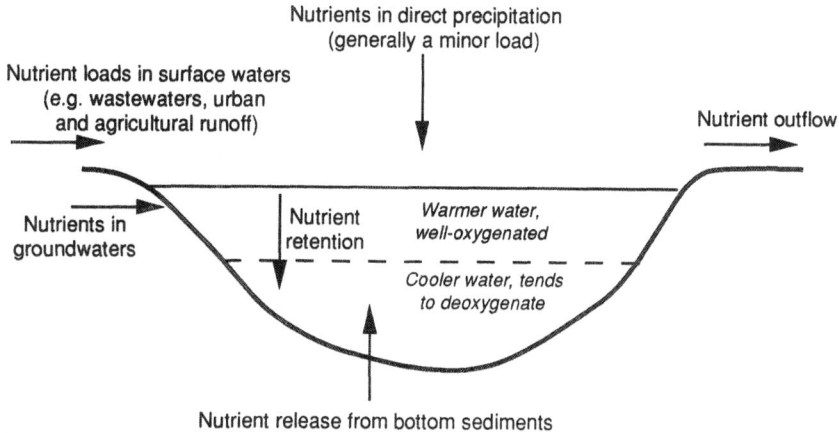

Fig. 5.12. Sources and sinks of plant nutrients to a hypothetical, thermally stratified lake. In summer the cooler bottom waters are isolated from the atmosphere and may deoxygenate, with an accompanying release of nutrients from the bottom sediments.

5.6.2 In-lake control of biomass

Techniques for in-lake control of phytoplankton biomass can be classified as follows (e.g. Howard-Williams 1987):

(1) chemical control, either poisons (algicides) or flocculants;
(2) physical control, including flushing, mixing, shading or harvesting;
(3) biological control, either grazing or control by pathogens.

Copper sulphate is a widely-used algicide (e.g. McKnight *et al.* 1983). It has been particularly useful in controlling phytoplankton in drinking water reservoirs, although its toxicity to certain aquatic animals has led to trials being undertaken on alternative compounds of copper, particularly copper oxychloride (Diatloff & Anderson 1984). It has been found that when barley straw is allowed to rot in a pond an unknown chemical is produced which inhibits phytoplankton growth (P.R.F. Barrett, Aquatic Weeds Research Unit, Reading, UK, personal communication). Studies are being conducted into the nature and action of this chemical at the time of writing.

Chemicals can be used to cause suspended particles such as phytoplankton to agglomerate and settle out of the water column. Aluminium salts such as aluminium sulphate ('alum') are commonly used coagulants (Section 5.5.3). As well as removing phytoplankton, aluminium coagulants can adsorb phosphorus from the water column, or inhibit its release from bottom sediments. In this way concentrations of this particular plant nutrient are also reduced which, in certain instances, can further limit phytoplankton densities.

The rate at which phytoplankton cells are flushed from a lake in the outflowing water can limit the phytoplankton density resulting from growth within the lake, particularly where the lake residence time is short. Pridmore & McBride (1984) showed that hydraulic residence times of about a fortnight or less are necessary to control phytoplankton populations by flushing. The flow of water needed to achieve such a short residence time is not usually available except in river impoundments. Control by water column mixing (which causes the phytoplankton to spend more time at depths where their growth is light-limited), shading or by harvesting are examples of further physical control techniques (e.g. Howard-Williams 1987). Their practical application is generally restricted to small ponds or reservoirs.

Phytoplankton grazing by planktivorous fish has been explored experimentally (e.g. Howard-Williams 1987). Silver carp (*Hypophthalmicthys molotrix*), for example, can control phytoplankton in small lakes or ponds, but is not clear whether control would be feasible in larger lakes. Although zooplankton also feed on phytoplankton, active management of their grazing is probably not feasible. Similarly, the practical details of controlling phytoplankton by deliberately exposing them to pathogenic organisms have only recently been addressed (e.g. Cooke *et al.* 1986).

In summary, of the various in-lake control techniques available, only chemical control (copper sulphate or aluminium sulphate) is likely to be generally applicable, and then only in quite small lakes (area <3 km^2) or bays of larger lakes. Generally, in-lake control should only be regarded as a short-term palliative. Only when plant nutrient loads are controlled at source is excessive phytoplankton growth addressed permanently.

5.6.3 Control by nutrient reduction
Nutrients and their bioavailability
Of all the chemical elements required by phytoplankton for growth, only nitrogen or phosphorus is usually in short supply; the others are typically present in excess of plant requirements. If phytoplankton growth is not limited by other factors, including flushing, light-limitation and grazing, then control of the supply of either N or P will generally control phytoplankton growth (Pridmore 1987). In some situations it may be easier to control the supply of one or other of these. Therefore it can be important to be able to determine which of N or P is limiting phytoplankton growth in a particular lake. Managing P-limited lakes is generally fairly straightforward (Sas 1989). However, even where lakes are N limited it will often be more practical to reduce the P load until P limitation is induced (e.g. as in the analysis of White (1989)).

A suite of techniques can be used to identify the limiting nutrient in particular lakes. These include determination of the concentrations of readily available or dissolved inorganic forms of N and P present at the times of peak phytoplankton growth, determination of the average concentrations of various dissolved and particulate forms of N and P, and laboratory and field experiments to examine the response of phytoplankton in samples of lakewater to nutrient addition.

Nitrogen and phosphorus are found in a variety of different forms in lakes and their inflowing waters. For example, N can be present as nitrate or ammonium ions, or as dissolved or particulate organic forms; similarly P may be present as phosphate ions, as other dissolved forms which may or may not react readily with certain laboratory reagents and as organic or inorganic particulates. Certain of these forms of both N and P, notably the dissolved inorganic forms, can be used more readily by phytoplankton than can others, particularly the dissolved organic and the particulate forms. As a result, in any consideration of the importance to phytoplankton of different nutrient sources it is necessary to take into account the differing 'bioavailability' of the various sources.

Broad guidelines have been established for assessing the likely bioavailability of various sources of nutrients to lakes. The nutrients in sewage wastewaters, for example, are readily available, while those from recent soils and subsoils are rather less so (Williamson & Hoare 1987).

Nutrient budgets and nutrient load control

Certain sources of nutrients will be more important than others, so it is necessary to establish the relative magnitude of these, that is, to derive a nutrient budget. Together with an assessment of the feasibility and costs of controlling particular sources, a budget allows priorities to be identified amongst various nutrient control options. Preparing a nutrient budget for a particular lake involves determining the rates of inflow of the different nutrient sources within the catchment (Fig. 5.12). These rates may be either estimated or measured directly (e.g. Vant & Hoare 1987). Estimation techniques provide results which are often only of low precision, and are therefore generally most useful when nutrient control is first being contemplated (i.e. at an early planning stage). Direct measurement of nutrient loads can be expensive and time consuming, and is usually only warranted when the decision has been made that nutrient control will be undertaken (i.e. when the precision of predictions about the results of nutrient control becomes important).

Controlling point sources of nutrients (e.g. sewage wastewaters) is, in principle, straightforward: the load must be reduced by treatment or be diverted away from the lake (e.g. Cooke et al. 1986). Control of non-point (diffuse) nutrient loads from land uses such as urbanization, forestry and agriculture is more difficult, however. While loads from undeveloped land (e.g. forests) are low (and essentially uncontrollable), those resulting from human landuse are usually substantially higher. Schueler (1987) describes how loads from urban runoff can be controlled by

(1) reducing stormwater generation by increasing infiltration,
(2) routing runoff to a treatment facility through a network which minimizes interference with the natural stream system, and
(3) treating runoff by filtering, settling and infiltration.

Further information on these topics and on the control of nutrient loads from agriculture, many of which involve controls on soil erosion, can be found in Novotny & Chesters (1981). Many of the techniques available for achieving reductions in diffuse nutrient loads from grazed pasture simply embody good farming practice (Williamson & Hoare 1987):

(1) matching farming type to land capability;
(2) accurate spreading of fertilizer at the appropriate rate and time;
(3) minimizing soil exposure, trampling damage and overgrazing, especially in areas where runoff is high;
(4) maintaining vegetation cover and infiltration;
(5) careful use of irrigation water and of artificial drainage.

An alternative approach is to seek to reduce diffuse nutrient runoff to streams, along with direct fertilizer and animal waste inputs, by protecting the contributing areas (Section 5.5.4). This can also help to reduce nutrient loads associated with streambank erosion. In certain situations the yield of P may be more than halved by riparian protection, although the reduction of N loads is likely to be more modest (e.g. <20% reduction for typical New Zealand pasture (Williamson & Hoare 1987)).

Nutrients may also be taken up and stored or transformed to potentially unavailable forms by instream plants (e.g. Howard-Williams *et al.* 1983; Cooke & Cooper 1988). Perhaps more importantly, aquatic N may be converted back to nitrogen gas by bacteria (denitrification) in swampy anaerobic areas adjacent to streams. This can substantially reduce a catchment's export of nitrate (e.g. by more than 90% (Cooper 1987)), so that the maintenance or enhancement of riparian wetlands is desirable over and above any riparian retirement which may be contemplated. Conversely, ill-considered drainage can destroy these wetlands, resulting in large increases in the catchment yield of nitrate.

Internal nutrient loads
Under certain circumstances part of the particulate nutrient flux which settles to the bottom of a lake may be recycled into the overlying water: an 'internal' nutrient load occurs (Fig. 5.12). This means a greater proportion of the catchment nutrient load will remain in the water column, thereby promoting higher phytoplankton biomasses than otherwise. Reducing the size of the internal load can thus reduce phytoplankton biomass. Nutrients are released from both oxidized and reduced (i.e. anoxic) lake sediments. However, release rates from oxidized sediments are very much slower and may generally be ignored.

Cooke *et al.* (1986) discuss major engineering remedies for controlling sediment nutrient release, including chemical treatment to oxidize the upper sediment layer (20 cm thick), and removal of nutrient-rich sediment altogether, generally by dredging. Alternatively, and more practically, especially in larger and deeper lakes, sediment nutrient release can be reduced by reducing the period for which the cooler bottom waters are anoxic (Fig. 5.12). This can be achieved by supplying additional dissolved oxygen to the bottom waters ('hypolimnetic aeration'). Alternatively, deoxygenation can be reduced by lowering the supply of oxidizable material to the bottom waters. As much of this material is often derived from the phytoplankton, practices which reduce phytoplankton biomass will thus reduce sediment nutrient release, thereby further reducing phytoplankton biomass—positive feedback occurs. As a result internal nutrient loads should decrease as external loads are controlled, although predicting the size of the decrease is usually difficult (but see, for example, Rutherford *et al.* (1989)).

Predicting resulting nutrient and phytoplankton levels
A reduction in catchment nutrient loads will produce a simple proportional reduction in nutrient concentrations in the lake, provided that the extent of nutrient loss or retention within the lake remains unchanged. However, in certain circumstances controlling catchment nutrient loads will increase nutrient retention in the lake, thus further reducing the final in-lake nutrient concentrations. This can happen when the magnitude of internal nutrient loads decreases as a result of reduced lake productivity (see above). Predicting the size of such a reduction is difficult, and is probably best ignored, the improvement predicted from other considerations therefore representing an expected minimum.

Reducing the volume of inflowing water is also expected to increase nutrient retention because the loss of particulate nutrients to the bottom sediments is enhanced when hydraulic residence time increases (e.g. Nurnberg 1984). The extent to which nutrient retention will increase depends both on the present retention characteristics of the lake and on the size of the proposed reduction in inflow volume. In practice only major reductions in inflow volume (e.g. >50%) can be expected to markedly increase retention, so that consideration of these effects will be limited to situations such as those where major stormflow diversions are proposed.

The final matter to be addressed is the expected response of a lake's phytoplankton to reduced lake nutrient concentrations. In brief, if a lake remains P deficient, then control of P, or P and N, should result in a predictable decrease in phytoplankton density (e.g. Pridmore 1987). However, since the atmosphere is potentially both a source and a sink for N in lakes, the outcomes of reducing catchment loads of N to an N-deficient lake are less certain. Phytoplankton biomass should decrease, but by how much will depend on whether or not fixation of atmospheric nitrogen gas or denitrification of inorganic N in the lake occurs and, if so, to what extent. Control of P sources alone are unlikely to reduce phytoplankton density in such a lake unless the reduction is sufficiently large for the lake to become P deficient.

6

Case studies

6.1 INTRODUCTION

During the time that we have been carrying out scientific investigations relevant to the optical quality of waters, we have made a number of in-depth studies of colour and clarity of particular surface water bodies. These studies have used a combination of established techniques for measuring various optical aspects of waters, together with some new methods that we have introduced. The studies illustrate many of the concepts, methods of measurement, features of different surface water environments, and management approaches that have been discussed in previous chapters of this book. Some of these studies have been selected for presentation below.

Lakes and rivers have featured about equally in our experience with analysing water colour and clarity and related management problems, but, unfortunately, we do not yet have any examples of estuarine optical water quality case studies 'under our belts' with the exception of predictions of the underwater light climate for phytoplankton in Manukau Harbour near Auckland, New Zealand (Vant 1991; see Section 4.4). In the remainder of this chapter we give brief accounts of six lake and river case studies.

Section 6.2 summarizes the study by Biggs & Davies-Colley (1990) of Lake Coleridge, a large alpine glacial lake in the South Island of New Zealand. This study was designed to quantify the effects on optical water quality, with emphasis on visual clarity, of turbid glacial rivers diverted to the lake to augment hydro-electric power generation flows.

Section 6.3 uses the findings of Vant (1987b), Wells *et al.* (1988) and the authors' unpublished data to link the optical water quality changes in Lake Whangape to catastrophic decline of macrophyte stands in this lake in the North Island of New Zealand. Both Lake Whangape and neighbouring Lake Waahi have been subjected in the past to turbid discharges of highly light-attenuating kaolinite clay from open-cast coal mines, which we believe may have so limited light penetration into the lakewaters that the macrophyte stands were largely eliminated, resulting in weed-choked wetlands of considerable value as wildlife habitat becoming turbid weed-free systems of little appeal to either aquatic life or recreational users.

Section 6.4 is a comparative study of two eutrophic lakes in the Rotorua volcanic zone of North Island, New Zealand. Lakes Rotorua and Okaro are similar in visual clarity, but differ in their optical quality and related water quality. Light attenuation in Okaro is mostly attributable, directly or indirectly, to the high algal biomass in this small, deep lake. In Lake Rotorua, by contrast, mineral suspensoids that are unrelated to the phytoplankton standing crop frequently dominate light attenuation. Thus there is some uncertainty about the optical water quality response of Lake Rotorua to nutrient control measures designed to reverse the progressive eutrophication in this lake.

In Lake Horowhenua, discussed in Section 6.5, sewage effluent has been diverted in order to reverse eutrophication as well as to improve sanitary quality. Phytoplankton biomass in this lake has declined in response to reduced nutrient loading as expected, but the visual water clarity is still poor and is likely to remain so because of high non-algal turbidity.

Section 6.6 discusses the problem with water appearance of the Tarawera River, North Island, New Zealand, which is arguably one of the country's most severely polluted water bodies. The main influence on the optical water quality of the river is the highly light-absorbing effluent discharged from a large Kraft pulp mill. This effluent changes the typically attractive blue–green and visually clear waters of the Tarawera River to a dark inky black and visual clarity drops by a factor of three. Panel studies of the threshold for detection of colour change in this river suggest that a 20-fold reduction in colour loading would be required before this water body would be suitable for recreational uses.

The impact of diversion of clear-water tributaries, for hydro-electric power generation in another catchment, on the optical water quality of the Whanganui River, is discussed in Section 6.7. The visual clarity of this river without diversion can be calculated from measured clarity in the diverted water and in the residual river, assuming a balance on light attenuation. This analysis permitted the impact of the diversions on the optical quality of the residual river to be defined quantitatively.

6.2 LAKE COLERIDGE: EFFECTS OF TURBID GLACIAL WATERS ON OPTICAL WATER QUALITY

6.2.1 Background

Lake Coleridge (43° 19' S, 171° 32' E) is a beautiful blue-water lake near the Southern Alps of inland Canterbury, South Island, New Zealand (Fig. 6.1, Plate 14). The lake occupies a glacial trough formed by the Wilberforce Glacier during previous glacial advances. The lake, at an altitude of 507 m, is 200 m deep, 17.8 km long, and relatively narrow (3.4 km at the widest point). In spite of its remote alpine location, Lake Coleridge is used for recreation, including salmon fishing. The lake is also used to store water for hydro-electric power generation at the Coleridge Power Station (Fig. 6.1).

To augment the small natural inflows of water (about 4 m^3 s^{-1}) from the lake catchment, water from two rivers with glaciated headwaters, the Harper River and the Wilberforce River, is diverted to the lake (Fig. 6.1), so increasing flow through Coleridge Power Station to 24 m^3 s^{-1}. Natural inflows transport relatively little suspended matter, but both the diverted rivers, and especially the Wilberforce River, are turbid owing to their

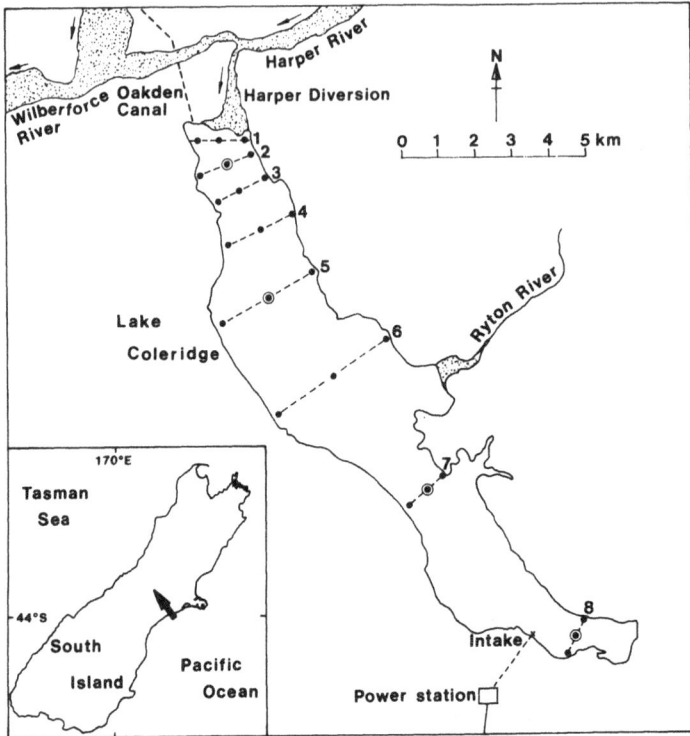

Fig. 6.1. Map of Lake Coleridge showing the locations of the inflows diverted to the lake and the sampling transects established by Biggs & Davies-Colley (1990) (with permission). Circled points are the 'main' stations, denoted 2/2, 5/2, 7/2 and 8/2, numbering from top to bottom.

loads of fine-grained suspended silt from glaciers, and they deliver about 130 000 t yr^{-1} of suspended material to the lake. This silt-sized material, often referred to as 'glacial flour', is efficiently light scattering.

The Harper River was diverted in 1921 before any water quality records were made on the lake. However, a major environmental concern regarding the Wilberforce diversion (commencing in 1977) was that glacial flour-laden water would reduce the clarity of the lake and change its colour. Increased fine suspended matter concentrations in the lake could change its visual clarity, and thus recreational value and habitat for sighted animals, and could also reduce light penetration into the lake water for plant growth. Also, increased light scattering by glacial flour would be expected to increase the reflectance and therefore the brightness of the water colour.

To determine whether changes in lake water colour and clarity had taken place, a one year study was carried out (Biggs & Davies-Colley 1990). Fortunately some measurements of Secchi disc clarity had been made prior to the Wilberforce diversion and these provided a baseline with which to compare the present optical quality. No study had been undertaken of the clarity impact of the diversions prior to this study, although a relatively small decline in visual clarity had been predicted based on a regression model (I.G. Jowett, NIWA, Freshwater Division, Christchurch, unpublished results).

6.2.2 Methods

Eight transects were established in a logarithmic progression down the lake from the northwestern end where the diverted water enters (Fig. 6.1). Three stations were sampled by boat on each transect: one at a central point (average 200 m depth) and two at 'near-shore' positions in about 20 m of water. Each station was sampled approximately monthly for a year, starting in September 1986.

At all stations, Secchi depth was measured on the sunny side of the boat using a viewing box, and 1 l water samples were obtained. These were used for measurement of nephelometric turbidity with a Hach 2100A instrument, and the absorption coefficient (filtered samples) at 270 nm (absorption at longer wavelengths, e.g. 440 nm, was too low for routine measurement). At four of the central stations (referred to as 'main stations') large volume (30 l) samples were collected for suspended solids and chlorophyll a analysis by standardized methods. Also, at these stations, profiles of irradiance were measured, using Li-Cor LI-192SB PAR sensors, for the estimation of PAR band average absorption and scattering coefficients following Kirk (1981b).

On one occasion (23 September 1986) water samples from central stations on transect 2 and transect 8 were analysed using a Pye-Unicam PU8800 spectrophotometer following the procedures of Davies-Colley (1983) for the measurement of water colour. Spectral absorption coefficients ($a(\lambda)$) and scattering coefficients ($b(\lambda)$) were estimated from the spectrophotometric measurements and water colour was inferred from the spectral relative reflectance $R_r(\lambda) = b(\lambda)/a(\lambda)$.

6.2.3 Optical character of Lake Coleridge

Average Secchi depths were fairly high (Table 6.1) and increased downlake from 7.5 m on transect 2 to 9.0 m on transect 8. Turbidity and SS, as expected, were low and followed a reverse trend (Table 6.1). Visual clarity and turbidity at the near-shore stations (data not shown) were usually similar to those at central stations, although on two occasions following windy weather the patterns of visual clarity and turbidity suggested that the shoreline and shallows can sometimes contribute suspensoids to the water column. Turbidity and suspended solids concentrations in the lake centre (transect 6) correlated with the suspended solids load from the Wilberforce diversion (Oakden Canal, Fig. 6.1) lagged by about one month ($r = 0.57$), suggesting that the diversion may indeed have affected the lakewater clarity.

The average irradiance attenuation coefficient, K_d, was low and the euphotic depth was correspondingly deep (about 31 m, Table 6.1). The reflectance coefficient for PAR averaged about 12%, a fairly high value consistent with the high ratio of scattering to absorption (b/a in Table 6.1). The absorption coefficient was low (0.073 m^{-1} average) whereas the scattering coefficient was higher (0.95 m^{-1} average—still 'low' as lakewaters go).

Yellow substance concentrations, as indicated by g_{440} values estimated from the absorption measurements at 270 nm, were low and more typical of marine waters (e.g. Kirk 1983, p. 57) than lakes. The average g_{440} value was about 0.04 m^{-1}, approximately the 3 percentile value on the distribution for New Zealand lakes given in Fig. 4.25.

Table 6.1. Optical water quality of Lake Coleridge

Variable		Central stations			
		Transect 2		Transect 8	
z_{SD}	(m)	7.5	(34)	9.0	(22)
SS	(g m^{-3})	0.99	(39)	0.74	(34)
Turbidity	(NTU)	0.80	(38)	0.69	(29)
Chl a	(mg m^{-3})	0.37	(92)	0.42	(130)
K_d	(m^{-1})	0.162	(25)	0.150	(15)
z_{eu}	(m)	30.1	(25)	31.4	(15)
R	(%)	12.5	(26)	11.4	(31)
a	(m^{-1})	0.075	(20)	0.072	(17)
b	(m^{-1})	1.03	(42)	0.85	(35)
b/a	(dimensionless)	13.5	(29)	12.1	(34)

Data are means with coefficients of variation in parentheses. (Adapted from Biggs & Davies-Colley (1990).)

The chlorophyll a content of the lake water was very low (Table 6.1), consistent with the ultra-oligotrophic status of Lake Coleridge. The absorption coefficient correlated fairly closely with chlorophyll a ($r = 0.61$), but the other optical properties were weakly, if at all, correlated with phytoplankton biomass. In contrast, the suspended solids content of the lakewater correlated with the visual clarity, light penetration, reflectance coefficient, and scattering coefficient ($r = 0.86$, 0.59, 0.76 and 0.56 respectively).

Fig. 6.2A shows spectral absorption and scattering coefficients measured by spectrophotometer on a sample from the central station on transect 2. Total light absorption was dominated by that of water itself at wavelengths greater than about 520 nm. Yellow substance was the most important blue-light-absorbing constituent ($\lambda < 440$ nm). Absorption by particulates was relatively unimportant. The scattering coefficient varied with wavelength according to a power law, and averaged about 0.3 m^{-1} in the visible range on this occasion, appreciably lower than the 0.95 m^{-1} average (Table 6.1). Fig. 6.2B shows the reflectance spectrum for this water, calculated from the spectral absorption and scattering coefficients. Also shown is the similarly calculated spectrum for the central station on transect 8, near the other end of the lake (Fig. 6.1). Both spectra peak in the blue region at about 500 nm. The resulting colour is basically the blue hue of optically pure water, slightly shifted towards green owing to increasing absorption of yellow substance at short visible wavelengths.

6.2.4 Effect of the diversion water on optical quality
The study demonstrated that the visual clarity of Lake Coleridge is strongly influenced by mineral suspended solids, mainly glacial flour derived from the Wilberforce and Harper Rivers. No visual clarity measurements were made before the Harper River was diverted in 1921, but some measurements were made near our transect 6 in 1965–67 (prior to the Wilberforce diversion) by M. Flain (personal communication). These mea-

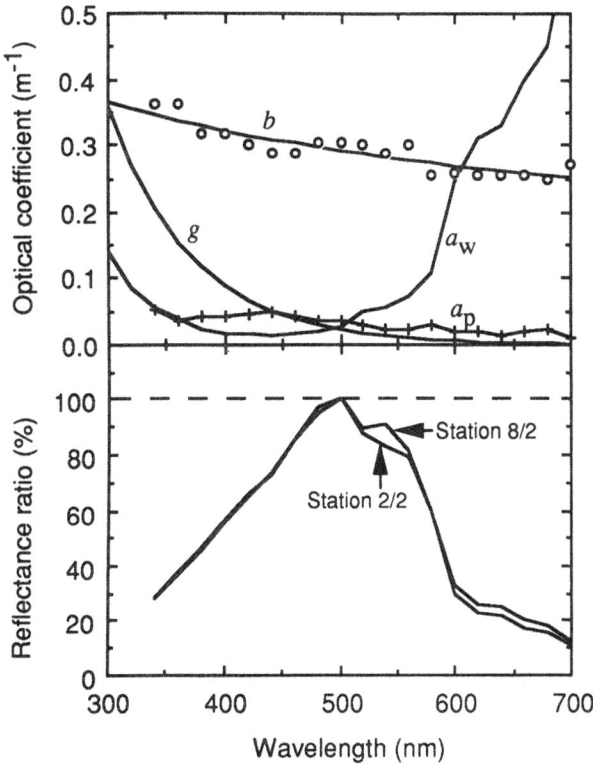

Fig. 6.2. (A) Absorption and scattering spectra for water from Station 2/2, Lake Coleridge, 23 September 1986. Absorption spectra are shown for yellow substance (g), suspensoids (a_p, +), and water itself. The line through the scattering coefficients represents a power law: $b = 4.5 \lambda^{-0.44}$ ($r^2 = 0.82$). (B) Reflectance spectra for Lake Coleridge at Stations 2/2 and 8/2 as calculated from the spectral absorption and scattering coefficients. The spectra are normalized to the maximum value. (Replotted after the data of Biggs & Davies-Colley (1990), with permission.)

surements are compared in Table 6.2 with those in the 1986–1987 study.

Mean Secchi depth changed appreciably from 13.4 to 8.6 m (Table 6.2), so visual clarity has been reduced by about 36% by the Wilberforce diversion. This change is outside the 20% guideline for maximum visual clarity change recommended in Section 5.3 to protect the aesthetic value of water. That is, the reduction in visual clarity attributable to the Wilberforce Diversion represents a significant aesthetic impact. However, an 8.6 m Secchi depth (corresponding roughly to a 6.5 m black disc range) still represents 'high' visual clarity and is consistent with most recreational water uses.

The question now arises, what effects has the Wilberforce diversion had on other optical attributes of the Coleridge water? Because Secchi depth (z_{SD}) and other apparent optical properties of water (K_d, R) can be related to the inherent optical properties (a, b), and thence to the water composition, it is possible to construct a model to address this question.

Table 6.2. Comparison of Secchi disc depth in Lake Coleridge with historical data
(from Biggs & Davies-Colley (1990))

Period	N	Range	Mean	Standard deviation	t	Degrees of freedom	P	Source
August 65–February 66	13	8.0–18.0	13.4	2.91				M. Flain personal communication
August 86–February 87	6	5.5–12.1	8.6	2.56	–3.67	11	0.0037	Biggs & Davies-Colley (1990)
January–October 67	8	6.0–15.5	13.4	3.22				M. Flain personal communication
January–October 87	7	5.5–12.2	8.0	2.54	–3.65	12	0.0033	Biggs & Davies-Colley (1990)

6.2.5 Optical model for Lake Coleridge—inferred changes in optical character

In order to construct an optical model for Lake Coleridge the PAR band absorption coefficient was taken to be constant at $a = 0.073$ m^{-1}, the average value measured in the monitoring programme. The variation of the other optical properties was then taken as a function of the scattering coefficient (and the suspended solids content). The theory underlying this optical model is discussed by Davies-Colley & Vant (1988).

Fig. 6.3 shows the apparent optical properties plotted as functions of the scattering coefficient, b. The recorded average Secchi depths in 1965 and 1967 (before diversion of the Wilberforce River) and in 1986–1987 (post diversion) are shown. The inferred values of the scattering coefficient corresponding to these measured visual clarities provide the link to the other optical properties of interest.

Table 6.3 gives the calculated values before and after diversion, together with the measured mean values after diversion. The estimated values of suspended solids and K_d (and therefore z_{eu}) after diversion are in reasonable agreement with the measured values, demonstrating the self-consistency of the model. However, the reflectance and scattering coefficients are too low, suggesting model bias. Biggs & Davies-Colley (1990) suggested that such bias might arise because the model depends heavily on the results of Monte Carlo simulations by Kirk (1981a) based on scattering with the same angular dependence as San Diego Harbour water (Fig. 2.3). The glacial flour particles in Lake Coleridge may be largely plate-shaped micaceous material (similar to that in the neighbouring glaciated Waitaki River (Graham 1990)) which, based on the work of Gibbs (1978), is expected to backscatter more efficiently than typical natural water suspensoids leading to higher reflectances at a given b/a ratio.

The optical model suggests that the measured 36% reduction in visual clarity due to the Wilberforce diversion has been associated with a 65% increase in the scattering coeffi-

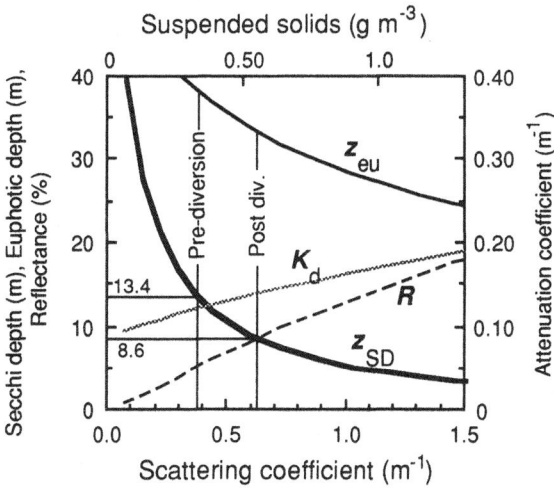

Fig. 6.3. Optical model for Lake Coleridge. The scattering coefficient, b, is assumed to be proportional to the suspended solids concentration and the absorption coefficient, a, is assumed constant at the average value of 0.073 m^{-1}. The other quantities are calculated from a and b using published relationships (see text). (Replotted after the data of Biggs & Davies-Colley (1990), with permission.)

cient and a similar increase in the average suspended solids content of the water (Table 6.3). The irradiance attenuation coefficient is calculated to have increased by 15% and the euphotic depth has decreased by 13%. The reflectance coefficient is calculated to have increased by 65%, an estimate which should be accurate in spite of the underestimation of

Table 6.3. Predictions from the optical water quality model for Lake Coleridge

Variable	Mean values			Change
	Pre-diversion	Post-diversion		
Measured z_{SD}	13.4	8.6		−36%
	(Calc.)	(Calc.)	(Meas.)	(Calc.)
b	0.38	0.625	0.95	+65%
SS	0.34	0.56	0.78	+65%
K_d	0.121	0.139	0.156	+15%
z_{eu}	38.0	33.1	30.7	−13%
R	5.09	8.39	12.3	+65%

Data are either calculated (Calc.) or measured (Meas.). (Adapted from Biggs & Davies-Colley (1990), with permission.)

the actual magnitude of the reflectance coefficient. Both these changes are (slightly) outside the maximum change guidelines recommended to protect optical quality of waters in Section 5.3. Thus, as well a sustaining an impact on its visual clarity, Lake Coleridge has suffered significantly reduced light penetration (with unknown ecological ramifications) and a change in the reflectance, and therefore the brightness of the water colour.

6.3 LAKE WHANGAPE: CHANGED OPTICAL QUALITY LINKED TO MACROPHYTE DECLINE

6.3.1 Background

Lake Whangape (37° 28' S, 175° 03' E, Fig. 6.4) is a shallow (mean depth 1.5 m), rapidly flushed (residence time about 2 months), lake (area = 14.5 km^2) in the Waikato Basin south of Auckland, New Zealand. The lake catchment is relatively large (about 300 km^2), and is used predominantly for grazing. Until mid-1987 the lake was notable for its extensive beds of aquatic macrophytes, present at about 95% coverage of the lake bed, and also its high wildlife values, with 6000 or more black swans (*Cygnus atratus*) feeding on the macrophytes.

In 1985–1986, when the main basin of the lake was still well covered by macrophytes, the Southern Arm of the lake (Fig. 6.4) was much more turbid than the main basin, and

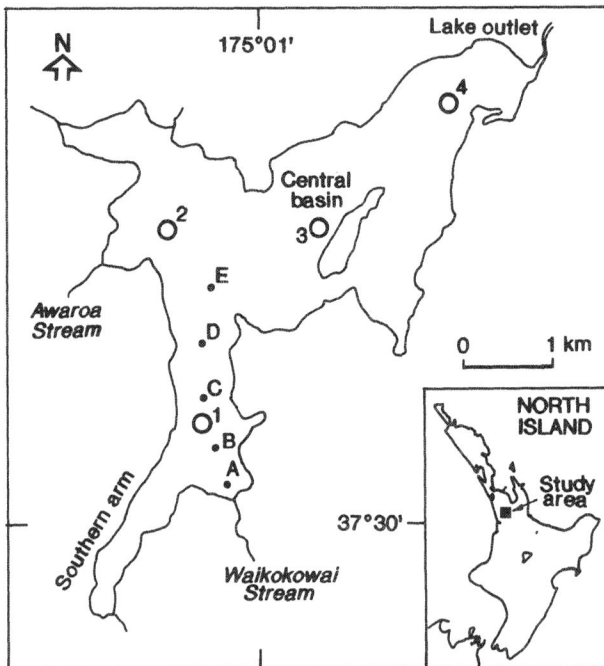

Fig. 6.4. Map of Lake Whangape showing sampling sites used by Wells *et al.* (1988) (with permission). Open-cast coal mines discharged turbid wastewaters to the Southern Arm via the Waikokowai Stream.

the macrophytes in the Southern Arm were much less abundant than elsewhere (Vant 1987b; Wells *et al.* 1988). At this time turbid runoff from areas of coal mining activity was discharging into this arm of the lake via the Waikokowai Stream (Fig. 6.4). Vant (1987b) and Wells *et al.* (1988) felt that there was a possibility that the localized macrophyte decline could initiate a catastrophic collapse of the macrophyte beds over the whole lake with consequent major changes in optical water quality. As it turned out, the macrophyte stands did indeed collapse over the whole lake, apparently around mid-1987.

Such abrupt declines of macrophyte stands, with attendant reduction in optical water quality and wildlife values, have been observed previously in other shallow lakes of New Zealand, including Lake Ellesmere near Christchurch (Gerbeaux & Ward 1991), and two neighbouring lakes in the Waikato Basin, Lakes Waikare and Waahi. Lakes Ellesmere and Waikare apparently lost their macrophyte stands in 1968 in a major storm (Gerbeaux & Ward 1991). These stands have never recovered because of the unfavourable light climate under a turbid water column and frequent disturbance of the lake bed by wind waves with re-suspension of bottom sediments.

In Lake Waahi, about 10 km to the south of Lake Whangape, the macrophyte collapse was more recent. Chapman (1981) described Lake Waahi as containing abundant and dense growths of macrophytes, particularly *Egeria densa*, in 1974–1975, but these plants had largely disappeared by 1984 (Kingett 1984) with an associated marked change in water quality (Plate 5). Lake Waahi receives discharges from several large coal mines, including open-cast operations, and in the past, prior to installation of wastewater treatment by coagulation/flocculation, some of these discharges have been markedly turbid with fine kaolinite clays associated with the coal beds. In 1977, turbid plumes were observed extending into the lake from the mouth of the mining-impacted major tributary, the Awaroa Stream (Kingett 1984). Much of the macrophyte biomass was subsequently removed, possibly during a large windstorm in June 1978. At several environmental consent hearings in 1985 and 1986 one of us (RJD-C) suggested how the turbid discharges could have been responsible for the macrophyte collapse and changed water quality. The plausibility of the suggested mechanism of macrophyte collapse (discussed below) has increased with the direct observation and documentation of just such a collapse in Lake Whangape.

6.3.2 Macrophytes

Four sites in Lake Whangape were monitored approximately quarterly from early 1985 to late 1986 by SCUBA divers (Clayton 1983, Wells *et al.* 1988). Three of these sites were in the main basin and one site was in the Southern Arm (Fig. 6.4). Three quadrats of 1 m^2 were harvested at each site, and the plant material was sorted into different species before being dried to constant weight at 80°C. Five sites, A to E, were located along the Southern Arm (Fig. 6.4) and sampled for suspended solids, turbidity and Secchi depth and (qualitatively) for macrophyte abundance. Exclosures 2.5 m × 4.0 m × 1.7 m high covered with 55 mm wire mesh were erected in the Southern Arm (Fig. 6.4), and at five sites in the main lake basin, to exclude browsing swans. These exclosures were assessed quarterly, and on 8 December 1986, after one year, a 1 m^2 quadrat was harvested inside, and one outside, each cage.

Macrophyte cover in 1985–1986 averaged 95% at sites 2, 3 and 4 compared with only 5% cover at site 1 in the Southern Arm. The biomass in the main lake basin averaged about 220 g dry weight m^{-2} but, at site 1, biomass was only about 1/10 of this value.

Egeria densa, Ceratophyllum demersum, and *Potomegaton ochreatus* were present at all sites and *Elodea canadensis* at all except site 1. Secchi disc clarity increased monotonically as suspended solids concentration and turbidity decreased from sites A to E, moving north along the Southern Arm away from the Waikokowai Stream (turbidity declined from 30 NTU to <3 NTU). Parallelling this gradient in clarity was an increase in macrophyte abundance from very low biomasses at site A near the Waikokowai Stream mouth to stands similar to the remainder of the lake at site E.

In the main basin of the lake, macrophytes inside exclusion cages grew to the water surface, whilst they were cropped by swans down to about 0.5 m below the surface outside the cages. This suggests that swan browsing removes plant biomass from the top of the water column and so, in turn, reducing the wave-damping effect of the macrophytes. The swan browsing was selective, with *Egeria* apparently being more palatable than *Ceratophyllum*. Exclusion of swans at site 1 in the Southern Arm did not result in recovery of the macrophytes: there was no significant difference in biomass or cover inside the cage compared with outside. Presumably regrowth of plants was discouraged by the unfavourable light climate in the Southern Arm, and swans may not have grazed the few macrophytes present there anyway. These observations suggest that swan grazing was not preventing recovery of the macrophytes in the Southern Arm. However, elsewhere in the lake the significantly higher macrophyte biomass within exclosures suggests that grazing pressure could have been a contributing factor responsible for the macrophyte decline (see Section 6.3.4).

6.3.3 Suspended solids and the light climate

On 28 May 1986 water samples were collected from sites throughout the lake and irradiance attenuation was measured with PAR sensors at a time when large gradients of turbidity and visual clarity occurred across the lake (Davies-Colley & Vant 1988). Inorganic suspensoids, measured as non-volatile suspended solids (NVSS), an index of non-algal matter, reached levels as high as 80 g m^{-3} in the lake waters. These mineral suspensoids were the main cause of light attenuation in the lake, and the irradiance attenuation coefficient increased approximately as the square root of the SS concentration (Fig. 6.5), consistent with the light penetration modelling studies of Kirk (1981a, 1984a). Light absorption in the lake water was fairly high, such that, even in the absence of mineral suspensoids, the irradiance attenuation coefficient was about 1.0 m^{-1}.

Based on the work of Vant *et al.* (1986) the light-limited maximum depth to which macrophytes can grow (z_c) is about 4.3/K_d, so, in Lake Whangape, the predicted z_c in the absence of mineral suspensoids is 4.3/(1.0 m^{-1}) = 4.3 m. At low mineral suspensoid concentrations, say <5 g m^{-3} (corresponding to a $z_c \sim$ 1.5 m), light penetration is probably adequate for macrophyte growth in this shallow lake. However, at the high suspensoid concentrations encountered in the Southern Arm (around 40 g m^{-3}, corresponding to $z_c \sim$ 0.8 m) the penetration of light would have been insufficient to promote regrowth of the macrophytes on bare areas of lake bed at greater depths, and the existing macrophytes were apparently only surviving because they had some biomass projecting up into the euphotic zone.

Elevated concentrations of suspensoids would be expected after heavy rainstorms in a lake like Whangape with a catchment disturbed by grazing (Section 5.5.4). During the winter of 1986 the turbidity in the outlet stream from Lake Whangape was recorded

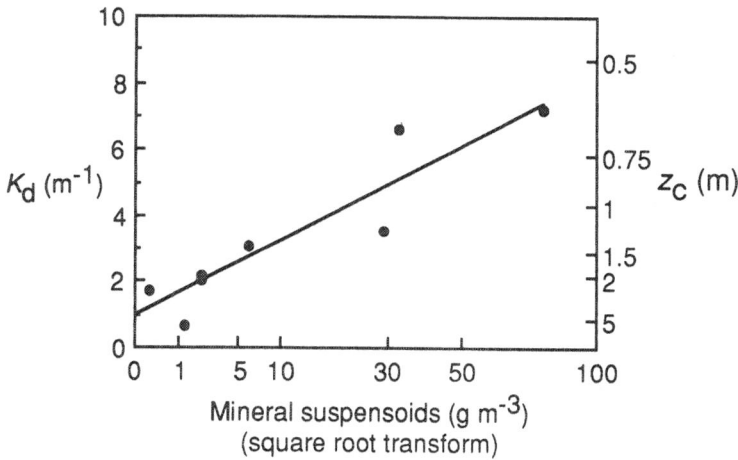

Fig. 6.5. Attenuation coefficient for irradiance (K_d) plotted against the concentration of mineral suspensoids in Lake Whanagape on 28 May 1986. The equation of the line is $K_d = 0.72M^{0.5} + 1.01$, where M is mineral (non-volatile) solids concentration. The corresponding maximum depths to which plants can grow are given as predicted from the equation of Vant *et al.* (1986): $z_c = 4.3/K_d$. (After Wells *et al.* 1988, with permission.)

during a flood of 10 year return period. Turbidity (in arbitrary units) was monitored at the lake outlet using a Monitek recording turbidimeter, and the irradiance penetration into the lakewater was estimated from this record (Fig. 6.6). The depth to which sufficient PAR penetrated for plant growth was in turn estimated from the irradiance attenuation and the known lake water level. For a period of about seven weeks most of the lake bed received less light than is typically required for plant growth (Fig. 6.6C); however, this period of shading from an extreme event apparently did not affect the plants in the main basin of the lake. Similar observations have been made in Lake Tutira (Johnstone & Robinson 1987), where shading by turbid floodwaters finally excluded macrophytes (*Elodea canadensis* and *Hydrilla verticillata*) after about three months (although regrowth from rhizomes of these particular species occurred rapidly once the lake waters had clarified).

Apparently, provided that macrophytes can store sufficient photosynthate during long periods of clear water, they can cope with occasional 'excursions' of low clarity of limited duration. Therefore, although catchment erosion delivers large quantities of light-attenuating suspensoids to Lake Whangape, the short residence time of the lake, particularly at times of high runoff, means that the exposure of plants in the lake to low light levels is limited. In contrast, the smaller mining loads produced much longer exposure times to an unfavourable light climate in the Southern Arm because discharge occurred over a range of states of flow including dry periods with long flushing times.

6.3.4 Collapse of the macrophyte stands
Following recognition of the risks posed by the discharge of turbid mining waters to the Southern Arm of Lake Whangape, the local statutory water authority was alerted and the suspensoid loadings from the mines were brought under control. Vant (1987b) suggested

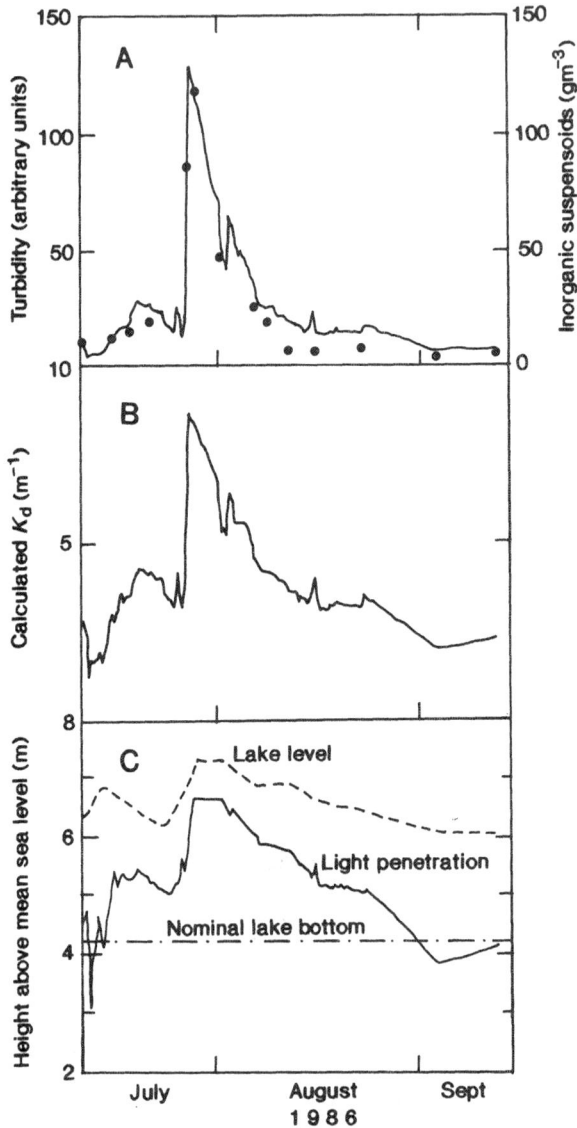

Fig. 6.6. The effect of a large flood (ten year return period) on the light climate in Lake Whangape:
(A) recorded turbidity (arbitrary units) and inorganic suspensoids (●); (B) calculated attenuation coefficients; (C) depth of penetration of light for macrophyte growth in relation to lake water level. (Average
lake bed level is about 4.2 m above mean sea level.) (From Vant (1976b), with permission.)

that with this control of mine loading the macrophytes in the Southern Arm should begin
to recover. There was apparently some evidence for such recovery with expansion of
Ceratophyllum to a virtual monoculture in the Southern Arm in summer, 1986/87 (Vant

1987b). However, this non-rooted species is more vulnerable to wave action than *Egeria*, and the recovery was unstable and short lived.

In winter 1987 a total macrophyte collapse in Lake Whangape did in fact occur (R. D. S. Wells, Ecosystems, NIWA, personal communication). It would appear that Lake Whangape has gone the way of Lake Waahi and for similar reasons. Macrophyte biomass in the lake is now very low and wave action is essentially unchecked. The lake is now very turbid (around 30 NTU) owing to frequent wind-wave disturbance of the exposed lake bed. The water column is now so light attenuating that macrophyte recovery, at least in deeper water (>1.0 m), is unlikely. Shallower areas of lake bed, although better lit, are no more favourable for macrophyte re-colonization because of exposure to wind-wave action. Thus, there is now little suitable habitat for aquatic macrophytes in Lake Whangape. Although recreational navigation in this lake is now possible in the absence of dense macrophyte stands, the lake water is aesthetically unpleasant owing to its low visual clarity (black disc range 0.12 m (Davies-Colley & Smith 1992)) and muddy appearance.

A plausible mechanism for the macrophyte collapse in Lake Whangape, and in neighbouring Lake Waahi (for similar reasons), is as follows (Fig. 6.7). The initial perturbation was discharge of turbid mining wastewaters into the lake via receiving streams. The resulting turbid streamflows would have dispersed only gradually in the lake because of the slow mixing in the presence of dense macrophyte stands. In restricted areas near impacted stream mouths, the light penetration into the water was so reduced that the macrophytes in those areas became light starved and moribund, and vulnerable to erosion by wind waves. Removal of plant biomass was probably achieved mainly by wave action during windstorms, although selective swan grazing on the dominant rooted *Egeria* probably contributed to the instability of the stands by favouring the less palatable and poorly anchored *Ceratophyllum*. Once reduction of plant biomass had occurred, particularly in the near-surface zone of swan grazing (to about 0.5 m depth), the damping of waves in windstorms would have been reduced, so destabilizing macrophytes in neighbouring areas as well as the remaining plants in the initially impacted areas. Furthermore, uprooting of plants would have exposed the bottom sediments to wind-wave erosion, with the resuspended sediments then exacerbating shading by the mine-derived turbidity. With the operation of these positive feedback mechanisms, the initial perturbation spread out as an expanding turbid plume. Ultimately the macrophyte decline and associated turbidity engulfed the whole lake.

Evidently turbid discharges to shallow, macrophyte-dominated lakes such as Whangape and Waahi can produce ecological and optical water quality changes out of all proportion to the (apparent) scale of the initial perturbation. Very restrictive conditions should therefore be placed on suspensoid concentrations in consents for even relatively minor discharges into wetlands with potentially vulnerable macrophyte stands.

6.4 LAKES ROTORUA AND OKARO: LIGHT ATTENUATION IN TWO CONTRASTING LAKES

6.4.1 Background
Lake Rotorua (38° 05' S, 176° 16' E) and Lake Okaro (38° 17' S, 176° 23' E) are two lakes of volcanic origin in the vicinity of Rotorua City, North Island, New Zealand.

A. Original State

B. Macrophyte Collapse

C. Present State

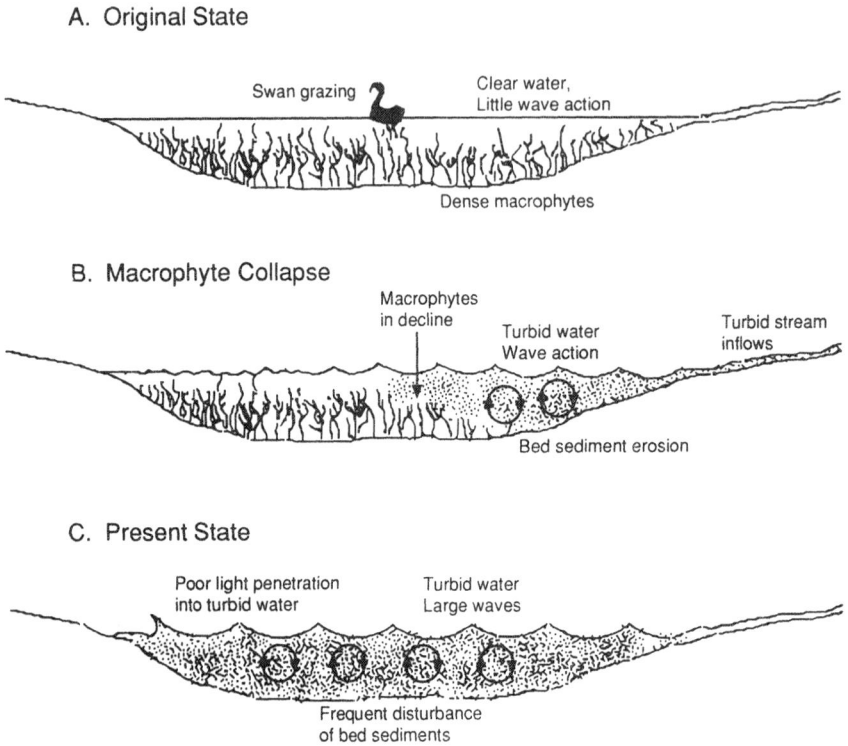

Fig 6.7. Conceptual model for the catastrophic collapse of dense macrophyte stands in shallow lakes such as Lake Whangape. (A) The original situation. The dense macrophyte stands restrict water circulation, damp wave action and protect the sediment bed from erosion. (B) Macrophyte collapse. Turbid wastewater enters the lake at the mouth of a small stream and the reduced light penetration leads to plant decline and eventual removal during windstorms. Increased water movement then permits erosion of further plants and erosion of bed sediments. (C) The present situation. Exposure to wave action and restricted light penetration prevents recovery of macrophytes.

Table 6.4, from Vant & Davies-Colley (1986), shows that these two lakes are very similar in some respects (e.g. mean depth) but Lake Rotorua is a large, mostly isothermal, lake, whereas Lake Okaro has a much smaller surface area and, being less wind exposed, stratifies strongly in summer.

Lake Rotorua is arguably the best-studied lake system in New Zealand. The lake is part of the scenic backdrop to the tourist centre of Rotorua City, and is much used for passive recreation, trout fishing, and general boating, although not so much for swimming—possibly because of the sometimes poor visual clarity (Vant & Davies-Colley 1988; Smith *et al.* 1991). Most (60%) of the catchment is grazed pasture, and drainage from this agricultural land, together with geothermal inflows to the lake, supplies a high background nutrient load (Hoare 1980). Until late 1990 the sewage from Rotorua (population = 60 000) was discharged to the lake and was a major nutrient source (50% of phosphorus and 27% of nitrogen (Rutherford *et al.* 1989)). The problem with progressive eutrophication of Lake Rotorua was recognized in the early 1970s (Fish 1975). Visual

clarity of the lake declined systematically with progressive eutrophication in the years
1978 to 1983 (Rutherford 1984), and unsightly cyanophyte scums have been observed on
the lake in calm, warm summers.

Table 6.4. Morphometric and limnological characteristics compared in Lake Okaro and
Lake Rotorua

Variable	Lake Okaro	Lake Rotorua
Mean depth (m)	11	11
Area (km^2)	0.3	81
Average Secchi depth (m)	1.8	2.3
Average chlorophyll a (mg m^{-3})	17	18
Average total phosphorus (mg m^{-3})	69	33

Lake Okaro data for 1970–1971 from McColl (1972); Lake Rotorua data for 1973–1983 from Rutherford (1984).

Since 1990 the sewage effluent from the city of Rotorua has been discharged to land
in *Pinus* plantations within the catchment. It is expected (Rutherford *et al.* 1989) that
both nitrogen (N) and phosphorus (P) from this effluent will be removed efficiently in
the soils and wetland areas of the catchment, so appreciably reducing the nutrient load on
the lake. Eventually, lake phytoplankton biomass should be reduced, with, it is hoped,
consequently improved recreational values, trout habitat, and optical water quality.

Lake Okaro has nothing like the value of Lake Rotorua as a water resource, but, as we
shall see, this lake provides a fascinating contrast as regards optical water quality and
likely response to nutrient control. (In contrast to Lake Rotorua, no such control is proposed.) The mean phytoplankton biomass, measured as chlorophyll *a* concentration, is
similar in Lakes Rotorua and Okaro (Table 6.4); however, the visual clarity of Lake
Rotorua is generally somewhat greater than that in Lake Okaro. McColl (1972) deduced
that the visual clarity of Lake Okaro is determined by phytoplankton biomass but, based
on analyses of long-term monitoring data, Rutherford (1984) suggested that the visual
clarity of Lake Rotorua is primarily dependent on constituents other than the phytoplankton standing crop. Here we have a puzzle: the mean phytoplankton biomass, as measured
by chlorophyll *a*, is similar in the two lakes (Table 6.4), implying similar phytoplankton
contributions to light attenuation, and yet Lake Rotorua is generally clearer, whilst light
attenuation in this lake depends on constituents other than the phytoplankton.

In order to explain this contrasting light-attenuating character in Lakes Okaro and
Rotorua, and to estimate the degree of visual clarity improvement to be expected from
nutrient control measures, Vant & Davies-Colley (1986) analysed a dataset comprising 16
monthly measurements of various optical characteristics and light-attenuating constituents.

6.4.2 Optical quality monitoring

Deep stations near the middle of Lakes Rotorua and Okaro were visited monthly from
July 1983 to October 1984 (Vant & Davies-Colley 1986). Secchi depth was measured on
the sunny side of the boat using a viewer, and water samples were collected for laboratory

analysis. Profiles of irradiance were measured (Appendix 2.1) using Li-Cor LI-192SB PAR sensors for the estimation of PAR band-average absorption, scattering, and attenuation coefficients (a, b, and c respectively) following Kirk (1981b).

Water samples were analysed in the laboratory for chlorophyll a and its phaeopigment degradation products, for suspended solids and volatile suspended solids, and for yellow substance (as g_{440}) (Section 3.6). Non-volatile suspended solids (ash weight at 400°C, Section 3.6), once corrected for the estimated ash content of phytoplankton, was used as an index of the concentration of mineral suspensoids, M, unrelated to the phytoplankton standing crop.

6.4.3 Water quality and light attenuation

Mean values and ranges for various optical and water quality measures are given in Table 6.5. Both average visual clarity and dispersion of clarity were rather similar in the two lakes. Note that the average Secchi depth measured by Vant & Davies-Colley (1986) in Lake Okaro (2.0 m, Table 6.5) is close to that reported earlier (1.8 m, Table 6.5) but Lake Rotorua was appreciably less clear (1.7 m compared with 2.3 m reported earlier). In each lake, Secchi depth and the beam attenuation coefficient were strongly inversely related, as expected (Fig. 2.5), so only c is considered further.

Table 6.5. Optical quality compared in Lake Okaro and Lake Rotorua

Variable	Lake Okaro Mean (range)	Lake Rotorua Mean (range)
Secchi depth (m)	2.0 (0.9–3.6)	1.7 (0.9–3.1)
Beam attenuation coefficient (m^{-1})	3.7 (1.2–8.5)	2.7 (1.4–4.6)
Chlorophyll a (mg m^{-3})	33.5 (2.9–92.6)	14.5 (3.7–34.2)
Phaeopigment (mg m^{-3})	1.7 (0.1–8.5)	9.9 (1.9–21.8)
Total pigment (mg m^{-3})	35.1 (2.9–92.6)	24.4 (11.6–42.6)
Suspended solids (g m^{-3})	5.3 (1.1–10.7)	6.7 (2.4–15.9)
Non-standing crop-related mineral suspensoids (g m^{-3})	0.8 (0–2.5)	3.3 (0–9.3)

Data for July 1983–October 1984 (from Vant & Davies-Colley (1986, Table 2), with permission).

Chlorophyll a and total pigment concentrations were comparable in the two lakes, although higher and more variable in Lake Okaro, which exhibited a pronounced seasonal cycle with spring maxima. A major difference between the lakes was that phaeopigment constituted a far greater proportion of total pigment in Lake Rotorua than in Lake Okaro (mean 41%, versus 6%). The two lakes had similar average and extreme suspended solids (SS) concentrations (Table 6.5), but the non-standing crop-related, mineral suspensoids differed markedly (average M = 0.8 g m^{-3}, M/SS = 15% in Okaro; versus M = 3.3 g m^{-3}, M/SS = 42% in Rotorua, Table 6.5).

6.4.4 Determinants of light attenuation

Total light beam attenuation in a water, c, can be regarded as the sum of contributions from each of the light-attenuating constituents, phytoplankton biomass (B measured as chlorophyll a), non-standing crop-related mineral suspensoids (M), detritus (D), yellow substance (Y), and water itself (w), thus:

$$c = c_B + c_M + c_D + c_Y + c_w$$

The last two terms, those contributions of yellow substance and water itself, were found to be small and together contributed approximately 0.3 m^{-1} to c in both lakes. Thus Vant & Davies-Colley (1986) assumed

$$c = C_B B + C_M M + c_D + 0.3$$

where C_B and C_M are the attenuation cross-sections (m^2 per unit concentration) of phytoplankton (B) and non-standing crop-related, mineral suspensoids (M) respectively. These cross-sections were evaluated by multiple linear regression (MLR) of beam attenuation versus B and M as the independent variables. Because detritus was not measured, its contribution to c (i.e. c_D) could not be estimated directly, but an average contribution over the 16 months of monitoring was estimated as the MLR intercept minus 0.3 m^{-1}.

Table 6.6. Optical cross-sections in Lakes Okaro and Rotorua, compared with average values for 25 New Zealand lakes

	C_B (m^2 g^{-1})	C_M (m^2 g^{-1})	Constant (m^{-1})	r^2
Lake Okaro	75 (11)	0.11[a] (0.35)	1.1 (0.49)	0.80
Lake Rotorua	28 (14)	0.32 (0.043)	1.2 (0.31)	0.80
25 NZ lakes (Vant & Davies-Colley 1984)	110 (16)	1.2 (0.12)	−0.48[a] (0.46)	0.92

The linear regression model is: $c = C_B B + C_M M +$ constant. Standard errors are shown in parentheses and the explained proportion of variance parameter (r^2) is also given.
[a] Value not significantly different from zero.

Table 6.6 shows the results of the MLR analysis. The regression equations accounted for about 80% of the variability in c in both lakes, and about 79% of this total variability was accounted for by variation in phytoplankton biomass (B) in Okaro and about 74% by that of M in Rotorua. The contribution of M to variation in light attenuation in Lake Okaro was so weak that the estimate of C_M was not significantly greater than zero for this lake (Table 6.6). The estimates of C_B and C_M are slightly biased by weak co-linearity between B and M ($r \sim 0.3$), and probably also by errors in measurement of these two variables.

Fig. 6.8 shows the contributions of the various light-attenuating constituents, as estimated from the MLR equations, to overall light beam attenuation in Lakes Rotorua and Okaro, plotted versus time. (Only average contributions could be given for water, yellow substance, and detritus.) Measured c values (not shown) compared closely with values calculated from the MLR equations ($\hat{c}$).

This partitioning of light beam attenuation shows that B was responsible for most (mean 68%) of $\hat{c}$ in Lake Okaro whereas M was the dominant light-attenuating constituent (mean 40%) in Lake Rotorua. Although c_M dominated $\hat{c}$ overall in Lake Rotorua, at times c_B was larger (January to March, and September 1984, Fig 6.8B). The role of

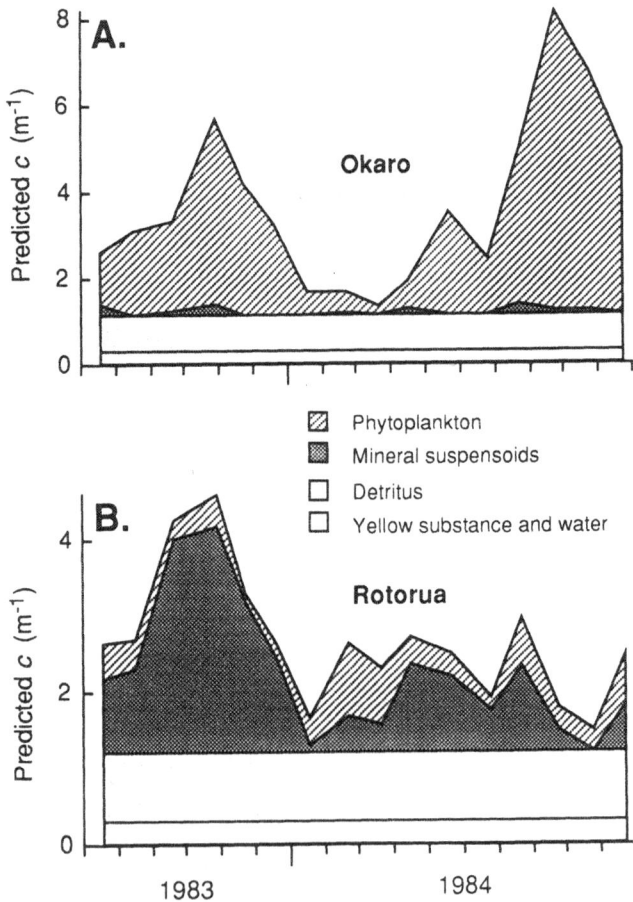

Fig 6.8. Contributions of chlorophyll a (an index of algal biomass, B) non-standing crop-related mineral suspensoids (M), and other factors (detritus, yellow substance and water) to predicted beam attenuation. (A) Lake Okaro data, partitioned as $0.075B + 0.11M + 1.1$. (B) Lake Rotorua data, partitioned as $0.028B + 0.32M + 1.2$. (After Vant & Davies-Colley 1986, with permission.)

detritus as a contributor to $\hat{c}$ is uncertain since only an average estimate of its value could be obtained. However, on occasions when clarity was relatively high (low c) detritus could have been relatively important or even dominant in Lake Rotorua (Fig. 6.8B).

6.4.5 Discussion

The attenuation cross-sections estimated from multiple linear regression for Lakes Okaro and Rotorua are compared with those for New Zealand lakes in general (Vant & Davies-Colley 1984) in Table 6.6. The value of C_B for Lake Okaro (75 m^2 g^{-1}) was similar to that found for New Zealand Lakes in general (110 m^2 g^{-1}), but C_B in Lake Rotorua was appreciably lower (28 m^2 g^{-1}). This indicates that a given phytoplankton biomass, measured as chlorophyll a, attenuates light with only 1/3 the efficiency of the same biomass in Lake Okaro and 1/4 the efficiency of New Zealand lakes in general. Bricaud et $al.$ (1983) and Davies-Colley et $al.$ (1986), among others, have reported C_B values for algal cultures encompassing these MLR values for C_B, and Morel (1987) has shown that such variation arises mainly from differences in cell size and chlorophyll a content. The algal flora of Lake Rotorua is usually dominated by large cylindrical diatoms of the genus $Melosira$ (Cassie 1975), which would be expected to attenuate light less efficiently than the much smaller greens and blue–greens dominating Lake Okaro (Flint 1975).

The value of C_M for Lake Rotorua (0.32 m^2 g^{-1}) was also much smaller than the MLR value reported by Vant & Davies-Colley (1984) for 27 New Zealand lakes (1.2 m^2 g^{-1}). (The value for Lake Okaro could only be determined with very low precision and does not warrant further discussion.) Vant & Davies-Colley's (1984) value is very high, close to the peak on Fig. 2.9, occurring for quartz-density particles with a diameter of about 1.2 μm, and we now know that this high value reflects weighting towards turbid clay-rich lake waters (e.g. Lake Waahi, Section 6.3) in their sample of 27 New Zealand lakes. Apparently the non-algal mineral suspensoids in Lake Rotorua are appreciably larger on average (of order 5 μm diameter), which would account for their attenuating with about one-third the efficiency of fine clays.

Vant & Davies-Colley (1986) suggested that the main source of the mineral suspensoids dominating the light attenuation in Lake Rotorua was wind-wave erosion of bottom sediments. Matthews (1979) and Rawlence (1984) have reported that the bottom sediments of Lake Rotorua are rich in algal remains, including silica (inorganic) frustules of diatoms (1–10 μm diameter) and phaeopigment (Matthews 1979; Rawlence 1984). A fairly close ($r = 0.49$) relationship was found between M in the lake water and wind stress over the previous five days, calculated (following Smith & Banke 1975) from meteorological records at nearby Rotorua Airport. Also, the (relatively high) phaeopigment content of the water samples correlated ($r = 0.46$) with the wind stress parameter ($P \sim 0.05$ in both cases). Thus wind-related resuspension of diatom-rich bottom sediments could account for the elevated concentrations of M and phaeopigments, and also the relatively high detritus content, of Lake Rotorua.

It is of interest to consider the likely changes in visual clarity in Lakes Rotorua and Okaro in response to reduced nutrient loading—which, as described above, actually has been implemented in Lake Rotorua, but is unlikely to be applied to Lake Okaro. In Lake Okaro, as we have seen, the light attenuation is dominated by phytoplankton. Therefore, if nutrient loading on Lake Okaro were to be reduced, the visual clarity would be expected to improve as the phytoplankton biomass fell, according to the Morel (1987) (Case I) relationship between light attenuation and biomass in Fig. 5.11. The optical quality response may be delayed because of internal nutrient loads (Section 5.6.3) in this strongly stratifying lake, but there is little doubt that it would occur eventually.

The situation is not so straightforward in Lake Rotorua. The sewage nutrient control measures described earlier are expected to decrease phytoplankton biomass gradually owing to the reduction in nutrient load. However, as reported by Vant & Davies-Colley (1986), both the average and worst visual clarity in Lake Rotorua are controlled by the light attenuation of mineral suspensoids (M) that are unrelated to the standing crop of phytoplankton, so it might appear that nutrient load reduction will not result in much improvement in visual clarity. However, as we have seen, M in Lake Rotorua water is, in fact, related to the (time-averaged) phytoplankton production in this lake, being composed largely of algal frustules entrained from the bottom sediments, along with detritus and phaeopigments. Therefore, the question is, how will the resuspendable bottom sediments in Lake Rotorua respond to the nutrient control measures?

Evidence that the visual clarity of Lake Rotorua will indeed improve with nutrient load reduction is to be found in the historical datasets compiled by Rutherford (1984). The period of most marked decline in Secchi depth (1978–1983) coincided with sharply increasing nutrient loading on the lake from the sewage of rapidly growing Rotorua City. It could, therefore, be inferred that the reduction in sewage nutrient loading with land disposal should improve water clarity after an unknown time required to flush the pool of suspendable algal-production-related material in the bottom sediments.

A further complication, however, relates to the role of aquatic macrophytes growing in the littoral of Lake Rotorua. As we saw in the previous section these plants can greatly modify optical quality of lake waters because of their role in restricting water mixing and wave action, and protecting bottom sediments. In Lake Rotorua the macrophytes growing to about 5.0 m (Vant et al. 1986) have declined somewhat in cover compared with previously (J. S. Clayton, Ecosystems Division, NIWA, Hamilton, NZ), possibly because of reduced light penetration with progressive eutrophication. If the macrophyte cover of littoral zones were to change in future, this effect could well outweigh the influence of nutrient control on visual clarity of Lake Rotorua; either nullifying any gains in clarity if further declines in cover occur, or appreciably improving clarity if the plants should return to their former luxuriance.

6.5 LAKE HOROWHENUA: CHANGES IN OPTICAL WATER QUALITY FOLLOWING SEWAGE EFFLUENT DIVERSION

6.5.1 Background

Lake Horowhenua (40° 37' S, 175° 15' E) is a small (2.9 km^2), shallow (mean depth 1.3 m) lake on the coastal plain of the lower North Island, New Zealand near the town of Levin (population 15 000). The lake waters are vertically well mixed and fairly rapidly flushed (mean residence time ~2 months). Plant nutrient concentrations and phytoplankton biomass are high and water clarity is low (Vant & Gilliland 1991). Historically a major nutrient source to the lake (about 20% of N and 90% of P, Table 6.7) was sewage effluent from the town of Levin. The lake is the focus for various recreational activities (but apparently not swimming) and is a major rowing venue.

Table 6.7. Annual water and nutrient inputs to Lake Horowhenua from 1978 to 1990 (after Vant & Gilliland 1991, with permission)

Source	1978–1987	1988–1990
Total P input ($t\ yr^{-1}$)		
Runoff and rainfall	0.7	0.7
Groundwater	0.1	0.1
Sewage	9.1 (92%)	0
Total N input ($t\ yr^{-1}$)		
Runoff and rainfall	204.2	204.2
Groundwater	41.5	41.5
Sewage	54.8 (18%)	0

Since late 1984, nutrients, chlorophyll *a*, turbidity and Secchi disc clarity have been monitored. In April 1987 sewage effluent from Levin was diverted away from the lake to a land disposal area out of the lake catchment. The main objective was to improve the sanitary quality of the lakewater, but it was also hoped that aesthetic quality of the lake-water, including visual clarity, would improve along with trophic status (Anon 1990). Vant & Gilliland (1991) examined the monitoring data up to April 1990 to see whether changes in water quality and optical character of the lake waters could be attributed to the diversion, and their dataset has been updated to the time of writing.

6.5.2 Monitoring
Since September 1984, Lake Horowhenua has been sampled eight times each year. Secchi disc depths are measured, and near-surface water samples are collected by boat at five different sites on the lake (Vant & Gilliland 1991). The samples are analysed in the laboratory, using standardized methods, for total and inorganic nitrogen (TN, TIN), total and dissolved reactive phosphorus (TP, DRP), chlorophyll *a*, turbidity (Hach 2100A nephelometer), and suspended solids. (The optical measurements have not been updated to more modern methods because it is desirable to maintain continuity of existing datasets.)

Seasonal variation in many characteristics was marked (Vant & Gilliland 1991), so results were summarized separately for 'summer' (December to April inclusive) and 'winter' (June to October). In the period immediately after diversion, some effluent would still have been present in the lake (flushing time ~2 months) so results for the first 6 months after diversion in April 1987 were omitted from the 'before' and 'after' comparisons.

6.5.3 Findings
Table 6.8 summarizes lake water quality data separately for summer and winter before and after diversion. Nutrient concentrations were markedly seasonal and the seasonal patterns suggested that phosphorus was controlling phytoplankton growth in winter, and

nitrogen in summer (Vant & Gilliland 1991). (There are also indications that light limitation of phytoplankton occurs in this lake on occasions.) Phytoplankton biomass, measured as chlorophyll *a*, generally peaked in the summer before sewage diversion, but winter peaks and summer troughs have occurred since then (Fig. 6.9A, Table 6.8). Secchi depth (Fig. 6.9B) was greater in winter than in summer (Table 6.8), reflecting a strong dependence on suspended solids concentrations (Fig. 6.9C).

Levels of dissolved reactive phosphorus (DRP) decreased markedly after diversion in both summer and winter. The summer total phosphorus (TP) also fell markedly, but the winter TP reduction was smaller (Table 6.8). In summer, total nitrogen (TN) and chlorophyll *a* decreased substantially (30–50%), but there was essentially no change in winter. Total phosphorus, TN, chlorophyll *a*, suspended solids and turbidity have all decreased since diversion, although some of the changes are not statistically significant (Table 6.8) in the presence of high variability.

Despite the substantial summer reduction in phytoplankton biomass, water clarity is still poor. Table 6.8 shows that mean Secchi depth in summer improved slightly from 0.28 to 0.42 m and the winter mean Secchi depth from 0.48 to 0.72 m.

The correlations between the two optical variables and suspended solids were examined. As expected, turbidity and Secchi depth were strongly (inversely) related (Spear-

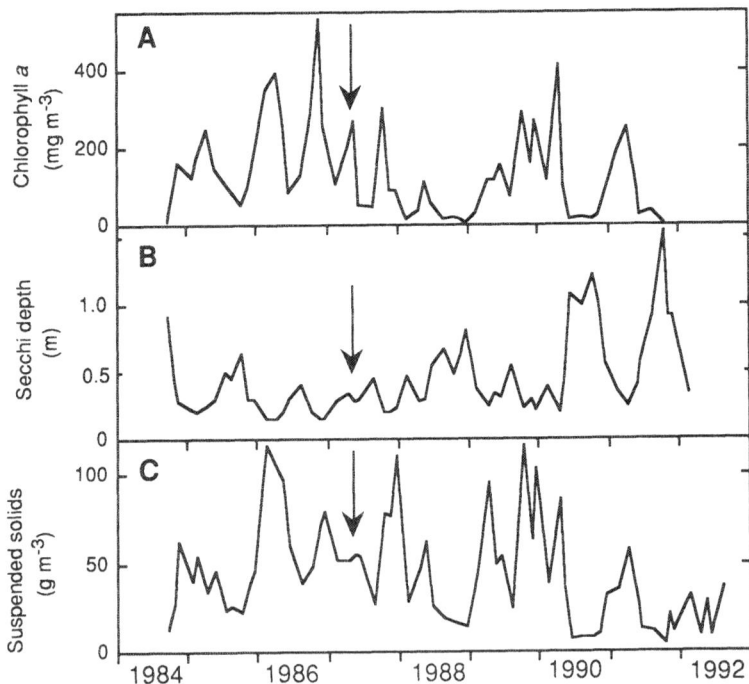

Fig. 6.9. Water quality in Lake Horowhenua, 1984–1992: (A) chlorophyll *a*; (B) Secchi depth; (C) suspended solids. The arrow on each graph indicates the time at which sewage effluent was diverted (April 1987).

man's rank correlation coefficient, $\rho = -0.87$) whilst both were related to suspended solids ($\rho = 0.82$ and -0.92 respectively). Evidently the high suspended solids content of the water in Lake Horowhenua controls the visual water clarity.

Table 6.8. Comparison of mean summer and winter conditions in Lake Horowhenua before and after diversion of sewage effluent from the lake in April 1987

	Summer (December–April)			Winter (June–October)		
	Before	After	P	Before	After	P
TP (mg m^{-3})	945	391	<0.001	225	181	ns
DRP (mg m^{-3})	535	111	<0.001	40	13	<0.06
TN (mg m^{-3})	4210	2877	<0.05	3890	3953	ns
TIN (mg m^{-3})	43	49	ns	1665	1736	ns
Chlorophyll a (mg m^{-3})	185	110	<0.08	90	81	ns
Secchi depth (m)	0.28	0.42	<0.06	0.48	0.72	<0.06
Suspended solids (g m^{-3})	55	63	ns	29	48	ns
Turbidity (NTU)	33.5	21.7	ns	15.6	12.1	ns
Non-algal turbidity	15.8	16.0	ns	8.3	5.9	ns

Differences not significant at $P < 0.1$: 'ns'. (Data from Vant & Gilliland (1991), with additional unpublished 'after' data (1990–1992).)

To analyse further the changes in visual clarity, the relationship between chlorophyll a, and turbidity was examined (Fig. 6.10). The lower bound to the data (line in Fig. 6.10 with a slope of unity, implying a linear relationship) was taken as indicative of the phytoplankton-specific turbidity in Lake Horowhenua. Turbidity in the lake increased as 0.08[chlorophyll a], similar to other New Zealand lakes (Vant & Davies-Colley 1984). The difference between this quantity and total turbidity was attributed to non-algal turbidity.

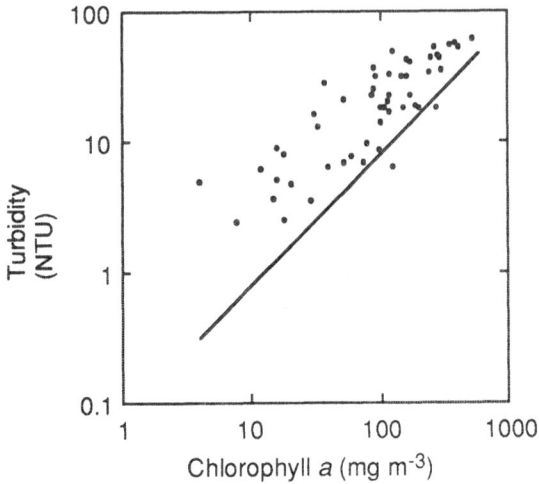

Fig. 6.10. Chlorophyll *a* plotted versus turbidity in Lake Horowhenua. The line is a lower bound to the data and corresponds to a phytoplankton-specific turbidity of 0.08 NTU (mg Chl *a* m^{-3})$^{-1}$.

Fig. 6.11 shows how the contributions to turbidity varied during 1984–1991. As summer chlorophyll *a* fell after diversion of sewage effluent, so did turbidity attributable to phytoplankton (Table 6.8). However, non-algal turbidity did not change (Table 6.8) because suspended solids concentrations remained high. Presumably, most of the suspended matter was sediment entrained from the lake bottom by wind waves in this shallow, exposed lake.

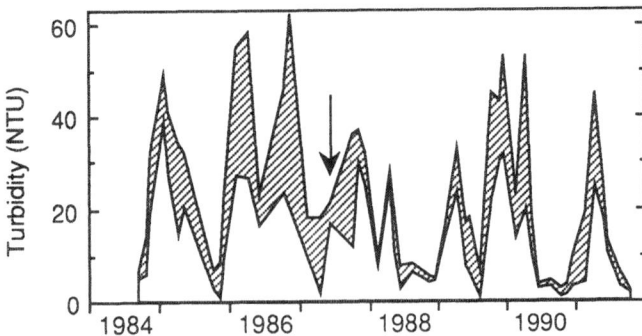

Fig. 6.11. Turbidity in Lake Horowhenua, September 1984–April 1991. The shaded area shows the calculated contribution of phytoplankton. The arrow indicates the time at which sewage effluent was diverted (April 1987).

6.5.4 Discussion

Vant & Gilliland (1991) presented data suggesting that, if lake nutrient concentrations in Lake Horowhenua were to fall further, phytoplankton biomass should also decrease. Lake TP should fall as the internal P load abates. In the winter, chlorophyll a levels should fall in parallel with TP, but summer TP would need to fall substantially (to about 120 mg m^{-3} (White 1989)) before chlorophyll a would respond by falling too. In contrast, a fall in TN would be expected to engender an immediate reduction in chlorophyll a, but this would require a reduction in the external N load from non-point sources (Table 6.7).

Even if chlorophyll a does decrease further, water clarity is not expected to improve much as a result because non-algal turbidity strongly influences total turbidity (Fig. 6.11). Vant & Gilliland (1991) suggested that, for the water to appear noticeably clearer, turbidity would need to be less than about 10 NTU, equivalent to a Secchi depth of more than 0.5 m. Table 6.8 suggests this is possible in the winter, provided that chlorophyll a concentrations fall by about four-fold (to less than 25 mg m^{-3}). However, noticeable improvements will occur in the summer only if the non-algal turbidity also falls substantially then. There seems no reason to expect a reduction in non-algal turbidity with a reduction in phytoplankton biomass, even though some fraction of the light-attenuating suspended matter is probably detritus derived from past algal production.

In Lake Horowhenua the improvement in visual clarity following nutrient control measures has been disappointing because of the high non-algal light attenuation. However, the observed small increase in visual clarity with reduced phytoplankton biomass is broadly consistent with Fig. 5.11. Additional evidence suggesting that Lake Horowhenua cannot be expected to further improve in visual clarity comes from a comparison of this lake with other dune lakes in the vicinity of Levin. The continuing low visual clarity of Lake Horowhenua is entirely consistent with the generally low clarity of other shallow lakes in the area, suggesting that this water body was never very clear.

6.6 TARAWERA RIVER: EFFECTS OF KRAFT EFFLUENT ON OPTICAL WATER QUALITY

6.6.1 Background

The source of the Tarawera River is oligotrophic Lake Tarawera in the Central Volcanic Zone of the North Island of New Zealand. At Kawerau (Fig. 6.12) the Tarawera River is an attractive water body, typically blue–green in colour and of moderate to high visual clarity (Wilcock & Davies-Colley 1986). Downstream of Kawerau the visual quality of the river water is degraded by a number of effluent discharges, including sewage effluent from Kawerau, geothermal condensates, effluent from a chemomechanical pulp mill and, most notably, effluent from a large Kraft pulp mill. Near the river mouth where the water is viewed by tourists on an otherwise scenic stretch of highway (Fig. 6.12), the water appears inky black in colour (Wilcock & Davies-Colley 1986). The river water discharges as a dark plume that is clearly visible from the air (Plate 12) because of colour contrast with the blue–green seawater of the Pacific Ocean.

Evidently the waters of the Tarawera River are severely visually degraded. The impacts on optical quality have been quantified by surveys of colour on the Pt–Co scale, and of light absorption and turbidity. This work was reported by Davies-Colley (1984) in

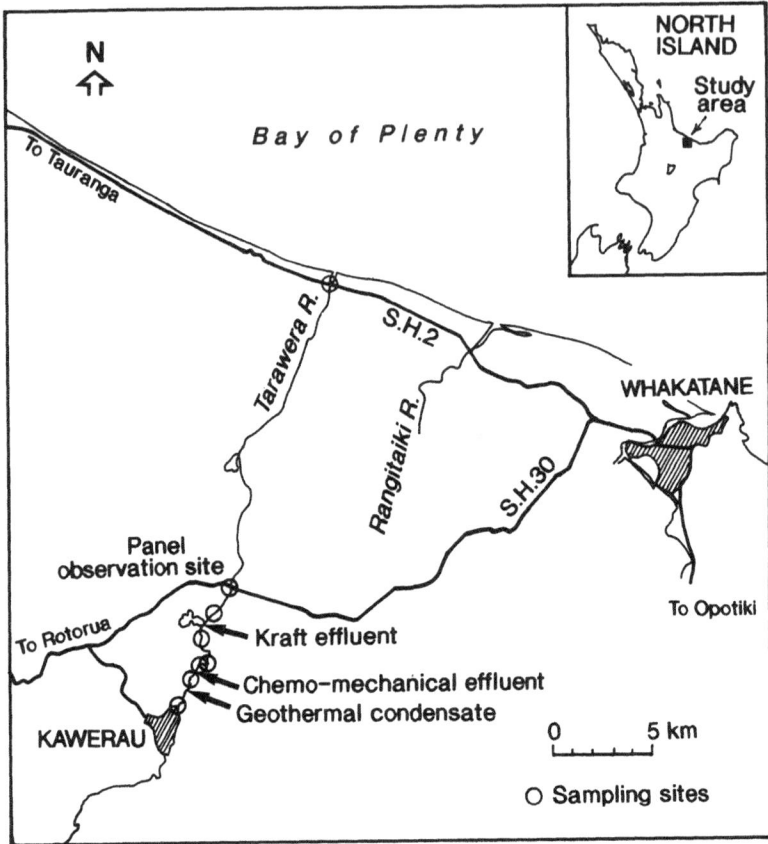

Fig. 6.12. Map of the lower Tarawera River showing the locations of effluent discharges and sampling sites.

a published 'Management Plan' for the Tarawera River. The Kraft effluent was shown to be by far the largest source of light-absorbing material in the river. This effluent greatly affects river water colour by changing the typically blue–green or green water to a very dark yellow or orange (Munsell hue = 20) and also significantly reduces visual water clarity and light penetration into the river water. By comparison, the other effluents cause relatively small optical quality impacts. A more recent study of the optical quality degradation in the Tarawera River (Davies-Colley 1990b) included black disc measurements of visual water clarity, transmissometer measurements of light beam attenuation, and light sensor measurements of irradiance attenuation in the river. The findings of this study are summarized below.

In the context of the management plan for the river, the question arose: what level of colour loading would be below the threshold for human detection in the river? This would be a lower bound to the colour loading that could be consented to if the river were to be restored for aesthetic and recreational uses. A special on-site perception study was carried out with a panel of human observers in order to define the detection threshold, and this study (by Wilcock & Davies-Colley 1986) is also summarized below.

6.6.2 Characterization of the optical quality impacts
Methods
On 10 January 1990, four effluents, including two geothermal flows and two paper industry effluents, were sampled together with one major tributary, the Rururanga Stream. Flow gaugings on effluents and river sites were made to permit attenuation balances to be calculated (Section 5.3.6). Visual water clarity was measured horizontally by the black disc method, in the river from the survey boat while drifting, and also on volumetrically diluted effluent samples in a trough (Davies-Colley & Smith 1992). Water colour (hue) as observed horizontally through the black disc viewer was matched directly to standards in the Munsell book of colour.

The beam attenuation coefficient was measured directly using a Martek XMS beam transmissometer. Measurements of irradiance (PAR) penetration were made (by wading) at two river sites: upstream (as a control) and downstream of the effluent discharges.

Within 48 h of collection of water samples, laboratory analyses were completed for turbidity (Hach 2100A), suspended solids (SS), and volatile suspended solids (VSS). Light absorption was measured at 440 nm using a Pye Unicam PU8800 spectrophotometer.

Findings
Longitudinal profiles for selected variables measured at sites on the Tarawera River are plotted in Fig. 6.13. Generally the optical character of the river was very similar to that observed in previous work (Davies-Colley 1984; Wilcock & Davies-Colley 1986), so the results can be regarded as typical.

Upstream of the industrial effluents the Tarawera River was fairly clear (black disc visibility = 1.9 m) and bluish-green in hue (Munsell hue = 47.5). The turbidity and beam attenuation coefficients were fairly low (about 1.8 NTU and 2.5 m^{-1} respectively), consistent with the fairly high visual clarity. The suspended matter concentration (SS) was low, and only a small fraction was organic (VSS/SS < 15%). Light absorption coefficients (a_{440}, g_{440}) were low, consistent with the observed bluish-green hue.

The visual clarity fell progressively downstream from Kawerau to a site immediately upstream of the Kraft effluent discharge (Fig. 6.13A). Associated with this fall in visual clarity was a rise in light attenuation (Fig. 6.13B), and also turbidity and SS. The hue of the water shifted slightly, but perceptibly, from bluish-green at Kawerau to green (Munsell hue = 42.5) (Fig. 6.13C). This hue shift was related to an increase in absorption of blue light, mainly by particulate constituents. Balances on light attenuation cross-section showed that these changes were attributable to effluent discharges: sewage from Kawerau, geothermal condensates and effluent from a chemo-mechanical pulp mill.

By far the largest influence on the visual quality and related optical character of the river was the inflow of the Kraft effluent (Fig. 6.13). This effluent tripled light beam attenuation in the river, resulting in a fall in visual clarity to less than one-half of the upstream value. The hue was markedly shifted from a green (Munsell hue 42.5) to a yellow–orange (Munsell hue 20) by the Kraft discharge, but the colour was very dark owing to intense light absorption (increasing five-fold for a_{440}, and nine times for g_{440}) such

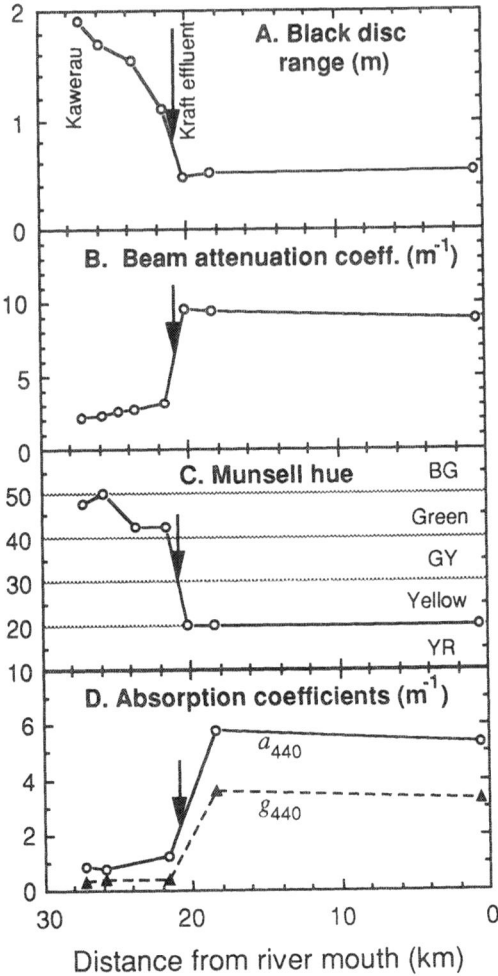

Fig. 6.13. Longitudinal profiles of various optical observations and related constituents in the Tarawera River on 10 January 1990: (A) black disc range; (B) beam attenuation coefficient calculated from transmissometer readings; (C) Munsell hue; (D) absorption coefficients. The arrow indicates the location of the Kraft effluent discharge.

that the above-water visual impression was of 'inky blackness'. The low saturation ('brownish') colour of the Tarawera River downstream of this discharge is generally only apparent in the wake of a boat or when seen against a light-coloured object at very shallow depths (Davies-Colley 1984).

Euphotic depth dropped markedly from about 8.5 m at Kawerau (considerably deeper than the river bed) to about 1 m (shallower than much of the river bed) near the river mouth owing to the very high absorption of light by the Kraft effluent. Benthic plants are not very abundant in the river, in part because of unsuitable substrate (mobile sands), so the inferred light limitation may not be very important, ecologically. However, the impacts on the river fauna of greatly reduced illumination of the water and bed features, together with compressed visual range, are possibly of greater ecological significance than reduction in primary production.

6.6.3 Detectability of change in river water colour

Wilcock & Davies-Colley (1986) reported the findings from an on-site panel study of the human visual detectability of Kraft effluent in the Tarawera River. The aim of the study was to determine the reduction in Kraft loading that would be required to reduce the colour change in the Tarawera River below a 'conspicuous' level (in the words of New Zealand water law). On 13 March 1984 a panel of 11 high school students (all of whom had verified normal colour vision and good eyesight) was located on the river bank a few kilometres downstream of the Kraft effluent discharge (Fig. 6.12). The panelists were told only that the river colour would change owing to upstream effluent loading changes.

Starting from a low effluent loading (practically as near to zero as could be achieved by stopping Kraft effluent flow at the Tasman outfall for 1 h) the effluent discharge was varied incrementally to produce an up-ramp followed by a down-ramp of colour loading as measured on the Pt–Co scale. On the day of the experiment (March 13 1984) the river water above the outfall was fairly turbid and the river bottom was not visible. The background water colour was a greyish yellow–green, so that hue changes in the river were indistinct. Therefore, the panelists were probably responding to changes in colour brightness rather than hue or water clarity (Wilcock & Davies-Colley 1986).

From the panel responses during the up-ramp of effluent loading, the median just-noticeable-difference (JND) in colour loading was found to be about 4 Pt–Co colour units over background. Four Pt–Co units corresponds to an effluent dilution in the river water of about 200-fold compared with the present 10-fold (at around 80 Pt–Co units), suggesting that a twenty-fold reduction in Kraft loading would be required if the visual quality of the Tarawera River is to be suitable for recreational uses. On the down-ramp of colour loading from a high of 32 Pt–Co units, the colour needed to be reduced to approximately 4 Pt–Co units before the change was recognized by a majority of the panel—that is, before there was a detectable colour change from the initial high condition. (It is coincidental—but noteworthy—that the point at which an improvement in colour was recognized was similar in magnitude to the point at which a change over background was recognized.)

These findings suggest (1) that only a small Kraft colour loading is acceptable in the Tarawera River (about 1/20 the present loading) if this water body is to be suitable for recreational uses, and (2) that a major colour reduction compared with the present severely degraded situation may be required before people will notice an improvement in river water appearance.

6.7 WHANGANUI RIVER: EFFECTS OF TRIBUTARY DIVERSION ON OPTICAL WATER QUALITY

6.7.1 Background

The Whanganui River is the second largest river in the North Island of New Zealand (median flow near the mouth = 140 m^3 s^{-1}). The river rises in the snows of Tongariro National Park (Fig. 6.14) and flows west and then south through rugged and remote

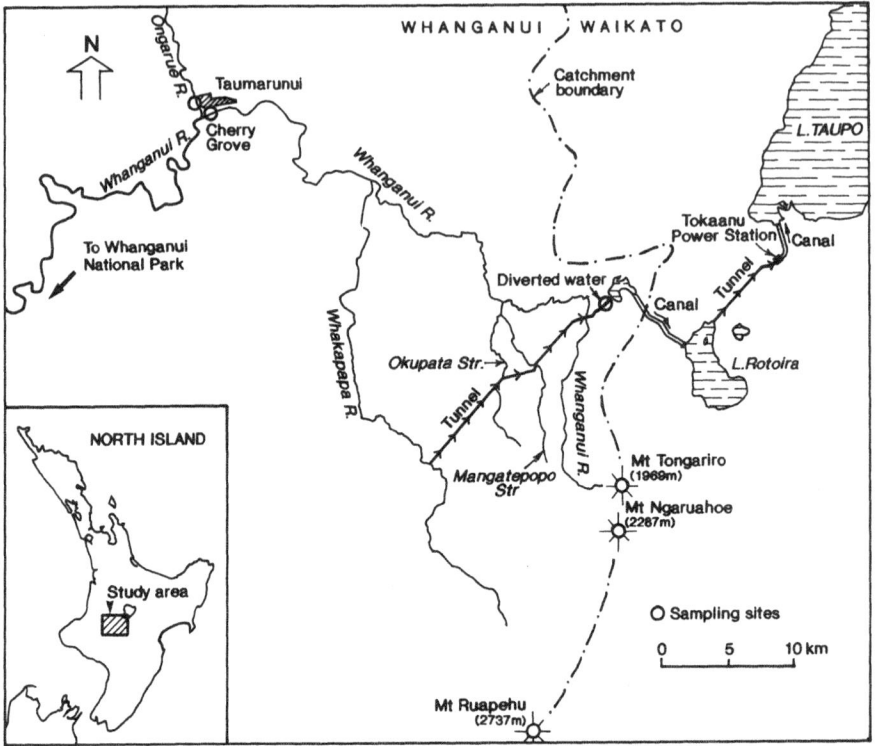

Fig. 6.14. Location map showing headwater tributaries of the Whanganui River that are diverted to another catchment (that of the Waikato River) to augment power generation flows. Sampling stations are shown on the diverted water (Te Whaiau Canal), and on the Ongarue River and the residual Whanganui River near Taumarunui.

mudstone country, including lands within the Whanganui National Park, before entering the Tasman Sea near the city of Wanganui. The river and its environs are increasingly used for a variety of 'wilderness' activities, notably hiking, canoeing, and jet boating. Most canoeists enter the river at Cherry Grove near the town of Taumarunui (Fig. 6.14) and make their way downstream through the National Park.

Headwater tributaries of the Whanganui River, particularly the Whakapapa River, Okupata Stream, Mangatepopo Stream, and the upper Whanganui River (in order moving north-east, Fig. 6.14), are diverted to the Waikato Catchment to augment flows for hydro-electric power generation at Tokaanu Power Station (Fig. 6.14) and in the Waikato River downstream of Lake Taupo. The tributary flows are normally diverted up to the capacity of the diversion structures, except for a residual flow of 0.6 m^3 s^{-1} left in the Whakapapa River according to an operating rule negotiated in 1983. The diversions according to the 1983 rule amount to about 84% of the mean aggregate flow at the diversion structures and about 41% of the natural mean annual flow in the Whanganui River at Cherry Grove, Taumarunui (Fig. 6.14). Fig. 6.15 shows the flow duration curve for the

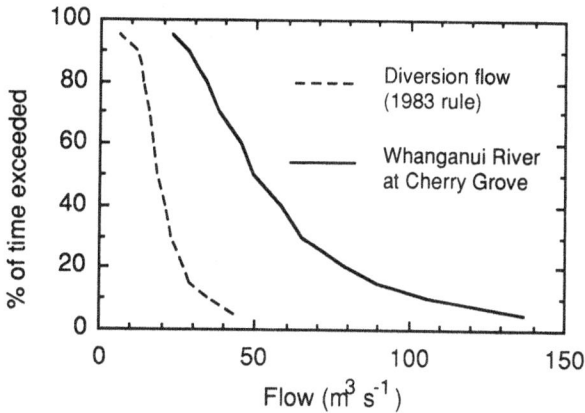

Fig. 6.15. Flow–frequency distributions (flow duration curves) for the Whanganui River at Cherry Grove, Taumarunui (natural flows) (——) and for the aggregate flow of water diverted from the headwater tributaries according to the 1983 operating rule (– – –).

total diversion flow according to the 1983 rule (which is fairly steady around a median of 19 m³ s⁻¹), together with that for natural flows in the Whanganui River at Taumarunui (median = 48 m³ s⁻¹).

As well as affecting the downstream flow-related activities such as recreational navigation and fishing, the diversions affect water quality, particularly optical water quality, of the residual Whanganui River. The headwater tributaries draining the hard andesite rocks of the Tongariro volcanoes are much clearer and more attractive in colour (blue to green hued) than the waters of other Whanganui tributaries such as the Ongarue River (Fig. 6.14), which is typically yellow coloured and turbid. Thus diversion of the headwater tributaries leaves the residual Whanganui River discoloured and less clear than in its natural condition.

No information is available on the optical water quality of the Whanganui River prior to commissioning of the diversions (in 1973). However, by assuming conservative behavour of the light-attenuating material in the different tributary waters following mixing with the Whanganui water, it is possible, at least in principle, to estimate the visual clarity in natural flows. That was the philosophy of a study by Davies-Colley (1990c). The visual clarity and water flows were measured in the diversion water and in the residual river, and from these data the flow and clarity were estimated for the mixed waters (i.e. no diversion flows), so that the impact on optical water quality of the diversions could be quantified.

6.7.2 Methods
Based on a jet boat reconnaissance, a programme of regular monitoring of optical quality was instigated. The study area is remote from suitable laboratory facilities so the monitoring was mainly confined to field observations of black disc clarity (Davies-Colley 1988a) and colour as observed through the black disc viewer (Davies-Colley & Close 1990).

Monitoring was carried out by Whanganui National Park (Department of Conservation) staff at three stations. The residual Whanganui River and a major tributary, the Ongarue River were monitored at the ranger station at Cherry Grove, Taumarunui (Fig. 6.14). Flows of both rivers at these stations could be estimated with good accuracy from nearby level recorder stations. The total diverted flow was measured routinely near the Whanganui diversion (Fig 6.14).

The monitoring was carried out for one year commencing on 21 April 1988. Usually measurements were made weekly at the two Cherry Grove stations and fortnightly for the diversion water. On four sampling occasions in January and February of 1989 natural flows occurred in the Whanganui River when the diversion structures were shut down for maintenance. Measurements on these occasions provided an opportunity to observe optical water quality in the absence of diversion.

6.7.3 Results of the monitoring programme
The monitoring record showed that the diversion waters were usually of high clarity

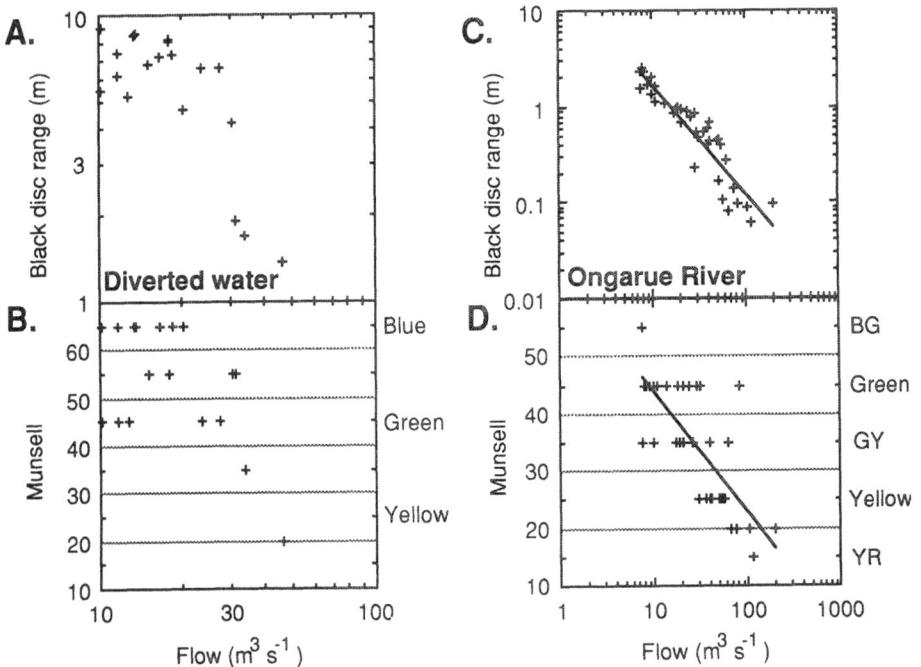

Fig. 6.16. Black disc clarity and hue (Munsell scale) plotted against flow in the water diverted from the headwater tributaries of the Whanganui River compared with similar plots for the Ongarue River, a turbid water tributary representative of the lower reaches of the Whanganui River system. (A) Black disc clarity and (B) hue on the Munsell scale, versus flow in the diverted water. (C) Black disc clarity and (D) hue on the Munsell scale versus flow in the Ongarue River. Linear regression lines are shown for the Ongarue river data ($r = -0.93$ for black disc range versus flow and -0.74 for hue versus flow).

(median = 6.6 m) and of blue to green hue (median Munsell hue number = 55). Figs 6.16A and 6.16B show that both clarity and hue were weakly related to diversion flow which, in turn, was dependent on tributary streamflow up to the capacity of the diversion structures. At high diversion flow, clarity was about 2 m, much lower than typical, and hue tended to be yellow.

In contrast, the tributaries of the Whanganui River downstream of Taumarunui (represented by the Ongarue River) had generally lower optical quality. In the Ongarue River, median black disc range was 0.61 m and median Munsell hue number was 35 (green–yellow). Both clarity and hue in this tributary were quite variable, mostly in relation to flow (Figs 6.16C and 6.16D). The residual Whanganui River at Cherry Grove (upstream of the Ongarue River confluence, Fig. 6.14) was intermediate in both clarity and colour (medians = 1.23 m for y_{BD} and 45—green—for Munsell hue number).

The black disc clarity at Cherry Grove for natural flow (i.e. had no diversion occurred) was calculated assuming conservation of light attenuation in the Whanganui River system (see Section 5.3.6). On each monitoring occasion, the clarity (symbol y) with no diversion (subscript ND) was calculated from the measured river water clarity with diversion (D) and the measured diversion water clarity (div), together with the measured flows (Q) using the following equation:

$$(Q/y)_{ND} = (Q/y)_{D} + (Q/y)_{div}$$

Fig. 6.17 shows the resulting calculated black disc ranges for no diversion plotted against natural flows, together with the measured black disc ranges plotted against residual flows, at Cherry Grove, Taumarunui. A fairly close log–log (power law) relationship between clarity and flow is indicated in Fig. 6.17, and linear regression on log-transformed variables was used to fit lines to the data. Comparison of the visibilities at Cherry Grove, with and without diversions, shows that the diversions have an appreciable effect on clarity of the residual river over a range of flow conditions. For example, on 21 April 1989, the measured black disc clarity was 3.90 m but would have been 5.22 m had water not been diverted (small arrow on Fig. 6.17).

Also shown in Fig. 6.17 are four measured clarities plotted against natural flows at times when no diversions were occurring owing to maintenance shutdowns at the diversion structures. The agreement of these measured values (solid points) with the predicted values (open points) is good, and confirms the validity of the analysis.

Colour (hue) observations are related to flow in the Whanganui River at Cherry Grove, as are those in the Ongarue River (Fig. 6.16D). At high flow the river was generally described as 'yellow' or 'yellow–red' whereas at low flow the water was a more attractive colour, usually described as 'blue–green'. Fig. 6.18A shows the inverse relationship between hue descriptions (assigned to the Munsell hue scale) and flow at Cherry Grove, and Fig. 6.18B shows the direct relationship between hue and clarity. Although there is some scatter, probably related to observer error with this necessarily subjective assessment, the relationships are sufficiently close to be considered predictive.

6.7.4 Effects of the diversions on river water clarity and colour

To quantify the magnitude of the impact over a range of flow conditions, clarity distribution curves with and without diversion were calculated from the relationships between

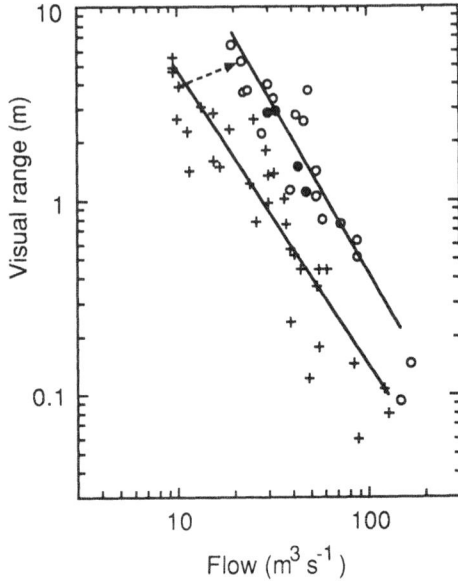

Fig. 6.17. Black disc clarity plotted against flow in the Whanganui River, North Island, New Zealand: +, measured values at times when flow was being diverted; ○, calculated values for natural flow (no diversion). The small arrow joins the value of 3.90 m observed on 21 April 1989 with the value of 5.22 m predicted for natural flow. Four observations at natural flow are also shown ●. Linear regression lines are shown ($r = 0.920$ for observed data and 0.934 for natural flow).

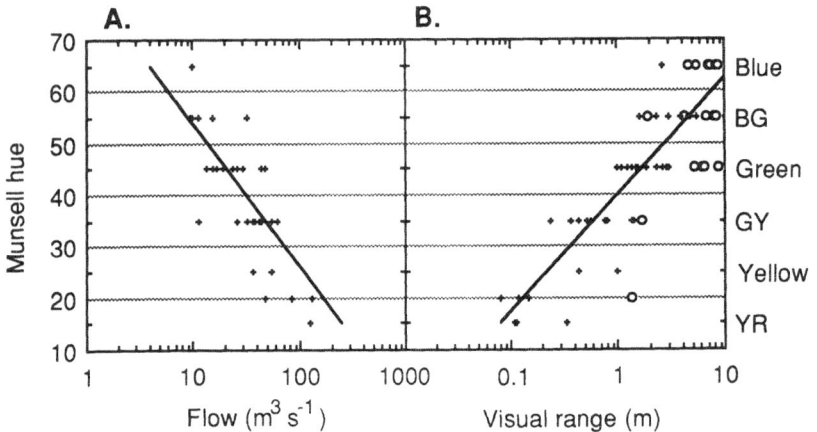

Fig. 6.18. Colour (hue) observations in the Whanganui River at Cherry Grove, Taumarunui, plotted versus (A) residual flow and (B) black disc range. Linear regression lines are shown ($r = 0.88$ in (A) and 0.85 in (B)). Also shown in (B) are colour versus clarity data for the diversion water (from Fig. 6.16A and 6.16B) (○).

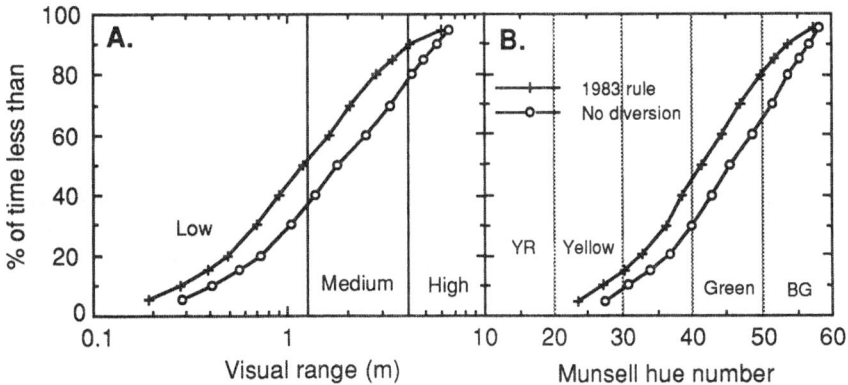

Fig. 6.19. Cumulative frequency distribution curves for clarity and colour in the Whanganui River at Cherry Grove, Taumarunui, in natural flows and with diversion according to the 1983 operating rule: (A) visual range of a black disc; (B) hue on the Munsell scale. Superimposed on the clarity distribution graph are boundaries of low–medium (1.2 m) and medium–high (4 m) clarity.

clarity and flow together with the known flow duration curves, using the method outlined in Section 4.3. Fig. 6.19A shows the resulting clarity distribution curves at Cherry Grove, Taumarunui. The median clarity of the residual river is about 1.19 m compared with about 1.79 m predicted for natural flows. The proportional decrease in clarity at this station is rather insensitive to state of flow, being about 33% over much of the range in flow.

To interpret further the impact on clarity, the clarity classification of Davies-Colley & Close (1990) is shown on Fig. 6.19A. This shows that the Whanganui River at Cherry Grove is not a particularly clear river at natural flow (although fairly 'average' as New Zealand rivers go), with clarity being 'low' ($y_{BD} < 1.2$ m) about 36% of the time and 'high' ($y_{BD} > 4.0$ m) 21% of the time. However, the diversions reduce the proportion of time the clarity is 'high' to 10% and 'low' clarity occurs 51% of the time.

Since colour (hue) in the Whanganui river at Cherry Grove is strongly related to visual clarity (Fig. 6.18B), the impact on clarity corresponds to an impact on the hue of the residual river (i.e. colours near the yellow end of the spectrum occur more frequently). Prediction of colour (hue) on a particular sampling occasion under different flow regimes is difficult because colour depends on spectral light attenuation, unlike clarity which, to a good approximation, depends only on visible waveband average light attenuation. To model colour under different flow regimes would require a balance on both absorption and scattering cross-sections across the spectrum. Fortunately, however, there is a simpler approach. This requires the assumption that the relationship between hue and clarity in residual water flow also applies to the diversion water. When the data for hue and clarity of the diversion water in Figs 6.16A and 6.16B is overlain on the plot for the residual river in Fig. 6.18B, the agreement is good, suggesting that the light-attenuating material in the Whanganui River system from diverse sources has rather similar optical properties.

With the assumption of a universal colour–clarity relationship in the upper Whanganui system, the distributions of hue given in Fig. 6.19B were calculated from the distributions of clarity in Fig. 6.19A. As expected, the hue is more frequently in the aesthetically preferred green-to-blue range in natural flows than when water is diverted. For example, under natural flows water colour is green to blue 70% of the time, but only 55% of the time when water is diverted. The change in colour averages about 4 Munsell hue units over most of the range of flow (Fig. 6.19B).

6.7.5 Discussion
The findings outlined above confirm what is obvious from casual observation: the waters diverted from the headwater streams of the Whanganui River are of much higher optical and visual quality than waters elsewhere in the Whanganui River system. Therefore, the diversion of these headwaters leaves the residual Whanganui River in a state of degraded optical quality, mainly because the diluting effect of the headwaters is no longer available. At Cherry Grove, Taumarunui, the 33% reduction of clarity over much of the flow range is outside the 20% maximum reduction in clarity guideline recommended in Section 5.3. The corresponding change in colour of about 4 Munsell hue units, while representing an undesirable impact, is probably acceptably small.

Further downstream in the Whanganui River we would expect the clarity-reducing effect of the diversions on the residual river to lessen as the diverted flow becomes a diminishing proportion of total flow. Data obtained at stations lower on the Whanganui River in New Zealand's National Water Quality Network (Smith & McBride 1990) bear out this expectation. For example, at Paetawa, the lowest monitoring station about 46 km from the river mouth, the average reduction in clarity is estimated to be 12%.

7

Future directions

7.1 OVERVIEW

In this book we have been concerned with an in-depth examination of the optical properties of water as they affect its use. We have examined the two main categories of water use that are affected by the water's optical characteristics: habitat for aquatic organisms and human recreational use. To enable the reader to gain a fuller understanding of some of the issues involved, we have addressed the physical concepts of aquatic optics and optical water quality (Chapter 2) and defined the separate attributes of clarity and colour. In Chapter 3 we discussed the measurement of optical properties of water, clarity and colour, and light-attenuating constituents, and we also briefly reviewed optical monitoring of water quality. Chapter 4 contrasted and compared clarity and colour in different aquatic environments: lakes, which are comparatively well studied, and rivers and estuaries, both of which are much less well characterized. Chapter 5 examined management of clarity and colour and recommended guideline values to protect optical water quality for a variety of water uses. It also considered sources of light-attenuating wastewaters, and the control of suspended solids from soil erosion and phytoplankton in lakes. Chapter 6 illustrated many of the concepts and approaches with a discussion of case studies.

Throughout the book we have been mindful of some shortcomings in the present state of scientific understanding and measurement technology. This brief chapter interprets where the information gaps lie and presents our recommendations for measurement and for further study of the optical properties of natural waters. For convenience what follows is in the same order as the text, and frequent cross-references are made to facilitate location of the full text.

7.2 CONCEPTS (CHAPTER 2)

The concepts of optical water quality and the underlying science of aquatic optics are not widely understood, let alone applied, by water managers. This book is a first attempt to redress this situation. Our hope is that, in future, optical water quality will be better appreciated and better managed than it has been in the past. The system of relationships (Fig. 2.24) between the water composition, the optical properties (both inherent and apparent) and the colour and clarity of water represents a powerful tool for the management of optical water quality, and warrants much wider adoption and application by water management professionals.

7.3 MEASUREMENT (CHAPTER 3)

7.3.1 Optical properties
The single most powerful optical measurement in water management is probably visual water clarity. Unfortunately, visual clarity, as such, cannot be measured continuously. However, beam transmittance, which is closely related to clarity, can be. Beam transmissometers are much used in oceanography, but have not been widely adopted for use in water management. Instruments suitable for spot measurement of the beam attenuation coefficient, c, would be very useful in water management applications (Section 3.3.1), particularly if suitable for continuous monitoring. An ideal instrument would have the following features: measurement of beam transmittance with negligible error due to unwanted detection of forward-scattered light, and low power consumption (preferably battery power) and compensation for window fouling (or window cleaning) so that remote monitoring for extended periods is feasible.

The absorption of light by filtered water samples is already measured routinely in water resources investigations, although usually only absorbances are reported rather than the (preferred) absorption coefficients. However, spectral absorption measurements and particulate matter absorption measurements are not often performed, possibly because of the lack of suitable instrumentation (a research-quality spectrophotometer with some means to collect most scattered light is required) and measurement complexity, but probably also because of lack of understanding of their significance (Section 3.3.2). More widespread measurement of absorption coefficients in water management is encouraged so that water colour can be related to water composition, and also to permit prediction of the colour of waters and spectral irradiance penetration.

Nephelometric turbidity measurements are common in water resources investigations, but these are not as appropriate for optical characterization of waters as measures of inherent or apparent optical properties in absolute physical units (Section 3.3.3). Transmittance measurements are to be preferred to nephelometric turbidity measurements, because turbidity measurements are highly instrument specific and very different values may be obtained from different nephelometers (despite identical calibration to formazin), because of differences in optical geometry, orientation, spectral power of the light source, and detector sensitivity. The sensible recommendations of Austin (1973) and McCluney (1975) for nephelometric measurements to be made by a fixed angle (preferably a forward angle, e.g. 15°) scattering meter which can be calibrated in absolute units of scattering have not yet been taken up (Section 3.3.3).

7.3.2 Water clarity
For the assessment of visual water clarity, we strongly recommend use of a black disc as a visual target (Section 3.4.2). Horizontal visibility measured with a black disc has the major advantage that it can be used in optically shallow waters including rivers. Other

virtues of the black disc technique in water resources work include its robust simplicity, ease of use and low cost. The horizontal black disc visibility is inversely proportional to the beam attenuation coefficient, which can thereby be calculated for the wavelength of peak sensitivity of the human eye (555 nm). If a vertical range is measured, this can be used, together with the horizontal range, to obtain a rough estimate of the illuminance attenuation coefficient, and thence the irradiance attenuation coefficient (Section 3.4.2). Widespread adoption of the black disc technique will lead to rapid advance in optical characterization of waters, especially rivers—which hitherto have been little studied (see Section 4.3.2, and Sections 5.3.2 and 5.3.6).

7.3.3 Water colour

We recommend that the platinum–cobalt colour measurement (Section 3.5.1) be discontinued. This variable bears little relationship to the appearance of natural waters, is of low precision for waters low in yellow substance (many natural waters), and blue–green and green waters are very difficult, or even impossible, to assess because of the difference in hue from the platinate standards. Measurement of platinum–cobalt colour by the visual comparison method should be replaced by measurement of absorption of yellow substance in filtered water samples at a standard wavelength (e.g. g_{440} (Kirk 1976)).

There is a need for more widespread measurement of spectral irradiance in water bodies, suitable for objective water colour specification (e.g. Smith *et al.* 1973; Davies-Colley *et al.* 1988) (Section 3.5.3) as well as for quantifying spectral attenuation of light (Section 3.4.3). However, it is recognized that the cost and complexity of the instrumentation (i.e. submersible spectroradiometers) will limit application in water management, except perhaps in special investigations of particularly valuable water bodies. Field matching of water hue to Munsell colour standards would seem to be a feasible technique for water colour specification (Section 3.5.4), although the field protocols need refinement.

7.3.4 Light-attenuating constituents

Suspended solids concentration is a traditional and very common water quality measurement (Section 3.6.2), but mass concentration of suspended matter is often not the main concern. We have observed that in many, if not most, situations the concern with suspended solids is actually with their optical effect, rather than being related directly to their mass. That is, the environmental degradation caused by suspensoids is linked to the attenuation of light by the suspended particles. Therefore, measurements of light attenuation by the suspensoids are often more appropriate than measurement of mass concentration. It follows that visual clarity or transmittance measurements, or even traditional turbidity measurement, could replace suspended solids measurement in many situations. Furthermore, these optical measurements, unlike suspended solids measurement, have the great advantage that they can be made *in situ*—and continuously in the case of the instrumental measurements. Water quality professionals should ask themselves, what is the real purpose of a given suspended solids mass concentration measurement, and could it be replaced by an optical measurement? Note, though, that measurement of suspended solids as an addendum to optical measurements is often valuable, since this permits optical cross-sections (optical coefficient divided by mass concentration) to be calculated—providing useful information on the character of the suspensoids (Section 3.6.2, Table 3.3).

Prediction of the light-attenuating contributions of phytoplankton in waters from indices of biomass, particularly chlorophyll *a*, is not very precise because the absorption and scattering cross-sections of different phytoplankters vary appreciably (Section 3.6.3). There is considerable information on the optical characteristics of marine species as a result of the work by Morel (1987) and coworkers, but much less on those of freshwater species. This may hamper management of optical quality in freshwaters, particularly lakes. Better data on absorption and scattering cross-sections of a greater number of species are desirable for the prediction of clarity changes in waters, particularly lakes, whose phytoplankton character may change (Section 3.6.3). Of course, this presupposes that the species composition of the future phytoplankton assemblage is predictable—which, in the absence of historical data, may be an intractable problem.

Detritus in natural waters is often an important light-absorbing and scattering constituent but its optical properties are poorly known and are expected to be rather variable (Section 3.6.4). Further work is required on optical characteristics of detrital particles and on means to distinguish the phytoplankton-related detritus (co-varying with chlorophyll *a*) from the detritus derived from other sources, including benthic plants.

7.4 DIFFERENT AQUATIC SYSTEMS (CHAPTER 4)

The optical characteristics of lakes have been generally well studied (Section 4.2), and optical knowledge of these waters is exceeded only by that of the deep sea. However, relatively little work has been carried out on the optical properties of rivers for a variety of reasons, including the fact that the water is often optically shallow, and there is typically much temporal variability, mainly with flow. As emphasized above (Section 7.3), we expect that progress on the optical characterization of rivers will be more rapid in future with the adoption of the black disc technique for direct measurement of visual clarity (Section 4.3.2), and probably also with increased use of beam transmissometers for measuring and monitoring water clarity.

If visual clarity of river waters has been little studied, light penetration (Section 4.3.3) has been almost totally neglected except in highly light-attenuating systems. It is possible that light limitation due to attenuation in the water column added to riparian shading, and self-attenuation by plants, could be more important for periphyton and macrophyte growth in streams than has been previously recognized. This warrants further study.

The optics of estuarine waters is also less well studied than the optics of lakewaters. In particular, light absorption and water colour in estuaries has been largely neglected. There is a need for better characterization of light absorption by various estuarine water constituents. More widespread use of *in situ* spectroradiometric measurements (Section 4.4.2) would contribute to a better understanding of light absorption and colour characteristics of estuaries.

The coagulation of river-derived particulates within estuaries affects their optical character. Further study of the optical ramifications of particle agglomeration in estuaries is required so that the effects of river loadings on estuarine optical properties can be predicted (Sections 4.4.1 and 4.4.3).

Comparison of the optical quality of different types of surface water environments in New Zealand has shown that light attenuation by phytoplankton is typically much more important in lakes than in estuaries, and is often almost negligible in rivers. In contrast, suspensoids other than phytoplankton dominate the optics of rivers and estuaries. The temporal variability of optical character also varies: lakes are less variable than either rivers (optical character varying with flow) or estuaries (varying with wind events and state of the tide). It would be interesting to know whether these patterns, recognized in New Zealand, occur globally.

The optics of spring waters deserves renewed research attention with modern optical instrumentation and methods. The clarity of many coldwater springs is remarkable, as is their pure water hue, and these optical attributes may be important considerations in their conservation and management. Geothermal springwaters, whilst typically turbid with colloidal silica and other fine precipitates, often display unusual and fascinating colours which contribute to their scenic appeal. Again, these optical attributes may be important management and conservation issues.

7.5 MANAGEMENT (CHAPTER 5)

The field of human perception of natural waters is a very neglected research area, possibly because it falls uncomfortably at the boundary of natural and social sciences. The paucity of relevant research is reflected in the fact that there are few well-founded guidelines for protection of the visual quality of waters for recreation and aesthetic purposes. In our opinion the general area of public perception of water quality, and of optical water quality in particular, deserves much greater research effort than it has received hitherto.

Likewise, the optical water quality requirements of aquatic fauna are, in general, poorly known, and again the lack of information is reflected in the lack of water management guidelines. For instance, although some information is available on the physiological aspects of the vision of aquatic organisms (Lythgoe 1979), little useful information appears to be available on their general 'optical habitat' requirements (i.e. visual clarity, illuminance, spectral quality of the light field). The whole area of sensitivity of aquatic animals to aspects of the optical character of water bodies requires much more research effort so that water managers will have the tools required to protect optical habitat.

In our experience, the two distinct aspects of water clarity, namely visual clarity and light penetration, are often confused or not distinguished by both scientists and water managers. In water management generally, and, in particular, when developing guidelines and standards for water clarity, it is important that these two aspects of clarity are kept distinct so that they can be separately protected.

7.5.1 Guidelines

For bathing waters a generally accepted safety philosophy is that the substrate (a low contrast target) should be visible. Over more than two decades, a common requirement has been that a Secchi disc (a high contrast target) should be visible on the bottom, or that it should be visible at 1.2 m. However, these requirements appear never to have been justified and recent studies by the authors indicate that a greater visual clarity is required—around 1.6 m horizontal black disc visibility, or 2.0 m Secchi depth. In addition little work has been carried out on the required clarity–depth relationships to ensure substrate visibility. This topic of clarity requirements for safe and aesthetically enjoyable contact recreation requires further study (Sections 5.2.1 and 5.3.2).

Likewise, little work has been conducted on the threshold for human detection of a reduction of clarity. Stipulations on not-to-be-exceeded maximum clarity (or corresponding turbidity) change often appear in water quality guidelines and standards documents, but their empirical basis is lacking (Sections 5.2.1 and 5.3.2). Based on admittedly limited research, we have recommended a 20% maximum reduction in visual clarity, but research to refine this guideline is desirable.

We are unaware of clarity criteria to protect visual requirements of aquatic fauna: this should be addressed so that defensible standards can be developed to protect sensitive, desirable species and aquatic ecosystems in general (Section 5.3.2).

To protect light penetration for photosynthesis, we have recommended a maximum relative change in euphotic depth of 10%, echoing an earlier recommendation in terms of the less-easily measured compensation depth (Section 5.3.3). However, we cannot at present justify this particular numerical value. Further work is therefore required to examine the sensitivity of aquatic plant communities, and, indeed, aquatic ecosystems generally, to reduced light penetration.

Further studies are desirable on the relationship of suitability for use of water to its hue (Section 5.3.4). The interaction of hue and other optical characteristics of water may confound such relationships. For example, Smith & Davies-Colley (1992) speculated that people's response to hue may be affected strongly by other visual attributes of water such as its brightness and visual clarity. Additionally, we have assumed that water colour requirements for bathing and aesthetics can be considered under one heading only, i.e. aesthetics. That is, the response of bathers to water colour is dictated by their aesthetic sense. Studies are desirable to examine this assumption.

In the absence of any scientific criteria, we have assumed that limiting maximum hue change in order to protect aesthetics will also protect the quality of the light field underwater as it may affect animals and plants. Studies of the sensitivity of aquatic biota to changes in light spectral quality are desirable (Section 5.3.4). Possibly some aquatic organisms may be more sensitive to change in spectral quality of the light field in waters than the human aesthetic sense.

7.5.2 Wastewaters
There is little information available on the optical characteristics of wastewaters and urban stormwaters, or their optical impact on receiving waters (Section 5.4). Waste treatment engineers and stewards of receiving water quality appear to have largely ignored such matters and devoted their attention in the main to oxygen-demanding materials, total solids concentration, pathogens, nutrients and toxic materials. Therefore there is little literature to draw upon when attempting to predict the optical character of a water receiving a wastewater discharge from a given industry. Quantitative data are needed by managers so that they can perform calculations in setting effluent standards (Appendix 3).

The optical properties of stormwaters are highly variable, both within and between catchments, and the optical characteristics of municipal and industrial wastewaters may be site specific and temporally variable. We have presented some data to give the reader an impression of the optical characteristics of some wastewaters (Section 5.4), but our data are often from only one (hopefully representative) effluent and are sometimes based on single grab samples. More information is needed on the optical character of wastewaters generally, and on strongly light-attenuating effluents in particular.

7.5.3 Soil erosion control

Most publications on erosion and sediment control emphasize suspended solids (SS) concentrations or catchment yields rather than downstream optical water quality. The optical impact of suspensoids is often not addressed by water managers or soil conservators, despite the fact that it is usually the most obvious feature of these constituents of waters. In particular, the significance of the fine-grained fraction does not always receive due recognition. Research emphasizing the fine-grained 'wash' load is required in order to develop control measures and guidelines.

Techniques exist for reducing mobilization of suspended matter in overland flow, but they tend to be more efficient for coarse particles than the strongly light-attenuating fine-grained particles (Section 5.5). From an optical water quality perspective (i.e. in order to prevent muddied waters) it is the fine material which is of most importance. Research is needed on techniques for reducing mobilization of the fine fraction of soils subject to erosion. The efficacy of erosion control measures needs to be measured in terms of the size distribution of solids yielded (or the yield of attenuation cross-section) as well as mass yields.

Once fine particles are mobilized, their removal is often difficult. Physicochemical treatment of storm waters with coagulants or flocculants is efficient for fines removal, but is generally only feasible in certain cases where storm flows can be contained and controlled, such as with some mining operations. However, 'natural' filtration by wetland vegetation or riparian cover of suspended matter eroded from soils has the potential for major optical quality benefits and deserves more research effort from this perspective.

7.5.4 Phytoplankton control
At present it is difficult to predict the optical quality improvements that will occur in a particular lake with a given reduction in phytoplankton biomass following control measures such as nutrient load reduction (Section 5.6). As we have seen, a major reason is that the absorption and scattering by a given phytoplankton biomass (as chlorophyll a) can vary widely. Eutrophication researchers need to give more attention to optical water quality aspects. Better information on optical cross-sections of phytoplankton, and phytoplankton-derived detritus, would be helpful in practical lake management, together with better predictability of the composition of the phytoplankton.

7.6 CONCLUSION

To conclude, we recommend that, in future, water managers should give greater consideration to the optical quality of waters. Once a reasonable level of familiarity with the concepts, methods of measurement, characteristics of particular water bodies, and guidelines regarding optical quality of waters has been achieved, a sharp increase in the level of activity in this under-studied area of water science, and water management, is to be expected. Our main purpose in writing this book will have been achieved if it has succeeded in encouraging a wider appreciation of, and concern for, the colour and clarity of natural waters.

References

Addis, M. C.; Simmons, I. G.; Smart, P. L. 1984: The environmental impact of an open cast operation in the forest of Dean, England. *Journal of environmental management 19*: 79–95.

AFS 1976. A review of the EPA Red Book: Quality Criteria for Water. American Fisheries Society, Bethesda, Maryland. 313 p.

Alpine, A. E.; Cloern, J. E. 1988: Phytoplankton growth rates in a light-limited environment, San Francisco Bay. *Marine ecology progress series 44*: 167–173.

Anderson, B.; Potts, D. F. 1987: Suspended sediment and turbidity following road construction and logging in Western Montana. *Water resources bulletin 23*: 681–690.

Anon 1966: The science of color. The Committee on Colorimetry of the Optical Society of America. Optical Society of America, Washington, District of Columbia. 385 p.

Anon 1979: The instrumental determination of total organic carbon, total oxygen demand and related determinands. Methods for the Examination of Waters and Associated Materials. Standing Committee of Analysts, Department of the Environment, HMSO, London. 24 p.

Anon 1984: Colour and turbidity of waters. 1981 tentative methods. Standing Committee of Analysts, Department of the Environment, HMSO, London. 25 p.

Anon 1987. Canadian water quality guidelines. Inland Waters Directorate, Environment Canada, Ottawa, Ontario.

Anon 1990: Levin scheme wins IPENZ Environmental Award. *New Zealand engineering 45*: 23–26.

ANZECC 1992: Australian water quality guidelines for fresh and marine waters. Australian and New Zealand Environment and Conservation Council, Canberra, ACT, Australia.

APHA 1989: Standard methods for the examination of water and wastewater, 17th Ed. America Public Health Association, American Water Works Association, Water Pollution Control Federation, Washington, District of Columbia.

ASCE 1975: Sedimentation engineering. Vanoni, V.A. (*ed.*) Manual No. 54. American Society of Civil Engineers, New York. 745 p.

ASTM 1980: Specifying color by the Munsell system. American Society for Testing and Materials. Standard Method D1535-68.

Atlas, D.; Bannister, T. T. 1980: Dependence of mean spectral extinction coefficient of phytoplankton on depth, water color, and species. *Limnology and oceanography 25*: 157–159.

Austin, R. W. 1973: Problems in measuring turbidity as a water quality parameter. USEPA seminar on methodology for monitoring the marine environment. Seattle, Washington, October 16–18, 1973.

Austin, R. W. 1974: The remote sensing of spectral radiance from below the ocean surface. Chapter 14 *In* Jerlov, N. G.; Steeman-Nielsen, E. (*eds*) Optical aspects of oceanography. Academic Press, London–New York. pp. 317–344.

Avnimelech, Y.; Troeger, B. W.; Reed, L. W. 1982: Mutual flocculation of algae and clay: evidence and implications. *Science 216*: 63–65.

Baker, K. S.; Smith, R. C. 1979: Quasi-inherent characteristics of the diffuse attenuation coefficient for irradiance. *Proceedings of the Society of Photo-optical Instrumentation Engineers 208*: 60–63.

Baker, K. S.; Smith, R. C. 1982: Bio-optical classification and model of natural waters. 2. *Limnology and oceanography 27*: 500–509.

Bannister, T. T. 1974: A general theory of steady state phytoplankton growth in a nutrient saturated mixed layer. *Limnology and oceanography 19*: 13–30.

Banoub, M. W. 1973: Ultraviolet absorption as a measure of organic matter in natural waters in Bodensee. *Archiv für Hydrobiologie 71*: 159–165.

Barfield, B. J.; Warner, R. C.; Haan, C. T. 1981: Applied hydrology and sedimentology of disturbed areas. Oklahoma Technical Press, Stillwater, Oklahoma. 603 p.

Bartz, R.; Zanevald, J. R. V.; Pak, H. 1978: A transmissometer for profiling and moored observations in water. *Society of Photo-optical Instrumentation Engineers, Ocean optics 5*: 102–108.

Beasley, R. P. 1972: Erosion and sediment pollution control. Iowa State University Press, Ames, Iowa. 320 p.

Beschta, R. L. 1978: Long-term patterns of sediment production following road construction and logging in the Oregon Coast Range. *Water resources research 14*: 1011–1016.

Beschta, R. L. 1979: Debris removal and its effects on sedimentation in an Oregon Coast Range stream. *Northwest science 53*: 71–77.

Biggs, B. J. B.; Davies-Colley, R. J. 1990: Optical properties of Lake Coleridge: the impacts of turbid inflows. *New Zealand journal of marine and freshwater research 24*: 441–451.

Biggs, B. J. B.; Duncan, M. J.; Jowett, I. G.; Quinn, J. M.; Hickey, C. W.; Davies-Colley, R. J.; Close, M. E. 1990: Ecological characterisation, classification and modelling of New Zealand's rivers: an introduction and synthesis. *New Zealand journal of marine and freshwater research 24*: 277–304.

Bilby, R. E.; Sullivan, K.; Duncan, S. H. 1989: The generation and fate of road-surface sediment in forested watersheds in Southwestern Washington. *Forest science 35*: 453–468.

Bioresearches 1987: The effects of effluent discharged from the Manukau Sewage Purification Works on the water quality and natural environment of the Manukau

Harbour: Review of existing data. Bioresearches, Auckland. 192 p.

Birge, E. A.; Juday, C. 1929: Transmission of solar radiation by the waters of the inland lakes. *Transactions of the Wisconsin Academy of Science, Arts and Letters 24*: 509–580.

Birge, E. A.; Juday, C. 1932: Solar radiation and inland lakes, fourth report. Observations of 1931. *Transactions of the Wisconsin Academy of Science, Arts and Letters 27*: 523–562.

Blackwell, H. R. 1946: Contrast thresholds of the human eye. *Journal of the Optical Society of America 36*: 624–643.

Bordas, M. P.; Walling, D. E. (*eds*) 1988: Sediment budgets. Proceedings of a symposium, Porto Alegre, Brazil. *IAHS Publication No 174*.

Bowling, L. C.; Steane, M. S.; Tyler, P. A. 1986: The spectral distribution and attenuation of underwater irradiance in Tasmanian inland waters. *Freshwater biology 16*: 313–335.

Brezonik, P.L. 1978: Effect of organic colour and turbidity on Secchi disc transparency. *Journal of the Fisheries Research Board, Canada 35*: 1410–1416.

Bricaud, A.; Stramski, D. 1990: Spectral absorption coefficients of living phytoplankton and nonalgal biogeneous matter: A comparison between the Peru upwelling area and the Sargasso Sea. *Limnology and oceanography 35*: 562–582.

Bricaud, A.; Morel, A.; Prieur, L. 1981: Absorption by dissolved organic matter of the sea (yellow substance) in the UV and visible domains. *Limnology and oceanography 26*: 43–53.

Bricaud, A.; Morel, A.; Prieur, L. 1983: Optical efficiency factors for some phytoplankters. *Limnology and oceanography 28*: 816–832.

Briers, M. G.; Wheeler, M. A. 1986: Aztec single stream colour monitor: an initial evaluation. WRc Environment, Medmenham Laboratory. 10 p.

Briggs, R.; Schofield, J. W.; Gordon, P. A. 1976: Instrumental methods for monitoring organic pollution. *Water pollution control 1976*: 47–57.

Bristow, M.; Nielsen, D.; Bundy, D.; Furtek, R. 1981: Use of water Raman emission to correct airborne laser fluorosensor data for effects of water optical attenuation. *Applied optics 20*: 2889–2906.

Brown, G. W.; Krygier, J. T. 1971: Clear-cut logging and sediment production in the Oregon Coast Range. *Water resources research 7*: 1189–1198.

Bukata, R. P.; Jerome, J. H.; Bruton, J. E.; Jain, S. C.; Zwick, H. H. 1981a: Optical water quality model of lake Ontario. 1. Determination of the optical cross-sections of organic and inorganic particulates in Lake Ontario. *Applied optics 20*: 1696–1703.

Bukata, R. P.; Bruton, J. E.; Jerome, J. A.; Jain, S. C.; Zwick, H. H. 1981b: Optical water quality model of Lake Ontario. 2. Determination of chlorophyll *a* and suspended mineral concentrations of natural waters from submersible and low altitude optical sensors. *Applied optics 20*: 1704–1714.

Bukata, R.P.; Jerome, J.H.; Bruton, J.E. 1981c: Validation of a five-component optical model for estimating chlorophyll *a* and suspended mineral concentrations in Lake Ontario. *Applied optics 20*: 3472–3474.

Bukata, R.P.; Bruton, J.E.; Jerome, J.H. 1983: Use of chromaticity in remote measurements of water quality. *Remote sensing of environment 13*: 161–177.

Cambridge, M. L.; McComb, A. J. 1984: The loss of seagrasses in Cockburn Sound,

Western Australia. I. The time course and magnitude of sea-grass declines in relation to industrial development. *Aquatic botany 20*: 229–243.

Campbell, D. E.; Spinrad, R. W. 1987: The relationship between light attenuation and particle characteristics in a turbid estuary. *Estuarine, coastal and shelf science 25*: 53–65.

Campbell, I. C.; Doeg, T. J. 1989: Impact of timber harvesting and production on streams: A review. *Australian journal of marine and freshwater research 40*: 519–539.

Canfield, D. E.; Langeland, K. A.; Linda, S. B.; Haller, W. T. 1985: Relations between water transparency and maximum depth of macrophyte colonization in lakes. *Journal of aquatic plant management 23*: 25–28.

Carder, K. L.; Steward, R. G.; Harvey, G. R.; Ortner, P. B. 1989: Marine humic and fulvic acids: Their effects on remote sensing of ocean chlorophyll. *Limnology and oceanography 24*: 68–81.

Carlson, R. E. 1977: A trophic state index for lakes. *Limnology and oceanography 22*: 361–369.

Cassie, U. V. 1975: Phytoplankton of Lakes Rotorua and Rotoiti (North Island). *In* Jolly, V. H.; Brown, J. M. A. (*eds*) New Zealand Lakes. Auckland University Press/ Oxford University Press, Auckland. pp. 193–205.

Chambers, P. A.; Kalff, J. 1985: Depth distribution and biomass of submersed aquatic macrophyte communities in relation to Secchi depth. *Canadian journal of fisheries and aquatic sciences 42*: 701–709.

Champ, M. A.; Gould, G. A.; Bozzo, W. E.; Ackleson, S. G.; Vierra, K. C. 1980: Characterization of light extinction and attenuation in Chesapeake Bay, August, 1977. *In* Kennedy, V. S. (*ed.*) Estuarine perspectives. Academic Press, New York. pp. 263–277.

Chapman, M. A. 1981: The summer limnology of Lake Waahi, New Zealand. *In* Barica, J.; Mur, L. R. (*eds*) Developments in hydrobiology, Vol 2. W. Junk, The Hague.

Chase, R. R. P. 1979: Settling behaviour of natural aquatic particulates. *Limnology and oceanography 24*: 417–426.

CIE 1957: Vocabulaire international de l'Eclairage, 2nd Ed. Commission International de l'Eclairage, Paris. 136 p.

Clark, E. H.; Haverkamp, J. A.; Chapman, W. 1985: Eroding soils: The off-farm impacts. The Conservation Foundation, Washington, District of Columbia. 252 p.

Clarke, R. T.; Marker, A. F. H.; Rother, J. A. 1987: The estimation of the mean and variance of algal cell volume from critical measurements. *Freshwater biology 17*: 117–128.

Clayton, J. S. 1983: Sampling aquatic macrophyte communities. *In* Biggs, B. J.; Gifford, J. S.; Smith, D. G. (*eds*) Biological methods for water quality surveys. *Water and soil miscellaneous publication No. 54*. Ministry of Works and Development, Wellington.

Cloern, J. E. 1987: Turbidity as a control of phytoplankton biomass and productivity in estuaries. *Continental shelf research 7*: 1367–1381.

Cloern, J. E.; Powell, T. M.; Huzzey, L. M. 1989: Spatial and temporal variability in South San Francisco Bay (USA). II. Temporal changes in salinity, suspended sediments, and phytoplankton biomass and productivity over tidal time scales. *Estuarine, coastal and shelf science 28*: 599–613.

Coffey, B. T.; Kar-Wah, C. 1988: Pressure inhibition of anchor root production in *L. major* (Ridl.) Moss: A possible determinant of its depth range. *Aquatic botany 29*: 289–301.

Colijn, F. 1982: Light absorption in the waters of the Ems-Dollard Estuary and its consequences for the growth of phytoplankton and microphytobenthos. *Netherlands journal of sea research 15*: 196–216.

Collier, K. J. 1987: Spectrophotometric determination of dissolved organic carbon in some South Island streams and rivers. *New Zealand journal of marine and freshwater research 21*: 349–351.

Cooke, G. D.; Welch, E. B.; Peterson, S. A.; Newroth, P. R. 1986: Lake and reservoir restoration. Butterworth Publishers, Boston, Massachusetts. 392 p.

Cooke, J. G.; Cooper, A. B. 1988: Sources and sinks of nutrients in a New Zealand hill pasture catchment. III. Nitrogen. *Hydrological processes 2*: 135–149.

Cooper, A. B. 1987: Nitrogen removal by denitrification in land treatment systems: implications for Rotorua. *In* New Zealand Water Supply and Disposal Association, Proceedings of 1987 annual conference. NZWSDA, Hamilton. pp. 278–294.

Cordery, I. 1977: Quality characteristics of urban storm-water in Sydney, Australia. *Water resources research 13*: 197–202.

Davies-Colley, R. J. 1979: River appearance. *In* Strachan, C. (*ed.*) The Waikato River, a water resources study. *Water and Soil technical publication 11*: 113–125. Ministry of Works and Development, Wellington.

Davies-Colley, R. J. 1983: Optical properties and reflectance spectra of 3 shallow lakes obtained from a spectrophotometric study. *New Zealand journal of marine and freshwater research 17*: 445–459.

Davies-Colley, R. J. 1984: Summary Report on water appearance studies — Tarawera River. *In* Tarawera River Management Plan. Appendix D5. Whakatane, Bay of Plenty Catchment Commission and Beca Carter Hollings and Ferner Ltd.

Davies-Colley, R. J. 1987a: Water appearance: survey and interpretation. *In* Vant, W. N. (*ed.*), Lake managers handbook. *Water and Soil miscellaneous publication 103*. Ministry of Works and Development, Wellington. pp. 66–78.

Davies-Colley, R. J. 1987b: Optical properties of the Waikato River, New Zealand. *Mitteilungen aus dem Geologisch-Paläontologischen Institut der Universität Hamburg, SCOPE/UNEP Sonderband, 64*: 443–460.

Davies-Colley, R. J. 1988a: Measuring water clarity with a black disc. *Limnology and oceanography 33*: 616–623.

Davies-Colley, R. J. 1988b: Mixing depths in New Zealand lakes. *New Zealand journal of marine and freshwater research 22*: 517–527.

Davies-Colley, R. J. 1990a: Frequency distributions of visual water clarity in 12 New Zealand rivers. *New Zealand journal of marine and freshwater research 24*: 453–460.

Davies-Colley, R. J. 1990b: Colour and clarity studies on the Tarawera River. New methods applied to an old problem. Water Quality Centre consulting report 7138/1 for the Bay of Plenty Regional Council. 15 p.

Davies-Colley, R. J. 1990c: Visual quality of the Whanganui River. The effects of headwater diversions to the Tongariro Power Scheme. Water Quality Centre Consulting Report 8011/2 to the Department of Conservation, Wanganui, New Zealand. 36 p.

Davies-Colley, R. J. 1992: Yellow substance in coastal and marine waters round the

South Island, New Zealand. *New Zealand journal of marine and freshwater research* 26: 311–322.

Davies-Colley, R. J.; Close, M. E. 1990: Water colour and clarity of New Zealand rivers under baseflow conditions. *New Zealand journal of marine and freshwater research* 24: 357–365.

Davies-Colley, R. J.; Smith, D. G. 1990: A panel study of the detectability of change in turbidity of water induced by discharge of inorganic suspensoids to a small stream. Water Quality Centre Publication No. 17, Water Quality Centre, DSIR, Hamilton, New Zealand.

Davies-Colley, R. J.; Smith, D. G. 1992: Offsite measurement of the visual clarity of waters. *Water resources bulletin 28*: 1–7.

Davies-Colley, R. J.; Vant, W. N. 1987: Absorption of light by yellow substance in freshwater lakes. *Limnology and oceanography 32*: 416–425.

Davies-Colley, R. J.; Vant, W. N. 1988: Estimation of optical properties of water from Secchi disc depths. *Water resources bulletin 24*: 1329–1335.

Davies-Colley, R. J.; Vant, W. N.; Latimer, G. J. 1984: Optical characterisation of natural waters by PAR measurement under changeable light conditions. *New Zealand journal of marine freshwater research 18*: 455–460.

Davies-Colley, R. J.; Pridmore, R. D.; Hewitt, J. E. 1986: Optical properties of some freshwater phytoplanktonic algae. *Hydrobiologia 133*: 165–178.

Davies-Colley, R. J.; Vant, W. N.; Wilcock, R. J. 1988: Lake water color: Comparison of direct observations with underwater spectral irradiance. *Water resources bulletin 24*: 11–18.

Davies-Colley, R. J.; Quinn, J. M.; Hickey, C. W.; Ryan, P. A. 1992: Effects of clay discharges on streams 1. Optical properties and epilithon. *Hydrobiologia 248*: 215–234.

De Haan, H.; De Boer, T.; Kramer, H. A.; Voerman, J. 1982: Applicability of light absorbance as a measure of organic carbon in humic lake water. *Water research 16*: 1047–1050.

Di Toro, D. M. 1978: Optics of turbid estuarine waters: Approximations and applications. *Water research 12*: 1059–1068.

Diatloff, G.; Anderson, T. 1984: A new safer copper algicide for ponds and dams. *In* Madin, R. W. (*ed.*) Proceedings of the seventh Australasian Weeds Conference. Weed Society of Western Australia, Perth. pp. 295–298.

Dobbs, R. A.; Wise, R. H.; Dean, R. B. 1972: The use of ultraviolet absorbance for monitoring the total organic carbon content of water and wastewater. *Water research 6*: 1173–1180.

Drew, E. A. 1979: Physiological aspects of primary production in seagrasses. *Aquatic botany 7*: 139–150.

Dring, M. J. 1990: Light harvesting and pigment composition in marine phytoplankton and macroalgae. Chapter 5 *In* Herring, P. J.; Cambell, A. K.; Whitfield, M.; Maddock, L. (*eds*) Light and life in the sea. Cambridge University Press. 357 p.

Droppo, I. G.; Ongley, E. D. 1992: The state of suspended sediment in the freshwater fluvial environment: A method of analysis. *Water research 26*: 65–72.

Duarte, C. M.; Kalff, J. 1989: The influence of catchment geology and lake depth on phytoplankton biomass. *Archiv für hydrobiologie 115*: 27–40.

Dubinsky, Z.; Berman, T. 1979: Seasonal changes in the spectral composition of downwelling irradiance in Lake Kinneret (Israel). *Limnology and oceanography 24*: 652–663.

Duncan, M. J. 1987: River hydrology and sediment transport. *In* Viner, A. B. (*ed.*) Inland waters of New Zealand. *DSIR Bulletin 241*, Wellington. pp. 113–137.

Dunne, T.; Leopold, L. B. 1978: Water in environmental planning. Freeman and Co., San Francisco, California. 818 p.

Duntley, S. Q. 1963: Light in the sea. *Journal of the Optical Society of America 53*: 214–233.

Duysens, L. M. 1956: The flattening of the absorption spectra of suspensions compared to that of solutions. *Biochemica et biophysica acta 19*: 1–12.

Dyer, K. R. 1973: Estuaries: a physical introduction. Wiley, London. 140 p.

Dyer, K. R. 1986: Coastal and estuarine sediment dynamics. Wiley-Interscience, Chichester. 342 p.

Dyer, K.R. 1988: Fine sediment particle transport in estuaries. *In* Dronkers, J.; van Leussen, W. (*eds*) Physical processes in estuaries. Springer-Verlag, Berlin. pp. 295–310.

EEC 1975. Council Directive 75/440/EEC, *Official Journal*, No. L 194/26.

EEC 1976. Council Directive 76/160/EEC, *Official Journal*, No. L 31/1.

EEC 1978. Council Directive 78/659/EEC, *Official Journal*, No. L 222/1.

EEC 1979. Council Directive 79/923/EEC, *Official Journal*, No. L 281/47.

EEC 1980. Council Directive 80/778/EEC, *Official Journal*, No. L 229/11.

Effler, S. W. 1985: Attenuation versus transparency. *Journal of environmental engineering 111*: 448–459.

Effler, S. W. 1988: Secchi disc transparency and turbidity. *Journal of environmental engineering 114*: 1436–1447.

Effler, S. W.; Schafran, G. C.; Driscoll, C. T. 1985: Partitioning light attenuation in an acidic lake. *Canadian journal of fisheries and aquatic sciences 42*: 1707–1711.

Elmore, W.; Beschta, R. L. 1987: Riparian areas: perspectives in management. *Rangelands 9*: 260–265.

EPAV 1983: Recommended Water Quality Criteria. EPA of Victoria Publication No. 165. East Melbourne, Victoria. 247 p.

Faust, M. A.; Norris, K. H. 1985: *In vivo* spectrophotometric analysis of photosynthetic pigments in natural populations of phytoplankton. *Limnology and oceanography 30*: 1316–1322.

Fischer, H. B.; List, E. J.; Koh, R. C. Y.; Imberger, J.; Brooks, N. H. 1979: Mixing in inland and coastal waters. Academic Press, New York. 483 p.

Fish, G. R. 1975: Lakes Rotorua and Rotoiti, North Island, New Zealand: Their trophic status and studies for a nutrient budget. New Zealand Ministry of Agriculture and Fisheries. *Fisheries research bulletin 8.*

Flint, E. A. 1975: Phytoplankton in New Zealand lakes. *In* Jolly, V. H.; Brown, J. M. A. (*eds*) New Zealand lakes. Auckland University Press/Oxford University Press, Auckland. pp. 163–192.

Foster, P.; Morris, A. W. 1974: Ultraviolet absorption characteristics of natural waters. *Water research 8*: 137–142.

Gabrielson, J. O.; Lukatelich, R. J. 1985: Wind-related resuspension of sediments in the Peel-Harvey estuarine system. *Estuarine, coastal and shelf science 20*: 135–145.

Gallegos, C. L.; Correll, D. L.; Pierce, J. W. 1990: Modeling spectral diffuse attenuation, absorption, and scattering coefficients in a turbid estuary. *Limnology and oceanography 35*: 1486–1502.

Gallene, B. 1974: Study of fine material in suspension in the estuary of the Loire and its dynamic grading. *Estuarine, coastal and shelf science 2*: 261–272.

Ganf, G. G.; Oliver, R. L.; Walsby, A. E. 1989: Optical properties of gas vacuolate cells and colonies of *Microcystis* in relation to light attenuation in a turbid, stratified reservoir (Mount Bold Reservoir, South Australia). *Australian journal of marine and freshwater research 40*: 595–611.

Gellman, I. 1982: Kraft effluent decolourization — a program for assessment of its need, technological capability and general consequences. National Council on Air and Stream Improvement (NCASI) of the paper industry, Special Report, 2nd Ed.

Gerbeaux, P.; Ward, J. C. 1986: The disappearance of macrophytes and its importance in the management of shallow lakes in New Zealand. *EWRS International Symposium on Aquatic Weeds 7*: 119–124.

Gerbeaux, P.; Ward, J. C. 1991: Factors affecting water clarity in Lake Ellesmere, New Zealand. *New Zealand journal of marine and freshwater research 25*: 289–296.

Gibbs, R. J. 1978: Light scattering from particles of different shapes. *Journal of geophysical research 83 C1*: 501–502.

Gieskes, W. W. C. 1987: Secchi disc visibility world record shattered. *EOS transactions of the American Geophysical Union 68*: 123.

Gjessing, E. T. 1976: Physical and chemical characteristics of aquatic humus. Ann Arbor Science. 120 p.

Gloyna, E. F.; Tischler, L. F. 1980: Recommendations for regulatory modifications: the use of waste stabilization pond systems. *Journal of the Water Pollution Control Federation 53*: 1559–1563.

Goldman, C. R. 1988: Primary productivity, nutrients and transparency during the early onset of eutrophication in ultra-oligotrophic Lake Tahoe, California-Nevada. *Limnology and oceanography 33*: 1321–1333.

Goldman, S. J.; Jackson, K.; Bursztynsky, T. A. 1986: Erosion and sediment control handbook. McGraw-Hill Inc.

Gordon, H. R. 1989: Can the Lambert-Beer law be applied to the diffuse attenuation coefficient of ocean water? *Limnology and oceanography 34*: 1389–1409.

Gordon, H. R.; Morel, A. Y. 1983: Remote assessment of ocean colour for interpretation of satellite visible imagery. A review. *Lecture notes in Coastal and Estuarine Studies 4*. Springer-Verlag, New York. 114 p.

Gordon, H. R.; Wouters, A. W. 1978: Some relationships between Secchi depth and inherent optical properties of natural waters. *Applied optics 17*: 3341–3343.

Gordon, H. R.; Brown, O. B.; Jacobs, M. M. 1975: Computed relationships between the inherent and apparent optical properties of a flat, homogeneous ocean. *Applied optics 14*: 417–427.

Graham, A. A. 1990: Siltation of stone-surface periphyton in rivers by clay-sized particles from low concentrations in suspension. *Hydrobiologia 199*: 107–115.

Graynoth, E. 1979: Effects of logging on stream environments and faunas in Nelson. *New Zealand journal of marine and freshwater research 13*: 79–109.

Green, J. D. 1975: Light penetration. *In* Jolly, V. H.; Brown, J. M. A. (*eds*) New Zealand lakes. Auckland University Press/ Oxford University Press. pp. 84–89.

Griffiths, G. A.; Glasby, G. B. 1985: Input of river-derived sediment to the New Zealand continental shelf: I. Mass. *Estuarine, coastal and shelf science 21*: 773–787.

Grobbelaar, J. U. 1985: Phytoplankton productivity in turbid waters. *Journal of plankton research 7*: 653–663.

Harm, W. 1980: Biological effects of ultraviolet radiation. International Union of Pure and Applied Biophysics, Biophysics Series. Cambridge. 216 p.

Hazen, A. 1892: A new colour standard for natural waters. *American chemical journal 14*: 300.

Herbich, J. B.; Brahme, S. B. 1991: Literature review and technical evaluation of sediment resuspension during dredging. Contract Report HL-91-1, US Army Corps of Engineers Waterways Experiment Station, Vicksburg, Mississippi.

Hewlett, J. D. 1961: Soil moisture as a source of baseflow from steep mountain watersheds. US Forest Service, *South-eastern Forest Experimental Station, Paper No. 132*.

Hewlett, J. D.; Nutter, W. L. 1970: The varying source area of streamflow from upland basins. *In* Symposium on interdisciplinary aspects of watershed management. The American Society of Civil Engineers. pp. 65–83.

Hickey, C. W.; Quinn, J. M.; Davies-Colley, R. J. 1989a: Effluent characteristics of dairy shed oxidation ponds and their potential impacts on rivers. *New Zealand journal of marine and freshwater research 23*: 569–584.

Hickey, C. W.; Quinn, J. M.; Davies-Colley, R. J. 1989b: Effluent characteristics of domestic sewage oxidation ponds and their potential impacts on rivers. *New Zealand journal of marine and freshwater research 23*: 585–600.

Hilton, J. 1984: Airborne remote sensing for freshwater and estuarine monitoring. *Water research 18*: 1195–1223.

Hoare, R. A. 1980: The sensitivity to phosphorus and nitrogen loads of Lake Rotorua, New Zealand. *Progress in water technology 12*: 897–904.

Hoare, R. A.; Spigel, R. H. 1987: Water balances, mechanics and thermal properties. *In* Vant, W.N. (*ed.*) Lake managers handbook. *Water and Soil miscellaneous publication 103*. Ministry of Works and Development, Wellington. pp. 41–58.

Højerslev, N. K. 1986: Visibility of the sea with special reference to the Secchi disc. *Proceedings of the Society of Photo-optical Instrumentation Engineers 637*: 294–305.

Howard-Williams, C. 1987: In-lake control of eutrophication. *In* Vant, W. N. (*ed.*), Lake managers handbook. *Water and soil miscellaneous publication 103*. Ministry of Works and Development, Wellington. pp. 195–202.

Howard-Williams, C.; Vincent, W. F. 1984: Optical properties of New Zealand lakes. I. Attenuation, scattering and a comparison between downwelling and scalar irradiances. *Archiv für Hydrobiologie 99*: 318–330.

Howard-Williams, C.; Vincent, W. F. 1985: Optical properties of New Zealand lakes. II. Underwater spectral characteristics and effects on PAR attenuation. *Archiv für hydrobiologie 104*: 441–457.

Howard-Williams, C.; Pickmere, S.; Davies, J. 1983: Decay rates and nitrogen dynamics of decomposing watercress (*Nasturium officiniale* R. Br.). *Hydrobiologia 99*: 207–214.

Howard-Williams, C.; Law, K.; Vincent, C. L.; Davies, J.; Vincent, W. F. 1986: Limnology of Lake Waikaremoana with special reference to littoral and pelagic primary producers. *New Zealand journal of marine and freshwater research 20*:

583–597.

Hughes, H. R.; McColl, R. H. S.; Rawlence, D. J. 1974: Lake Ellesmere: a review of the lake at its catchment. *DSIR information series No. 99*. DSIR, Wellington.

Hutchinson, E. G. 1980: Stormwater quality in urban and rural catchments — October 1980. Upper Waitemata Harbour Catchment Study, Working Report No 11A. Unpublished report to Auckland Regional Authority.

Hutchinson, G. E. 1957, 1967, 1975: A treatise on limnology. I. Geography, physics and chemistry. II. Introduction to the lake biology and the limnoplankton. III. Limnological botany. John Wiley & Sons, New York.

Idso, S. B. 1982: Discussion: 'Secchi disc relationships'. French, R. H.; Cooper, J. J.; Vigg, S. *Water resources bulletin 18*: 1053.

Idso, S. B.; Gilbert, R. G. 1974: On the universality of the Poole and Atkins Secchi disc — light extinction equation. *Journal of applied ecology 11*: 399–401 .

Inglis, C. C.; Allen, F. H. 1957: The regimen of the Thames estuary as affected by currents, salinities and river flow. *Proceedings of the Institution of Civil Engineers 7*: 827–878.

Iturriaga, R.: Siegel, D. A. 1989: Microphotometric characterization of phytoplankton and detrital absorption properties in the Sargasso Sea. *Limnology and oceanography 34*: 1706–1726.

Jackson, R. H.; Williams, P. J. le B.; Joint, I. R. 1987: Freshwater phytoplankton in the low salinity region of the River Tamar estuary. *Estuarine, coastal and shelf science 25*: 299–311.

James, H. R.; Birge, E. A. 1938: A laboratory study of the absorption of light by lake waters. *Transactions of the Wisconsin Academy of Science, Arts and Letters 31*: 1–154.

Jerlov, N. G. 1968: Optical oceanography. *Elsevier Oceanography Series 5*. Amsterdam.

Jerlov, N. G. 1976: Marine optics. *Elsevier Oceanography Series 14*. Amsterdam. 231 p.

Jerlov, N. G.; Steeman-Nielsen, E. (*eds*) 1974: Optical aspects of oceanography. Academic Press, London–New York. pp. 177–219.

Jewson, D. H. 1977: Light penetration in relation to phytoplankton content of the euphotic zone of Lough Neagh, N. Ireland. *Oikos 28*: 74–83.

Jewson, D. H.; Talling, J. F.; Dring, M. J.; Tilzer, M. M.; Heaney, S. I.; Cunningham, C. 1984: Measurement of photosynthetically available radiation in freshwater: comparative tests of some current instruments used in studies of primary production. *Journal of plankton research 6*: 259–273.

Johnstone, I. M.; Robinson, P. W. 1987: Light level variation in Lake Tutira after transient sediment inflow and its effect on the submerged vegetation. *New Zealand journal of marine and freshwater research 21*: 47–54.

Jones, D.; Wills, M. S. 1956: The attenuation of light in sea and estuarine waters in relation to the concentration of suspended solid matter. *Journal of the Marine Biological Association UK 35*: 431–444.

Jordon, M. B. 1988: A new submersible recording scalar light sensor array. *Deep sea research 35*: 1411–1423.

Jorgensen, B. B.; Des Marais, D. J. 1988: Optical properties of benthic photosynthetic communities: fibre-optic studies of cyanobacterial mats. *Limnology and oceanography 33*: 99–113.

Josselyn, M. N.; West, J. A. 1985: The distribution and temporal dynamics of the estuarine macroalgal community of San Francisco Bay. *Hydrobiologia 129*: 139–152.

Judd, B.; Kokich, D. 1986: Water quality problems in Lake Omapere: a report by the Lake Omapere Task Force to the Northland Catchment Commission, May 1986. NCC, Whangarei. 35 p.

Kalle, K. 1966: The problem of Gelbstoff in the sea. *Oceanography and marine biology annual review 4*: 91–104.

Karr, J. R.; Schlosser, I. J. 1978: Water resources and the land-water interface. *Science 201*: 229–234.

Kiefer, D. A.; Austin, R. W. 1974: The effect of varying phytoplankton on submarine light transmission in the Gulf of California. *Limnology and oceanography 19*: 55–64.

Kiefer, D. A.; Olson, R. J.; Wilson, W. H. 1979: Reflectance spectroscopy of marine phytoplankton Part 1. Optical properties as related to age and growth rate. *Limnology and oceanography 24*: 664–672.

Kingett, P. D. 1984: Lake Waahi, an environmental history. Report by Kingett & Associates for Mines Division, Ministry of Energy, NZ. 210 p.

Kirk, J. T. O. 1975a: A theoretical analysis of the contribution of algal cells to the attenuation of light within natural waters 1. General treatment of suspensions of pigmented cells. *New phytologist 75*: 11–20.

Kirk, J. T. O. 1975b: A theoretical analysis of the contribution of algal cells to the attenuation of light within natural waters 2. Spherical cells. *New phytologist 75*: 21–36.

Kirk, J. T. O. 1976: Yellow substance (Gelbstoff) and its contribution to the attenuation of photosynthetically active radiation in some inland and coastal south-eastern Australian waters. *Australian journal of marine and freshwater research 27*: 61–71.

Kirk, J. T. O. 1977: Use of a quanta meter to measure attenuation and underwater reflectance of photosynthetically active radiation in some inland and coastal south-eastern Australian waters. *Australian journal of marine and freshwater research 28*: 9–21.

Kirk, J. T. O. 1979: Spectral distribution of photosynthetically active radiation in some south-eastern Australian waters. *Australian journal of marine and freshwater research 30*: 81–91.

Kirk, J. T. O. 1980: Spectral absorption properties of natural waters: Contribution of the soluble and particulate fractions to light absorption in some inland waters of South-eastern Australia. *Australian journal of marine and freshwater research 30*: 287–296.

Kirk, J. T. O. 1981a: Monte-Carlo study of the nature of the underwater light field in, and the relationships between optical properties of, turbid yellow waters. *Australian journal of marine and freshwater research 32*: 517–532.

Kirk, J. T. O. 1981b: Estimation of the scattering coefficient of natural waters using underwater irradiance measurements. *Australian journal of marine and freshwater research 32*: 533–539.

Kirk, J. T. O. 1982: Prediction of optical water quality. *In* O'Loughlin, E. M.; Cullen, P. (*eds*) Prediction in water quality. Proceedings of a symposium sponsored by the Australian Academy of Science and the Institution of Engineers, Australia, held in Canberra, November 30–December 2, 1982. Australian Academy of Science, Canberra. pp. 30–26.

Kirk, J. T. O. 1983: Light and photosynthesis in aquatic ecosystems. Cambridge

University Press, Cambridge. 401 p.

Kirk, J. T. O. 1984a: Dependence of relationship between inherent and apparent optical properties of water on solar altitude. *Limnology and oceanography 29*: 350–356.

Kirk, J. T. O. 1984b: Attenuation of solar radiation in scattering-absorbing waters: a simplified procedure for its calculation. *Applied optics 23*: 3737–3739.

Kirk, J. T. O. 1985: Effects of suspensoids (turbidity) on penetration of solar radiation in aquatic ecosystems. *Hydrobiologia 125*: 195–208.

Kirk, J. T. O. 1986: Optical limnology — a manifesto. *In* De Dekker, P.; Williams, W. D. (*eds*) Limnology in Australia. CSIRO/Dr W. Junk, Melbourne/Dordrecht. pp. 33–62.

Kirk, J. T. O. 1988: Optical water quality — what does it mean and how should we measure it? *Journal of the Water Pollution Control Federation 60*: 194–197.

Kirk, J. T. O. 1989: The upwelling light stream in natural waters. *Limnology and oceanography 34*: 1410–1425.

Kirk, J. T. O. 1991: Volume scattering function, average cosines, and the underwater light field. *Limnology and oceanography 36*: 455–467.

Kishino, M.; Booth, C. R.; Okami, N. 1984: Underwater radiant energy absorbed by phytoplankton, detritus, dissolved organic matter, and pure water. *Limnology and oceanography 29*: 340–349.

Kishino, M.; Takahashi, M; Okami, N; Ichimura, S. 1985: Estimation of the spectral absorption coefficients of phytoplankton in the sea. *Bulletin of marine science 37*: 634–642.

Krause, G.; Ohm, K. 1984: A method to measure suspended load transport in estuaries. *Estuarine, coastal and shelf science 19*: 611–618.

Kuittinen, R.; Sucksdoroff, Y. 1984: Determination of water depth using aerial photography. *Publication of the Water Research Institute (Finland) 60*: 22–34.

Larson, D. W. 1972: Temperature, transparency, and phytoplankton productivity in Crater Lake, Oregon. *Limnology and oceanography 17*: 410–417.

Latimer, P.; Eubanks, C. A. H. 1962: Absorption spectrophotometry of turbid suspensions: a method for correcting for large systematic distortions. *Archives of biochemistry and biophysics 98*: 274–285.

Leopold, L. B. 1968: Hydrology for urban land planning — a guidebook for the hydrologic effects of urban land use. *US Geological Survey, Circular 554*. 18 p.

Leopold, L. B.; Wolman, M. G.; Miller, J. P. 1964: Fluvial processes in geomorphology. Freeman, San Francisco. 522 p.

Leppard, G. G. 1984a: The ultrastructure of lacustrine sedimenting materials in the colloidal size range. *Archiv für hydrobiologie 101*: 521–530.

Leppard, G. G. 1984b: Organic coatings on suspended particles in lake water. *Archiv für hydrobiologie 102*: 256–269.

Lerman, A. 1979: Geochemical processes. Wiley-Interscience, New York. 481p.

Lewis, D. W. 1984: Practical sedimentology. Hutchinson-Ross, Stroudsburg, Pennsylvania. 229 p.

Likens, G. E.; Bormann, F. H.; Pierce, R. S.; Easton, J. S.; Johnson, N. M. 1977: Biogeochemistry of a forested ecosystem. Springer-Verlag, New York.

Lind, O. T. 1977: Handbook of common methods in limnology. 2nd Ed. C.V. Mosby.

Livingston, M. E.; Biggs, B. J.; Gifford, J. S. 1986: Inventory of New Zealand lakes. *Water and soil miscellaneous publication 80, 81*. DSIR Publishing, Wellington. 199 p. and 192 p.

Lloyd, D. S. 1987: Turbidity as a water quality standard for salmonid habitats in Alaska. *North American journal of fisheries management 7*: 34–45.

Lloyd, D. S.; Koenings, J. P.; La Perriere, J. D. 1987: Effects of turbidity in fresh waters of Alaska. *North American journal of fisheries management 7*: 18–33.

Lorenzen, M. W. 1980: Use of chlorophyll-Secchi disk relationships. *Limnology and oceanography 25*: 371–372.

Lumb, C. M. 1990: Algal depth distributions and long-term turbidity changes in the Menai Strait, North Wales. *Progress in underwater science 15*: 85–99.

Lyle, E. S. 1987: Surface mine reclamation manual. Elsevier, New York. 268 p.

Lythgoe, J. N. 1979: The ecology of vision. Oxford University Press. 244 p.

Malone, T. C. 1977: Environmental regulation of phytoplankton productivity in the lower Hudson River. *Estuarine, coastal and marine science 5*: 157–171.

Marker, A. F. H.; Nusch, E. A.; Rai, H.; Riemann, B. 1980: The measurement of photosynthetic pigments in freshwaters and standardization of methods. Conclusions and recommendations. *Archiv für Hydrobiologie Beihefte Ergebnisse der Limnologie 14*: 91–106.

Matthews, R. J. 1979: Chemical analysis of Lake Rotorua sediments. Unpublished MSc thesis, University of Waikato, New Zealand.

McCave, I. N. 1979: Suspended sediment. *In* Dyer, K. R. (*ed.*) Estuarine hydrography and sedimentation. Cambridge University Press, Cambridge. pp. 131–185.

McCluney, W. R. 1975: Radiometry of water turbidity measurement. *Journal of the Water Pollution Control Federation 47*: 252–266.

McColl, R. H. S. 1972: Chemistry and trophic status of seven New Zealand lakes. *New Zealand journal of marine and freshwater research 6*: 399–447.

McColl, R. H. S.; Hughes, H. R. 1981: The effects of land use on water quality — a review. *Water and soil miscellaneous publication 23*. Ministry of Works and Development, Wellington. 59 p.

McComb, A. J.; Lukatelich, R. J. 1990: Inter-relations between biological and physico-chemical factors in a database for a shallow estuarine system. *Environmental monitoring and assessment 14*: 223–238.

McDowell-Boyer, L. M.; Hunt, J. R.; Sitar, N. 1986: Particle transport through porous media. *Water resources research 22*: 1901–1921.

McGirr, D. J. 1974: Interlaboratory quality control study No. 10. Turbidity and filterable and non-filterable residue. Canada Centre for Inland Waters Report Series No. 37. Burlington Ontario. 9 p.

McKnight, D. M.; Chisholm, S. W.; Harlemann, D. R. F. 1983: $CuSO_4$ treatment of nuisance algal blooms in drinking water reservoirs. *Environmental management 7*: 311–320.

McPherson, B. F.; Miller, R. L. 1987: The vertical attenuation of light in Charlotte Harbour, a shallow, subtropical estuary, south-western Florida. *Estuarine, coastal and shelf science 25*: 721–737.

Megahan, W. F. 1975: Sedimentation in relation to logging activities in the mountains of Central Idaho. *In* Present and prospective technology for predicting sediment yields and sources. USDA Agricultural Research Service ARS-S-40. pp. 74–82.

Megard, R. O.; Settles, J. C.; Boyer, H. A.; Combs, W. S. 1980: Light, Secchi disks, and trophic states. *Limnology and oceanography 25*: 373–377.

Metcalfe & Eddy Inc. 1979: Wastewater engineering: treatment, disposal, reuse. 2nd Ed. McGraw-Hill, New York. 920 p.

Michaelis, F. B. 1974: The ecology of Waikoropupu Springs. Unpublished PhD thesis, University of Canterbury, Christchurch, New Zealand. 158p.

Mitchell, J. K.; Bubenzer, G. D. 1980: Soil loss estimation. *In* Kirkby, M. J.; Morgan, R. P. C. (*eds*) Soil erosion. Wiley. pp 17–62.

Mook, D. H.; Hoskins, C. M. 1982: Organic determinations by ignition: Caution advised. *Estuarine, coastal and shelf science 15*: 697–699.

Morel, A. 1974: Optical properties of pure water and pure sea water. *In* Jerlov, N. G.; Steeman-Nielsen, E. (*eds*) Optical aspects of oceanography. Academic Press, London–New York. pp. 1–24.

Morel, A. 1980: In-water and remote measurements of ocean-colour. *Boundary-layer meteorology 18*: 177–201.

Morel, A. 1987: Chlorophyll-specific scattering coefficient of phytoplankton. A simplified theoretical approach. *Deep sea research 34*: 1093–1987.

Morel, A. 1988: Optical modelling of the upper ocean in relation to its biogenous matter content (Case I waters). *Journal of geophysical research 93*: 10749–10768.

Morel, A.; Bricaud, A. 1988: Inherent properties of algal cells including picoplankton: Theoretical and experimental results. *Canadian bulletin of fisheries and aquatic sciences 214*: 521–559.

Morel, A.; Prieur, A. 1977: Analysis of variations in ocean color. *Limnology and oceanography 22*: 709–722.

Mosley, M. P. 1980: The impact of forest road erosion in the Dart Valley, Nelson. *New Zealand journal of forestry 25*: 184–198.

Munday, J. C.; Alfoldi, T. T. 1979: Landsat test of diffuse reflectance models for aquatic suspended solids measurement. *Remote sensing of the environment 8*: 169–183.

Newhall, S. M.; Nickerson, D.; Judd, D. B. 1943: Final report of the OSA subcommittee on the spacing of Munsell colors. *Journal of the Optical Society of America 33*: 385–418.

Newton, J. R. 1975: Factors affecting slick formation at domestic sewage outfalls. *In* Helliwell, R. R.; Bossanyi, J. (*eds*) Pollution criteria for estuaries. Poutech Press, London. pp. 12.1–12.6.

Novitski, R. P. 1978: Hydrological characteristics of Wisconsin's wetlands and their influence on floods, streamflow and sediment. *In* Greeson, P. E.; Clark, J. R.; Clark, J. E. (*eds*) Wetland functions and values: the state of our understanding. American Water Resources Association, Minneapolis. pp. 377–388.

Novotny, V.; Chesters, G. 1981: Handbook of nonpoint pollution: sources and management. Van Nostrand Reinhold, New York. 555 p.

NTAC 1968: Water *quality criteria*. Report of the National Technical Advisory Committee to the Secretary of the Interior, Washington, District of Columbia, 234 p.

Nurnberg, G. K. 1984: The prediction of internal phosphorus loads in lakes with anoxic hypolimnia. *Limnology and oceanography 29*: 111–124.

Nusch, E. A. 1980: Comparison of different methods for chlorophyll and phaeopigment determination. *Archiv für Hydrobiologie Beihefte Ergebnisse der Limnologie 14*: 14–36.

Odum, H.T. 1957: Trophic structure and productivity of Silver Springs, Florida.

Ecological monographs 27: 55–112.

OECD 1982: Eutrophication of waters: monitoring, assessment and control. Organisation for Economic Co-operation and Development, Paris. 154 p.

Orth, R. J.; Moore, K. A. 1983: Chesapeake Bay: an unprecedented decline in submerged aquatic vegetation. *Science 222*: 5153.

Paerl, H. W. 1988: Nuisance phytoplankton blooms in coastal, estuarine, and inland waters. *Limnology and oceanography 33*: 823–847.

Paerl, H. W.; Payne, G. W.; Mackenzie, A. L.; Kellar, P. E.; Downes, M. T. 1979: Limnology of nine Westland beech forest lakes. *New Zealand journal of marine and freshwater research 13*: 47–52.

Pain, S. 1987: After the goldrush. *New scientist 20*: 36–43.

Pearsall, W. H.; Ullyott, P. 1933: Light penetration into freshwater. I. A thermionic potentiometer for measuring light intensity with photoelectric cells. *Journal of experimental biology 10*: 293–305.

Pearsall, W. H.; Ullyott, P. 1934: Light penetration into freshwater. III. Seasonal variations in the light conditions in Windermere in relation to vegetation. *Journal of experimental biology 11*: 8–93.

Pejrup, M. 1986: Parameters affecting fine-grained suspended sediment concentrations in a shallow microtidal estuary, Ho Bugt, Denmark. *Estuarine, coastal and shelf science 22*: 241–254.

Pelevin, V. N. 1965: Measurement of the true absorption coefficient of light in the sea. *Izvestiya, Academy of Sciences, USSR, Atmospheric and oceanic physics series 1*: 539–545.

Pennock, J. R. 1985: Chlorophyll distributions in the Delaware estuary: regulation by light-limitation. *Estuarine, coastal and shelf science 21*: 711–725.

Peres, J. M.; Picard, J. 1975: Causes de la raréfaction et de la disparition des herbiers de *Posidonia oceanica* sur les Côtes Françaises de la Méditerranée. (Causes of decrease and disappearance of the seagrass *Posidonia oceanica* on the French Mediterranean coast.) *Aquatic botany 1*: 133–139.

Peterson, L. L. 1974: The propagation of sunlight and the size distribution of suspended particles in a municipally polluted ocean water. Unpublished PhD thesis, California Institute of Technology.

Petersen, R. C.; Madsen, B. L.; Wilzbach, M. A.; Magadza, C. H. D.; Paarlberg, A.; Kullberg, A.; Cummins, K. W. 1987: Stream management: emerging global similarities. *Ambio 16*: 166–179.

Petzold, T. J. 1972: Volume scattering functions for selected ocean waters. Visibility Laboratory, Scripps Institution of Oceanography Technical Report, Scripps Institute of Oceanography Ref. 72–78. 79 p.

Petzold, T. J.; Austin, R. W. 1968: An underwater transmissometer for ocean survey work. Scripps Institution of Oceanography Reference 68–9. 5 p.

Pickrill, R. A.; Mitchell, J. S.; Hill, P. J. 1986: Calibration and use of an optical transmissometer to determine concentrations of suspended particulate matter. *NZ Oceanographic Institute oceanographic field report 23*. 8 p.

Pierce, J. W.; Correll, D. L.; Goldberg, B.; Faust, M. A.; Klein, W. H. 1986: Response of underwater light transmittance in the Rhode River estuary to changes in water-quality parameters. *Estuaries 9*: 169–178.

Pilgrim, D. A; Redfern, T. A.; MacLauchlan, G. S.; Marsh, R. I. 1989: Estimation of optical coefficients from diver observations of visibility. *Progress in underwater science 14*: 33–52.

Pitt, R.; Field, Q. 1977: Water quality effects from urban runoff. *Journal of the American Water Works Association 67*: 432–436.

Poole, H. H.; Atkins, W. R. G. 1926: On the penetration of light into seawater. *Journal of the Marine Biological Association of the United Kingdom 14*: 177–198.

Poole, H. H.; Atkins, W. R. G. 1929: Photoelectric measurements of submarine illumination throughout the year. *Journal of the Marine Biological Association of the United Kingdom 16*: 297–324.

Postma, H. 1960: Einige Bemerkungen über den Sinkstofftransport in Ems-Dollard-Gebiet. (Some observations on the sediment transport in the Ems-Dollard area.) *Verhandelingen Koninklijk Nederlands Geologisch Mÿn bouwkundig Genootschap Geologische Serie 19*: 103–110.

Postma, H. 1961: Suspended matter and Secchi disc visibility in coastal waters. *Netherlands journal of sea research 1*: 359–390.

Preisendorfer, R. W. 1958: Some practical consequences of the asymptotic radiance hypothesis. Scripps Institute of Oceanography Ref. S10 58–60.

Preisendorfer, R. W. 1961: Application of radiative transfer theory to light measurements in the sea. *International Union of Geodesy and Geophysics Monographs, 10*: 11–29.

Preisendorfer, R. W. 1976: Hydrologic optics, 6 V. US Department of Commerce, Washington, District of Columbia.

Preisendorfer, R. W. 1986: Secchi disc science: Visual optics of natural waters. *Limnology and oceanography 31*: 909–926.

Pridmore, R. D. 1987: Phytoplankton response to changed nutrient concentrations. *In* Vant, W. N. (*ed.*) Lake managers handbook. *Water and soil miscellaneous publication 103*. Ministry of Works and Development, Wellington. pp. 183–194.

Pridmore, R. D.; McBride, G. B. 1984: Prediction of chlorophyll *a* concentrations in impoundments of short hydraulic retention time. *Journal of environmental management 19*: 343–350.

Pritchard, D. W. 1967: What is an estuary: physical viewpoint. *In* Lauff, G. H. (*ed.*) Estuaries. American Association for the Advancement of Science, Washington. pp. 3–5.

Privoznik, K. G.; Daniel, K. J.; Incropera, F. P. 1978: Absorption, extinction and phase function measurements for algal suspensions of *Chlorella pyrenoidosa*. *Journal of quantitative spectroscopy and radiative transfer 20*: 345–352.

Rawlence, D. J. 1984: A study of pigment and diatoms in a core from Lake Rotorua, North Island, New Zealand, with emphasis on recent history. *Journal of the Royal Society of New Zealand 14*: 119–132.

Rebhun, M.; Manka, J. 1971: Classification of organics in secondary effluents. *Environmental science and technology 5*: 606–609.

Reid, J. M.; Cresser, M. S., MacLeod, D. A. 1980: Observations on the estimation of total organic carbon from UV absorbance for an unpolluted stream. *Water research 14*: 525–529.

Reid, L. M.; Dunne, T. 1984: Sediment production from road surfaces. *Water resources research 20*: 1753–1761.

Reynolds, J. B.; Simmons, R. C.; Burkholder, A. R. 1989: Effects of placer mining discharge on health and food of arctic grayling. *Water resources bulletin 25*: 625–635.

Rieger, W. A.; Olive, L. J.; Gippel, C. J. 1988: Channel sediment behaviour as a basis for modelling delivery processes. Sediment Budgets. *International Association of Hydrological Sciences publication 174*: 541–548.

Rutherford, J. C. 1984: Trends in Lake Rotorua water quality. *New Zealand journal of marine and freshwater research 18*: 355–365.

Rutherford, J. C.; Pridmore, R. D.; White, E. 1989: Management of phosphorus and nitrogen inputs to Lake Rotorua, New Zealand. *Journal of water resources planning and management 115*: 431–439.

Sand-Jensen, K. 1988: Minimum light requirements for growth in *Ulva lactuca*. *Marine ecology progress series 50*: 187–193.

Sand-Jensen, K.; Moller, J.; Olesen, B. H. 1988: Biomass regulation of microbenthic algae in Danish lowland streams. *Oikos 53*: 332–340.

Sas, H. 1989: Lake restoration by reduction of nutrient loading: expectations, experiences, extrapolations. Academia Verlag, Richarz. 497 p.

SCB 1986: Lower Oreti River gravel extraction. Southland Catchment Board Publication No. 126. Invercargill, New Zealand.

Schanz, F. 1985: Vertical light attenuation and phytoplankton development in Lake Zurich. *Limnology and oceanography 23*: 247–259.

Schindler, D. A. 1981: Studies of eutrophication in lakes and their relevance to the estuarine environment. *In* Neilson, B. J.; Cronin, L. E. (*eds*) Estuaries and nutrients. Humana Press, New Jersey. pp. 71–82.

Schubel, J. R. 1968: Suspended sediment of the northern Chesapeake Bay. *Chesapeake Bay Institute technical report 35*. John Hopkins University.

Schueler, T. R. 1987: Controlling urban runoff: a practical manual for planning and designing best management practices. Metropolitan Washington Council of Governments.

Selby, M. J. 1982: Hillslope materials and processes. Oxford University Press, Oxford. 264 p.

Shibata, K. 1959: Spectrophotometry of translucent biological materials: opal glass transmission method. *Methods of biochemical analysis 7*: 77–109.

Sholkovitz, E. R. 1976: Flocculation of dissolved organic and inorganic matter during mixing of river water and seawater. *Geochimica et cosmochimica acta 40*: 831–845.

Simpson, W. R. 1982: Particulate matter in the oceans — sampling methods, concentration, size distribution and particle dynamics. *Oceanography and marine biology annual review 20*: 119–172.

Singer, J. K.; Anderson, J. B.; Ledbetter, M. T.; McCave, I. N.; Jones, K. P. N.; Wright, R. 1988: An assessment of analytical techniques for the size analysis of fine-grained sediments. *Journal of sedimentary petrology 58*: 534–543.

Skovlin, J. M. 1984: Impacts of grazing on wetlands and riparian habitat: a review of our knowledge. *In* Development strategies for rangeland management. National Research Council/National Academy of Sciences, Westview Press, Boulder, Colorado. pp. 1001–1104.

Slovacek, R. E.; Hannan, P. J. 1977: *In vivo* fluorescence determinations of phytoplankton chlorophyll *a*. *Limnology and oceanography 22*: 919–925.

Smith, C. M. 1987: Sediment, phosphorus and nitrogen in channelised surface runoff from a New Zealand pastoral catchment. *New Zealand journal of marine and freshwater research 21*: 627–639.

Smith, C. M. 1989: Riparian pasture retirement effects on sediment, phosphorus and nitrogen in channelised surface runoff from pastures. *New Zealand journal of marine and freshwater research 23*: 139–146.

Smith D. G. 1986: Heavy metals in the New Zealand aquatic environment: a review. *Water and soil miscellaneous publication 100.* Ministry of Works and Development, Wellington.

Smith, D. G.; Davies-Colley, R. J. 1992: Perception of water clarity and colour in terms of suitability for recreational use. *Journal of environmental management 36*: 225–235.

Smith, D. G.; McBride, G. B. 1990: New Zealand's national water quality monitoring network — Design and first year's operation. *Water resources bulletin 26*: 767–775.

Smith, D. G.; Roper, D. S.; Williams, B. L. 1987: Implications for the marine environment. *In* Coastal outfalls. Proceedings of a seminar, Christchurch, New Zealand, December 4, 1985. *Water and soil miscellaneous publication 107.* Ministry of Works and Development, Wellington.

Smith, D. G.; Cragg, A. M.; Croker, G. F. 1991: Water clarity criteria for bathing waters based on user perception. *Journal of environmental management 33*: 285–299.

Smith, R. C.; Baker, K. S. 1978: Optical classification of natural waters. *Limnology and oceanography 23*: 260–267.

Smith, R. C.; Baker, K. S. 1981: Optical properties of the clearest natural waters (200–800 nm). *Applied optics 20*: 177–184.

Smith, R. C.; Tyler, J. E. 1967: Optical properties of clear natural water. *Journal of the Optical Society of America 57*: 589–595.

Smith, R. C.; Tyler, J. C.; Goldman, C. R. 1973: Optical properties and color of Lake Tahoe and Crater Lake. *Limnology and oceanography 18*: 189–199.

Smith, S. D.; Banke, E. G. 1975: Variations of the sea surface drag coefficient with wind speed. *Quarterly journal of the Royal Meteorological Society 101*: 665–673.

Spence, D. H. N. 1976: Light and plant response in fresh water. *In* Evans, G. C.; Bainbridge, R.; Rackham, O. (*eds*) Light as an ecological factor: II. Sixteenth symposium of the British Ecological Society. Blackwell Scientific, Oxford.

Spence, D. H. N. 1982: The zonation of plants in freshwater lakes. *Advances in ecological research 12*: 37–125.

Spinrad, R. W. (*ed.*) 1989: Hydrologic optics. *Limnology and oceanography 34*: (special issue).

Stein, R. 1985: Rapid grain size analyses of clay and silt fractions by sedigraph 5000 D: comparison with Coulter Counter and Atterberg methods. *Journal of sedimentary petrology 55*: 590–615.

Sterner, R. W. 1990: Lake morphometry and light in the surface layer. *Canadian journal of fisheries and aquatic sciences 47*: 687–692.

Stramski, D.; Kiefer, D. A. 1990: Optical properties of marine bacteria. *Society of Photo-optical Instrumentation Engineers, Ocean optics 10*: 250–268.

Strickland, J. D. H.; Parsons, T. R. 1972: A practical handbook of seawater analysis. Bulletin 167 (2nd Ed.), Fisheries Research Board of Canada, Ottawa. 310 p.

Stross, R. G.; Sokol, R. C. 1989: Runoff and flocculation modify underwater light environment of the Hudson River estuary. *Estuarine, coastal and shelf science 29*: 305–316.

Stumm, W.; Morgan, J. J. 1981: Aquatic chemistry, 2nd Ed. Wiley-Interscience, New York. 780 p.

Swanson, F. P.; Swanson, M. M.; Woods, C. 1981: Analysis of debris avalanche erosion in steep forest land. *International Association of Hydrological Sciences publication 132*: 67–75.

Talling, J. F. 1957: Photosynthetic characteristics of some freshwater planktonic diatoms in relation to underwater radiation. *New phytologist 56*: 29–50.

Talling, J. F. 1971: The underwater light climate as a controlling factor in the production ecology of freshwater phytoplankton. *Mitteilungen der internationalen vereinigung für theoretische und angewandte limnologie 19*: 214–243.

Tam, A. C.; Patel, C. K. N. 1979: Optical absorptions of light and heavy water by laser opto-acoustic spectroscopy. *Applied optics 18*: 3348–3358.

Taylor, J. H. 1964: Use of visual performance data in visibility prediction. *Applied optics 3*: 562–569.

Thompson, M. J.; Gilliland, L. E.; Rosenfeld, L. K. 1979: Light scattering and extinction in a highly turbid coastal inlet. *Estuaries 2*: 164–171.

Thornes, J. B. 1980: Erosional processes of running water. *In* Kirkby, M. J.; Morgan, R. P. C. (*eds*) Soil erosion. Wiley. pp 129–182.

Thurman, E. M. 1985: Organic geochemistry of natural waters. Dordrecht.

Tilzer, M. M. 1983: The importance of fractional light absorption by photosynthetic pigments for phytoplankton productivity in Lake Constance. *Limnology and oceanography 28*: 833–846.

Tilzer, M. M. 1988: Secchi disk–chlorophyll relationships in a lake with highly variable phytoplankton biomass. *Hydrobiologia 162*: 163–171.

Timofeeva, V. A. 1974: Optics of turbid waters (Results of laboratory studies). *In* Jerlov, N. G.; Steeman-Nielsen, E. (*eds*) Optical aspects of oceanography. Academic Press, London–New York. 177–219 pp.

Tipping, E. 1981: The adsorption of aquatic humic substances by iron oxide. *Geochimica et cosmochimica acta 45*: 191–199.

Toy, T. J.; Hadley, R. F. 1987: Geomorphology and reclamation of disturbed lands. Academic Press Inc., Orlando, Florida. 480 p.

Tressler, W. L.; Wagner, L. G.; Bere, R. 1940: A limnological study of Chautauqua Lake. II. Seasonal variations. *Transactions of the American Microbiological Society 59*: 12–30.

Truper, H. G.; Yentsch, C. S. 1962: Use of glass fibre filters for the rapid preparation of *in vivo* absorption spectra of photosynthetic bacteria. *Journal of bacteriology 94*: 1255–1256.

Tyler, J. E. 1968: The Secchi disc. *Limnology and oceanography 13*: 1–6.

Tyler, J. E. 1973: Applied radiometry. *Oceanography and marine biology annual review 11*: 25.

Tyler, J. E.; Smith, R. C. 1970: Measurements of spectral irradiance underwater. Gordon and Breach, New York. 102 p.

UNESCO 1982: Sedimentation problems in river basins. White, W. R. (*ed.*) Project 5.3 of the International Hydrological Programme. United Nations Educational, Scientific and Cultural Organisation, Paris. 152 p.

USEPA 1973a: Water Quality Criteria 1972. US EPA Report No. EPA.R3.73.033. EPA, Washington, District of Columbia. 594 p.

USEPA 1973b: Methods for identifying and evaluating the nature and extent of nonpoint sources of pollutants. US Environmental Protection Agency Report EPA-4030/9-73-014, 261 p.

USEPA 1976: Quality Criteria for Water. US EPA, Washington, District of Columbia. 256 p.

USEPA 1979: Livestock grazing management and water quality protection. US Environmental Protection Agency Report, Seattle, Washington. EPA-910/9-79-67, 147 p.

USEPA 1983: Results of the nationwide urban runoff program, Volume 1. Final Report, Office of Regulations and Standards, US Environmental Protection Agency, Washington, District of Columbia. EPA 440/5-86/001.

USEPA 1986: Quality Criteria for Water 1986. US EPA Report No. EPA.440/5-86-001. EPA, Washington, District of Columbia.

USEPA 1988: Development document for final effluent guidelines and new source performance standards for the ore mining and dressing point source category. Gold placer mining subcategory. US Environmental Protection Agency, Washington, District of Columbia. EPA 440/1-88/061.

Utterback, C. L.; Phifer, L. D.; Robinson, R. J. 1942: Some chemical, planktonic and optical characteristics of Crater Lake. *Ecology 23*: 97–103.

Vale, C.; Sundby, B. 1987: Suspended sediment fluctuations in the Tagus estuary on semi-diurnal and fortnightly time scales. *Estuarine, coastal and shelf science 25*: 495–508.

van de Hulst, H. C. 1957: Light scattering by small particles. John Wiley & Sons, New York–London. 470 p.

van Olphen, H. 1977: An introduction to clay colloid chemistry. Wiley-Interscience, New York. 318 p.

van Roon, M. R. 1983: Water quality of the Upper Waitemata Harbour and catchment. Auckland Regional Authority, Auckland. 430 p.

Van Sickle, J.; Beschta, R. L. 1983: Supply-based models of suspended sediment transport in streams. *Water resources research 19*: 768–778.

Vant, W. N. 1987a: Lake managers handbook. *Water & Soil miscellaneous publication 103*. Ministry of Works and Development, Wellington.

Vant, W. N. 1987b: Lake Whangape — muddy water and macrophytes. *Soil and water 23*: 20–24.

Vant, W. N. 1990: Causes of light attenuation in nine New Zealand estuaries. *Estuarine, coastal and shelf science 31*: 125–137.

Vant, W. N. 1991: Underwater light in the northern Manukau Harbour, New Zealand. *Estuarine, coastal and shelf science 33*: 291–307.

Vant, W. N.; Davies-Colley, R. J. 1984: Factors affecting clarity of New Zealand lakes. *New Zealand journal of marine and freshwater research 18*: 367–377.

Vant, W. N.; Davies-Colley, R. J. 1986: Relative importance of clarity determinants in Lakes Okaro and Rotorua. *New Zealand journal of marine and freshwater research 20*: 355–363.

Vant, W. N.; Davies-Colley, R. J. 1988: Water appearance and recreational use of 10 lakes of the North Island (New Zealand). *Verhandlungen der internationale verein-igung für theoretische und angewandte limnologie 23*: 611–615.

Vant, W. N.; Gilliland, B. W. 1991: Changes in water quality of Lake Horowhenua fol-lowing sewage diversion. *New Zealand journal of marine and freshwater research 25*: 57–61.

Vant, W. N.; Hoare, R. A. 1987: Determining input rates of plant nutrients. *In* Vant, W. N. (*ed.*) Lake managers handbook. *Water and Soil miscellaneous publication 103*. Ministry of Works and Development, Wellington. pp. 158–166.

Vant, W. N.; Davies-Colley, R. J.; Clayton, J. S.; Coffey, B. T. 1986: Macrophyte depth limits in North Island (New Zealand) lakes of differing clarity. *Hydrobiologia 137*: 55–60.

Vincent, W. F. 1983a: Fluorescence properties of the freshwater phytoplankton: three algal classes compared. *British phycological journal 18*: 5–21.

Vincent, W. F. 1983b: Phytoplankton production and winter mixing. *Journal of ecology 71*: 1–20.

Vincent, W. F.; Howard-Williams, C.; Downes, M. T.; Dryden, S. J. 1989: Underwater light and photosynthesis at three sites in Pelorus Sound, New Zealand. *New Zealand journal of marine and freshwater research 23*: 79–91.

Viner, A. B.; White, E. 1987: Phytoplankton growth. *In* Viner, A. B. (*ed.*) Inland waters of New Zealand. DSIR Publishing, Wellington. pp. 191–223.

Walker, T. A. 1980: A correction to the Poole and Atkins Secchi disc/light attenuation formula. *Journal of the Marine Biological Association of the UK 60*: 769–771.

Walker, T. A. 1982: Use of a Secchi disc to measure attenuation of underwater light for photosynthesis. *Journal of applied ecology 19*: 539–544.

Walling, D. E. 1977: Assessing the accuracy of suspended-sediment rating curves for a small basin. *Water resources research 13*: 531–538.

Ward, L. G. 1985: The influence of wind waves and tidal currents on sediment resuspen-sion in middle Chesapeake Bay. *Geo-marine letters 5*: 71–75.

Weber, W. J. 1972: Physico-chemical processes for water quality control. Wiley-Interscience, New York. 640 p.

Weidemann, A. D.; Bannister, T. T. 1986: Absorption and scattering coefficients in Irondequoit Bay. *Limnology and oceanography 31*: 567–583.

Weidemann, A. D.; Bannister, T. T.; Effler, S. W.; Johnson, D. L. 1985: Particulate and optical properties during $CaCO_3$ precipitation in Otisco Lake. *Limnology and oceano-graphy 30*: 1078–1083.

Weilenmann, U.; O'Melia, C. R.; Stumm, W. 1989: Particle transport in lakes: models and measurements. *Limnology and oceanography 34*: 1–18.

Wells, R. D. S; Vant, W. N.; Clayton, J. S. 1988: Inorganic suspensoids and submerged macrophytes in Lake Whangape, New Zealand. *Verhandlungen der internationale vereinigung für theoretische und angewandte limnologie 23*: 1969–1972.

Westlake, D. F. 1966: The light climate for plants in rivers. *In* Bainbridge, R.; Evans, G.C.; Rackham, O. (*eds.*) Light as an ecological factor. Blackwell, Oxford. pp. 99–119.

Westlake, D. F. 1986: Measurement of underwater light. *In* The direct determination of biomass of aquatic macrophytes and measurement of underwater light. Standing Committee of Analysts, DOE, London, HMSO. 45 p.

White, E. 1989: Utility of relationships between lake phosphorus and chlorophyll *a* as predictive tools in eutrophication control studies. *New Zealand journal of marine and freshwater research 23*: 33–41.

Wilcock, R. J. ; Davies-Colley, R. J. 1986: Panel tests for evaluating the appearance and odour of the lower Tarawera River. *New Zealand journal of marine and freshwater research 20*: 699–708.

Williamson, R. B. 1985: Urban stormwater quality I. Hillcrest, Hamilton, New Zealand. *New Zealand journal of marine and freshwater research 19*: 413–427.

Williamson, R. B. 1986: Urban stormwater quality II. Comparison of three New Zealand catchments. *New Zealand journal of marine and freshwater research 20*: 315–328.

Williamson, R. B.; Hoare, R. A. 1987: Controlling nutrient loads and predicting resulting lake nutrient concentrations. *In* Vant, W.N. (*ed.*) Lake managers handbook. *Water and soil miscellaneous publication 103*. Ministry of Works and Development, Wellington. pp. 172–182.

Wischmeier, W. H. 1976: Use and misuse of the universal soil loss equation. *Journal of soil and water conservation 31*: 5–9.

Wrigley, T. J.; Toerien, D. F. 1990: Limnological aspects of small sewage ponds. *Water research 24*: 83–90.

Appendix 1: Radiation quantities and description of the light field in water

'Light' is normally taken to refer to electromagnetic radiation visible to the human eye. This radiation ranges in wavelength (symbol λ) between 400×10^{-9} and about 760×10^{-9} m (Fig. A1.1), and amounts to roughly half of the total energy radiated through our atmosphere from the sun. The unit 10^{-9} m will henceforth be called a nanometre (symbol nm). The 'visible' range is often taken as 400 nm < λ < 700 nm, since sensitivity of the

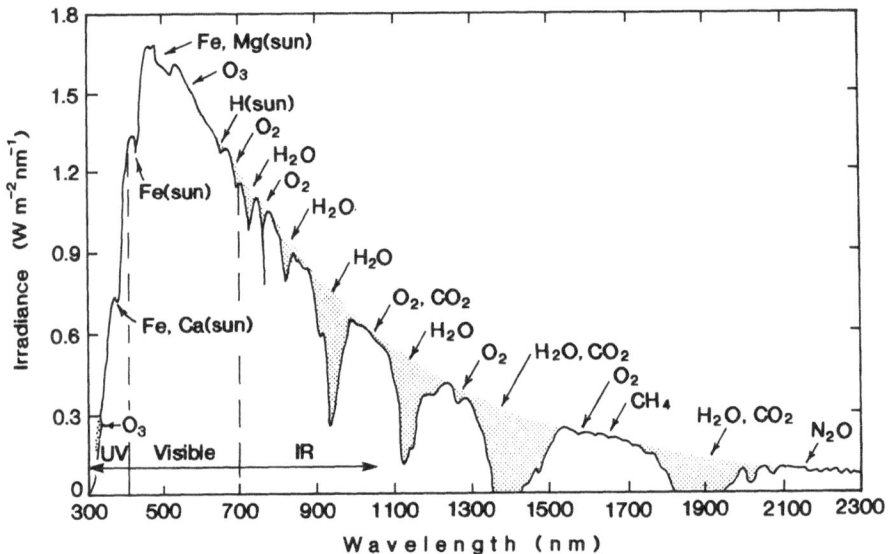

Fig. A1.1. The spectrum of solar radiation at sea level with spectral absorption features identified. The spectrum was recorded on 18 July 1980 at 12:36 local standard time, at Bedford, Massachussetts. (From Bird *et al.* (1982), with permission.)

human eye is low to 'ultraviolet' (UV) radiation shorter than 400 nm, and to 'infrared' (IR) radiation beyond 700 nm. This range also turns out to be that useful to green plants, including aquatic plants, for photosynthesis.

Light, as is well known, has a 'particle character' as well as a 'wave character'. The particles are referred to as photons and the energy content (quantum) associated with a given photon depends on wavelength:

$$E = h\nu = hc/\lambda \qquad (A1.1)$$

where h is Planck's constant, ν is the frequency of the electromagnetic radiation and c is the speed of light.

Light can be quantified as *radiant flux* (Φ) which has the units of energy per unit time (joules per second $\equiv$ watts) or photons per unit time. The following basic quantities are defined in terms of radiant flux.

Radiant intensity is the flux from a given direction per unit solid angle, ω, in steradians (abreviated sr):

$$I = d\Phi/d\omega \qquad \left(\text{units: } W \, sr^{-1}\right) \qquad (A1.2)$$

Radiance is the flux per unit area (S) from a given direction (θ, ϕ) (see Fig. A1.2).

$$L(\theta,\phi) = d^2\Phi/(dS\cos\theta \, d\omega) \qquad (A1.3)$$

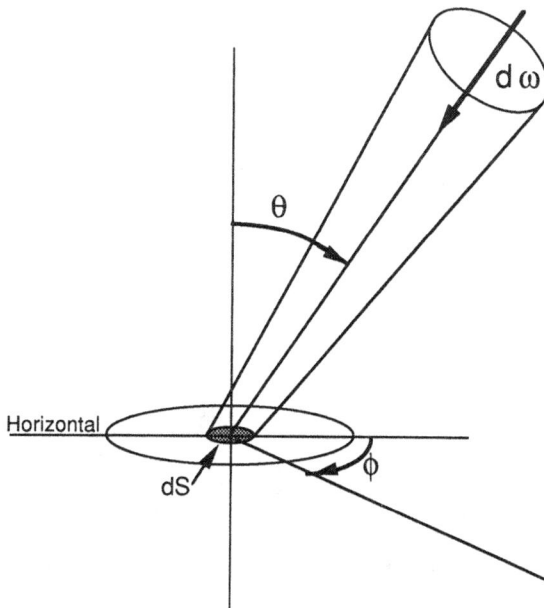

Fig. A1.2. Definition of radiance as radiant flux per unit surface per unit solid angle. Radiance is a function of direction, specified by zenith angle, θ, and azimuthal angle, ϕ.

Given this definition of radiance, we can define a quantity termed the *irradiance* as the radiance integrated over all directions (angles) (see Fig. A1.2). The irradiance is thus the radiant flux per unit area and has the units W m^{-2}. Fig. A1.3 indicates, schematically, how radiance, and also the integrated quantities defined below in terms of radiance, are measured.

We distinguish vector irradiance from scalar irradiance.

Downward irradiance (*vector irradiance*) is defined

$$E_d = \int_{2\pi} L(\theta,\phi)\cos\theta\, d\omega = \int_0^{\pi/2}\int_0^{2\pi} L(\theta,\phi)\cos\theta\sin\theta\, d\phi\, d\theta \qquad (A1.4)$$

(since dω = sin θ dϕ dθ; see Fig. A1.2) and can be conceptualized as all the radiance collected from the whole upper hemisphere by a flat plate detector—which is precisely how it is measured (Fig. A1.3).

Upward irradiance (*vector irradiance*) is defined similarly:

$$E_u = -\int_{-2\pi} L(\theta,\phi)\cos\theta\, d\omega = -\int_{\pi/2}^{\pi}\int_0^{2\pi} L(\theta,\phi)\cos\theta\sin\theta\, d\phi\, d\theta \qquad (A1.5)$$

and is all the radiance collected by a flat plate detector from the whole lower hemisphere.

Net vector irradiance is simply the downward vector irradiance minus the upward vector irradiance:

$$E = E_d - E_u = \int_{4\pi} L(\theta,\phi)\cos\theta\, d\omega \qquad (A1.6)$$

Scalar irradiance is defined as

$$E_o = \int_{4\pi} L(\theta,\phi)\, d\omega \qquad (A1.7)$$

and can be conceptualized as the total radiant intensity at a point in space. Scalar irradiance can be measured with a spherical collector (Fig. A1.3), although allowance has to be made for the fact that, in general, light rays strike such a finite-sized collector at oblique angles. It turns out that the geometric conversion factor from irradiance, E_s

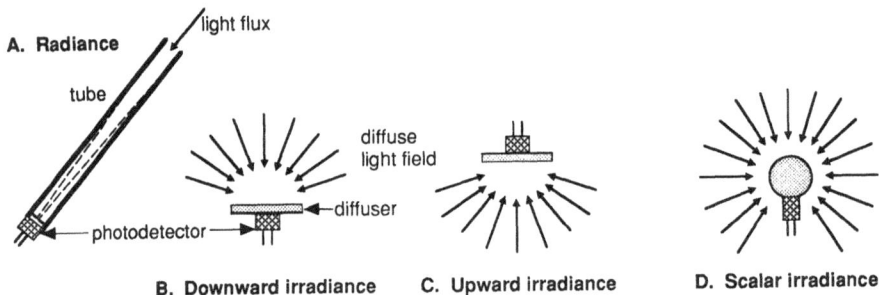

Fig. A1.3. Schematic indicating measurement of some radiometric quantities according to their definitions. A spherical sensor, as indicated in D, actually measures spherical irradiance which is 1/4 of scalar irradiance.

('spherical irradiance'), received by the spherical collector, to scalar irradiance at a point, is simply 4, being the ratio of the cross-sectional area of a sphere to its surface area, i.e. $E_o = 4E_s$ (Tyler & Preisendorfer 1962).

A complete specification of the radiance as a function of azimuthal (ϕ) and zenith (θ) angles completely describes the light field at a point in water. However, the complete directional structure is difficult and time consuming to measure, and is not required for most practical purposes. Very often it is sufficient to characterize the light field by simple indices of directional structure (angular distribution) or radiance. One such index is the ratio of net vector irradiance to scalar irradiance (Kirk 1983):

$$\mu = \vec{E}\big/E_o \qquad (A1.8)$$

which is referred to as the *average cosine* (e.g. Kirk 1983) for reasons which become clear on consideration of the definitions in equations (A1.6) and (A1.7). The average cosine, μ, can be thought of as the average vertical component of the radiance vectors in all directions at a point, and is thus a measure of the degree of diffusion of the light field. For a completely diffuse light field $\mu = 0$ while for a completely collimated, nadir-directed (vertically downwards), light field, $\mu = 1$. Equation (A1.8) can be seen as following directly from the definitions of $\vec{E}$ and E_o in equations (A1.6) and (A1.7).

It is sometimes useful to refer also to the *downward average cosine* defined by

$$\mu_d = \frac{E_d}{E_{od}} \qquad (A1.9)$$

and the *upward average cosine* defined by

$$\mu_u = \frac{E_u}{E_{ou}} \qquad (A1.10)$$

where E_{ou} and E_{od} are the upward- and downward-directed portions respectively of the total scalar irradiance. These average cosines, μ, μ_d and μ_u, are the reciprocals of the *distribution functions* (referring to angular distribution of the light field) introduced by Preisendorfer (Tyler & Preisendorfer 1962) and are classified together with the irradiance attenuation coefficients and the reflectance coefficient as *apparent* optical properties (AOPs).

All of the defining relations given above hold for the whole visible waveband or, for that matter, for the whole electromagnetic spectrum. However, the various quantities, e.g. the irradiance, E_d, may vary greatly with wavelength, and it is often necessary to consider this wavelength dependence. If these quantities are measured as a function of wavelength using essentially monochromatic light we obtain *spectral* quantities, for example: spectral irradiance $E_d(\lambda)$ (W m^{-2} nm^{-1}), spectral radiance $L(\lambda)$ (W m^{-2} sr^{-1} nm^{-1}), and spectral radiant flux $\Phi(\lambda)$ (W nm^{-1}).

In the context of human vision in (or into) water the spectral irradiances and spectral radiance must be weighted by the sensitivity function for the light-adapted human eye, the so-called photopic function, $\bar{y}$, which peaks at 555 nm. Table A1.1 shows the correspondence between the physical quantities and the associated psychophysical quantities weighted by $\bar{y}$. The correspondence is established by integrations with wavelength, for example the illuminance is calculated by integrating spectral irradiance:

$$I = K_{max} \int_0^\infty \bar{y} E(\lambda) d\lambda \qquad (A1.11)$$

where K_{max} is the maximum luminosity (683 lm W^{-1}—for 555 nm green light), and illuminance has the units lm m^{-2} or lux. Likewise, luminance is defined by integrating radiance weighted by $\bar{y}$.

Table A1.1. Correspondence between physical, physiological and psychophysical quantities

Physical	Physiological (e.g. plant photosynthesis)	Psychophysical (e.g. human vision)
Radiant flux Φ (W)	Photon flux $\Phi(PAR)$ (photons s^{-1})	Luminous flux Φ_I (lm)
Radiance L (W m^{-2} sr^{-1})	Quantum radiance $L(PAR)$ (photons s^{-1} m^{-2} sr^{-1})	Luminance B (lm m^{-2} sr^{-1})
Irradiance E (W m^{-2})	Quantum irradiance $E(PAR)$ (photons s^{-1} m^{-2})	Illuminance I (lm m^{-2} = lux)

Similarly, in the context of plant photosynthesis, the photon flux is required rather than energy flux since the light reactions occurring in plant photosynthetic structures have the same quantum efficiency for all photons in the visible (400–700 nm) spectrum. The PAR band quantum irradiance, for example is calculated as

$$E(PAR) = \frac{1}{hc} \int_{400}^{700} \lambda E(\lambda) d\lambda \qquad (A1.12)$$

REFERENCES

Bird, R. E.; Hulstrom, R. L.; Kliman, A. W.; Eldering, H.G. 1982: Solar spectral measurements in the terrestial environment. *Applied optics 21*: 1430–1436.

Kirk, J. T. O. 1983: Light and photosynthesis in aquatic ecosystems. Cambridge University Press, Cambridge. 401 p.

Tyler, J. E.; Preisendorfer, R. W. 1962: Transmission of energy in the sea. 8. Light. *In* Hill, M. N. (*ed.*) The sea. Wiley–Interscience, New York. pp. 397–451.

Appendix 2: Measurement recipes

A2.1 MEASUREMENT OF UNDERWATER IRRADIANCE PROFILES

A2.1.1 Principle

A light (irradiance) sensor of appropriate spectral and spatial sensitivity is lowered into a water body in daylight hours, and readings of irradiance (both up- and down-welling) are taken at different depths (Kirk 1977). From the profiles of irradiance versus depth in the water column an irradiance attenuation coefficient can be calculated. The reflectance coefficient, a weak function of depth, is calculated as the ratio of upwelling to down-welling irradiance (Kirk 1977).

Optically deep water is assumed (i.e. bottom reflectance negligible): it may not be possible to measure meaningful irradiance profiles in optically shallow water. Access by boat is assumed, although sometimes useful measurements can be made from a structure such as a jetty or bridge, or by wading in very light-attenuating waters. Readings at very low solar altitudes, say <20° above the horizon, should be avoided if possible.

A2.1.2 Checklist

Essential
  (1) Irradiance sensor(s), as appropriate, with underwater housings and connector system.
  (2) Support frame (e.g. Westlake 1986) and support system (e.g. graduated line).
  (3) Underwater cable (30 m is usually adequate for inland waters).
  (4) Readout instrument. Ratioing or logging instruments are particularly useful under changeable ambient lighting. Absolute calibration is desirable but not necessary.
  (5) Field notebook or clipboard and pencils (yes, we've forgotten these too!).

Desirable
  (1) Hydrological gauging winch and derrick/crane (for accurate depth measurement).

Alternatively, a depth transducer can be used.
(2) Four-cycle semilog paper for direct convenient plotting and checking of data on site in the field (see Fig. 3.3).
(3) Second irradiance sensor and cable, with 2 channel ratio readout meter, chart recorder or datalogger or field computer. A second sensor and appropriate instrumentation are probably necessary for the estimation of reflectance and irradiance attenuation if highly unstable lighting conditions prevail.

A2.1.3 Procedure
(1) Proceed in a boat to a suitable open water site (meaningful measurements are difficult to obtain in weed beds or near shadowing structures).
(2) Anchor boat if in standing water or slight currents (sensor support will need to be weighted). Drift with water current in rivers or other flowing water.
(3) Take air irradiance readings to check function, batteries etc. Ensure readout instrument is switched to 'AIR' position rather than 'WATER' (the signal in water is affected by the so-called 'immersion effect'). Under clear skies near noon, ambient irradiance is typically about 2000 μmol m^{-2} s^{-1} (illuminance ~100 klx).
(4) Suspend sensor(s) on sunny side of boat (to avoid boat shadow) preferably from a derrick or crane. Switch to 'WATER' calibration.
(5) Take at least five, or, preferably, ten or more, downwelling irradiance readings at constant depth intervals through the euphotic zone (i.e. down to the euphotic depth, z_{eu}. By definition $E_d(z_{eu}) = 0.01E_d(0-)$). Take care with depth measurements particularly in turbid water under wave action.
(6) Retrieve sensor(s) and check that near-surface irradiance levels have not changed during profile measurement. If they have, subsequent readings will not be comparable, making it difficult to estimate reflectance or irradiance attenuation.
(7) Reverse sensor orientation and repeat steps (4)–(6) for upwelling irradiance.
(8) Before moving to another site, check that the up- and down-welling irradiance profiles are log–linear and parallel, as is theoretically expected.

A2.1.4 Reporting of results
The irradiance attenuation coefficient for downwelling irradiance, K_d, is calculated for average depth $z_{av} = (z_2+z_1)/2$ as follows:

$$K_d\left(z_{av}\right) = \frac{\ln\left[E\left(z_1\right)/E\left(z_2\right)\right]}{z_2 - z_1} \qquad (A2.1)$$

where z_2 and z_1 are depths at which irradiances $E_d(z_2)$ and $E_d(z_1)$ are measured. K_d is a weak function of depth, z, and typically approaches a constant value with increasing depth in homogeneous water. Usually it is possible to estimate an average value of K_d by linear regression of $\ln E_d$ versus z, with a few near-surface readings excluded from the regression. The upwelling irradiance attenuation coefficient (K_u) is calculated similarly from the profile of upwelling irradiance (E_u) and the net vector irradiance (K_E) is calculated from the profile of E_d-E_u.

The reflectance coefficient is simply calculated:

$$R(z) = \frac{E_u(z)}{E_d(z)} \tag{A2.2}$$

Like K_d and K_u, R is a weak function of depth z. Usually it will be sufficient to report the attenuation coefficients and reflectance at the midpoint of the euphotic zone only ($z_m = z_{eu}/2$), i.e. $K_d(z_m)$, $K_u(z_m)$, $K_E(z_m)$ and $R(z_m)$. The surface reflectance, $R(0-)$, the reflectance coefficient at null depth below the water surface (in practice, usually within a few tens of centimetres), is of particular interest as a measure of the brightness of the water in the context of water colour, and should also be reported.

A2.1.5 Notes

(1) The above procedure applies to measurements under clear skies or stable overcast conditions. Under patchy cloud with unstable lighting, relative irradiance measurements should be taken with two identical sensors (e.g. Davies-Colley *et al.* 1984). For example, irradiance reflectance can be measured directly, independent of ambient light, as the ratio of the signal from an 'upward-looking' sensor to that from a 'downward-looking' sensor.

(2) Once $K_E(z_m)$ and $R(z_m)$ are obtained, Kirk's (1981a,b) nomograms can be used to estimate the PAR band average absorption and scattering coefficients (see the worked example in Table 3.1).

(3) The appropriate depth interval for irradiance measurements depends on water clarity, and may range from 5 cm in very highly light-attenuating water ($K_d \sim 10$ m^{-1}), to 5 m in very clear water ($K_d \sim 0.1$ m^{-1}).

(4) The quality of irradiance data can be checked before moving off-site by plotting directly on 4 cycle semilog paper (Fig. 3.3). Alternatively, if data have been logged electronically, a field computer can be used to check data graphically in the field.

A2.2 MEASUREMENT OF UNDERWATER BEAM TRANSMITTANCE FOR THE CALCULATION OF BEAM ATTENUATION COEFFICIENT

A2.2.1 Principle

The transmittance, T_c, of a highly collimated light beam is measured over a short path of light in water (e.g. Petzold & Austin 1968). The method given here assumes that an *in situ* (field) transmissometer of suitable optical design is available. However, laboratory transmittance meters, in which a water sample is contained in a glass cuvette, can also be used for estimating the beam attenuation coefficient, c, on samples no more than a few hours old.

The transmissometer used must have a very small angle of view so that virtually no scattered light is detected. In practice, because scattering in water is typically strongly biased in the forward direction (i.e. at low angles to the incident beam), this means that the half-angle of the cone of accepted (detected) light must be less than 1° and preferably less than 0.1°. Beam transmittance cannot be measured with a standard laboratory spectrophotometer unless specially modified to reduce greatly detection of forward-scattered light (e.g. Bricaud *et al.* 1983).

Access to the water by boat is assumed in standing bodies of water. Profiling of transmittance through the water column is desirable if density stratification is likely to be present. In rivers, the transmissometer can be deployed by wading or by lowering from a structure such as a jetty or bridge.

A2.2.2 Checklist

Essential
(1) Transmissometer suitable for measuring beam transmittance with cabling and readout instrument.
(2) Graduated support line (for profiling where water may be stratified, as in lakes).
(3) Field notebook or clipboard.

Desirable (for profiling in lakes and estuaries)
(1) Hydrological gauging winch or a depth (pressure) sensor, derrick/crane.
(2) Temperature and conductivity sensors.
(3) X–Y chart recorder (for direct plotting of per cent beam transmittance, T_c, temperature and conductivity versus depth).
(4) Datalogger or field computer.

A2.2.3 Procedure
(1) Calibrate the instrument. This is more conveniently done in air than in filtered or distilled water, since preparation of large volumes (tens of litres) of sufficiently clear calibration water is inconvenient. Transmittance will be lower than 100% in air owing to Fresnel reflection losses of about 4% at each air–glass interface (e.g. the Martek XMS instrument has a folded light path with four water-or-air–glass interfaces, so the air reading is set at $100(0.96)^4 = 85\%$).
(2) Lower the transmissometer into the water and simply record the transmittance. If density stratification is likely to be present, as in lakes or estuaries, record the transmittance profile, $T_c(z)$, down to the maximum depth of interest. Transmittance gradients are to be expected near the surface and near any density interfaces in the water column related to temperature or conductivity gradients.

A2.2.4 Reporting of results
The beam attenuation coefficient (in per metre units) is calculated:

$$c = \frac{\ln(100/T_c)}{r} \qquad (A2.3)$$

where T_c is the beam transmittance (in per cent) and r is path length of the light beam through water. The peak transmission wavelength of the instrument should be recorded. In stratified water bodies a plot of T_c (or c) versus depth should be made.

A2.2.5 Note
(1) Transmissometers with a chopped beam are to be preferred, other features being equal, because they are insensitive to ambient light which can affect even the best-collimated steady beam instruments.

A2.3 ABSORPTION AND SCATTERING MEASUREMENT BY SPECTROPHO-TOMETER. PART 1: ABSORPTION OF DISSOLVED ORGANICS ('COLOUR' MEASUREMENT) (YELLOW SUBSTANCE, *GELBSTOFF*, GILVIN)

A2.3.1 Principle

The absorbance or transmittance of a membrane-filtered water sample is measured using a laboratory spectrophotometer. The absorption coefficient, $a_f(\lambda)$ (= g_λ, g denoting *Gelbstoff* or gilvin (Kirk 1976)), is calculated from the measured absorbance, suitably corrected for any residual (filter-passing) scattering, and taking into account the path length of light in the cuvette.

Spectrophotometers are standard laboratory instruments which are designed to measure the absorbance, D, or transmittance, T, of a monochromatic light beam by non-turbid liquids.

A2.3.2 Checklist

Essential
   (1) Clean sample containers, preferably glass rather than plastic (which can leach or adsorb light-absorbing organics).
   (2) UV–visible laboratory spectrophotometer with cuvettes of 10 mm or greater path length (e.g. Anon 1984).
   (3) Filtration assembly, preferably glass.
   (4) Glass fibre filters (e.g. Whatman GF/C) and 0.2 μm pore size membrane filters.
   (5) High quality distilled or double-distilled water.

Desirable
   (1) Double-beam spectrophotometer with matched optical silica cuvettes of relatively long path length (40 to 100 mm).
   (2) Chart recorder or printer output from spectrophotometer (for scanning through the spectrum).
   (3) Baseline memory facility.
   (4) Five-figure absorbance readout.
   (5) Selectable integration times (long integration times at each wavelength, e.g. 10 s or more, can be used to smooth photometer noise).

A2.3.3 Procedure
   (1) Rinse sample containers twice with sample water before collection. In standing waters, collect water samples from at least 100 mm below the water surface so as to avoid sampling any floatables or organics in the surface microlayer. Chill samples to 4°C and keep dark. Filter samples as soon as possible, preferably within 24 h (Davies-Colley & Vant 1987).
   (2) Prefilter water samples through glass fibre filters then filter through membrane filters. Take care not to exceed about 0.5 atm differential pressure since greater pressures might disrupt cells of planktonic organisms. The first 50 ml of filtrate at each step should be discarded to minimize contamination. Filter assemblies and filtrate containers must be scrupulously clean.

(3) Set the wavelength at which absorption (or transmittance) is to be measured and zero the spectrophotometer (set transmittance = 100%) using distilled water as a blank in the cuvette (both cuvettes in a double-beam instrument).

(4) Rinse the sample cuvette twice with each sample before filling. Fill the cuvette (sample cuvette only in a double-beam instruments) with each sample filtrate in turn and read absorbance (transmittance). Always insert cuvettes with the same orientation.

(5) Check the absorbance (transmittance) of the blank for drift.

(6) Repeat steps (3) to (5) for all wavelength settings of interest. At least two wavelengths should be measured, one being an IR wavelength (e.g. 740 nm) for residual scattering correction. Recommended discrete wavelengths for measurement are 340 and 440 nm and at 740 nm to permit correction for residual scattering.

A2.3.4 Reporting of results

The absorption coefficient of yellow substance is calculated from the measured absorbance:

$$g_\lambda = \frac{2.303}{r} \left[D(\lambda) - D(\Lambda) \frac{\Lambda}{\lambda} \right] \tag{A2.4}$$

where D is absorbance, λ is wavelength and Λ is the IR wavelength used for scattering correction (e.g. 740 nm), and r is cuvette path length. The factor $2.303 = \ln(10)$ arises because absorbance is $-\log_{10}$(transmittance), whereas absorption coefficients are reported in $\log_e$ units.

The corrected absorption coefficients at 340 and 440 nm (i.e. g_{340}, g_{440}) should be reported (Kirk 1976, 1983; Davies-Colley & Vant 1987).

A2.3.5 Notes

(1) Glass containers and laboratory ware can be cleaned by oxidizing any adsorbed organic material in chromic acid. Cuvettes should be cleaned by soaking overnight in chromic acid. Plastic water sample containers can be cleaned by leaching with 1 M NaOH followed by exhaustive rinsing with tap, and finally, distilled, water.

(2) The glass fibre prefilter and membrane filter can be sandwiched together in order to replace two separate filter runs for each sample—if the filter residue is not to be used, e.g. for chlorophyll a measurement.

(3) Filtrates appear to be stable for at least one year if refrigerated to 4°C and kept in the dark in airtight glass bottles.

(4) If a chart recorder or printer output is available, scanning through the whole or part of the visible spectrum is feasible. Before scanning, a blank spectrum is run (or stored in machine memory if a baseline memory facility is available) and then sample absorbance (or transmittance) spectra are overlaid or printed as desired. In order to correct for residual scattering, the spectrum must be discretized by reading off (or printing out) absorbances (or transmittances) at intervals for example of 10 or 20 nm. Refer to Section A2.4 for more details on spectral scanning of water samples.

(5) Absorption coefficients of yellow substance (e.g. g_{340}, g_{440}), correlate passably

well with visual 'colour' of waters measured in platinum–cobalt units, and can be regarded as superseding the latter test.

A2.4 ABSORPTION AND SCATTERING MEASUREMENT BY SPECTROPHOTOMETER. PART 2: SPECTRAL ABSORPTION OF TOTAL WATER CONSTITUENTS

A2.4.1 Principle
The spectral absorption coefficient of the light-absorbing constituents (both dissolved and particulate) of a water sample is calculated from measurements made with a high quality laboratory UV–visible spectrophotometer.

Since natural water samples (unfiltered) are typically highly scattering it is imperative that measurements be made using a system for increasing the efficiency of detection of scattered light in the spectrophotometer (Kirk 1980, 1983; Davies-Colley 1983; Bricaud et al. 1983). Otherwise a diffuse light attenuance (rather than the absorbance) will be measured, and accurate estimation of absorption coefficients is not feasible.

A2.4.2 Checklist

Essential
 (1) Clean sample containers, preferably glass.
 (2) High precision, double-beam, UV–visible spectrophotometer with grating monochromator capable of 4 nm or less half-power bandwidth.
 (3) Matched, optical silica cuvettes, usually of 10 mm pathlength as a compromise between sensitivity and the need to ensure most scattered light is collected.
 (4) System for increasing the detected proportion of total scattered light, i.e. one of a turbid sample accessory (second sample position), an integrating sphere accessory, or light-diffusing ('opal glass') filters.
 (5) High quality distilled or double-distilled water.

Desirable
 (1) Chart recorder output (for scanning).
 (2) Baseline memory facility.
 (3) Five-figure absorbance readout.
 (4) Computer interfacing of spectrophotometer (to facilitate analysis of spectral data).
 (5) Long integration times at each wavelength (e.g. 10 s or more) to smooth photometer noise.

A2.4.3 Procedure
 (1) Rinse sample containers twice with sample water before collection. In standing waters, collect water samples from at least 100 mm below the water surface to avoid sampling any floatables or organics in the surface microlayer. Chill samples to 4°C and keep dark. Process samples as soon as possible, certainly within 48 h, unless optical stability of the samples over longer time periods has been demonstrated.

(2) Set the baseline (baseline memory facility) or scan the spectrum, with distilled water (blanks) in both sample and reference cuvettes. Wavelength scan should normally be from 740 nm in the near-IR to 340 nm in the near-UV. To minimize machine noise the widest available slit width (up to a maximum half-power bandwidth of 4 nm) should normally be used.

(3) Fill the sample cuvette with water subsampled from the (well-mixed) contents of the sample container.

(4) Scan the sample spectrum. The scanning speed will be a compromise between precision, which calls for slow scanning or long integration times at discrete wavelengths, and the effect of settling of the particulates, which calls for fast scanning speeds.

(5) Check for suspensoid settling immediately after completion of the sample scan by comparing absorbance at the initial wavelength before and after the scan. If settling is indicated by a lower absorbance after the scan, determine the rate of settling by timed absorbance measurements.

(6) If necessary, repeat the sample scan in steps (e.g. every 50 nm, dependent on settling rate), stopping at the end of each step to agitate the sample cuvette. This can be done with a piece of plastic laboratory film stretched over the cuvette mouth.

(7) Check the absorbance of the blank for drift.

A2.4.4 Reporting of results

The absorption coefficient of water constituents is calculated at all wavelengths (λ) of interest (cf. equation (A2.4)), or over the complete spectrum:

$$a_c(\lambda) = \frac{2.303}{r} \left[D_c^*(\lambda) - D_c^*(\Lambda)\frac{\Lambda}{\lambda} \right] \qquad (A2.5)$$

where D_c^* is the absorbance, in which the asterisk indicates absorbances measured using a device to improve collection efficiency of scattered light, and the subscript, c, indicates all constituents other than water itself.

The total absorption coefficient of the water can be calculated at discrete wavelengths by adding the absorption of water itself, $a_w(\lambda)$, to that of the constituents:

$$a(\lambda) = a_c(\lambda) + a_w(\lambda) \qquad (A2.6)$$

A2.4.5 Notes

(1) In principle the sample cuvette could be stirred continuously with a miniature paddle stirrer to obviate the need for periodic agitation of settleable particulates. This has not been tried by the authors. Care would need to be taken to ensure that the paddle did not obscure the light path.

(2) A more accurate correction for scattering than that in equation (A2.4) can be applied if the scattering spectrum is actually measured as detailed in Section A2.6.

(3) Accuracy of the spectrophotometer's monochromator should be checked periodically, e.g. with a lanthanide element filter (lanthanides have sharp, well-defined absorption peaks in the visible region). Photometric accuracy of spectrophotome-

ters can be checked using standard solutions of acid potassium dichromate which has a well-known molar absorptivity.

A2.5 ABSORPTION AND SCATTERING MEASUREMENT BY SPECTROPHOTOMETER. PART 3: SPECTRAL ABSORPTION OF PARTICULATE CONSTITUENTS

A2.5.1 Principle
The spectral absorption coefficient of the particulate constituents is estimated from measurements made with a high quality laboratory spectrophotometer on concentrated suspensions of particulate material captured by membrane filtration of water samples (Kirk 1980) (method A). Alternatively, the spectral absorption coefficient is estimated from measurements made directly on the particulate material captured on a glass fibre filter (Truper & Yentsch 1967) (method B).

In relatively highly absorbing waters (highly coloured waters), say with $a_c(440) >$ 2.3 m^{-1} ($D_c(440) = 0.010$ absorbance units in 10 mm cuvettes), the particulate absorption coefficient can be calculated as the difference between the total constituent absorption coefficient (Section A2.4) and the filtrate absorption coefficient (Section A2.3), i.e. $a_p(\lambda) = a_c(\lambda) - g_\lambda$. In less light-absorbing waters there is insufficient precision for direct measurement of a_c and the particulate absorption coefficient, a_p, is best measured directly on the suspensoids isolated by filtration.

A2.5.2 Checklist

Essential
Equipment as for Section A2.4, except that a filter assembly is required and either membrane (0.2 μm— method A) or glass fibre (GF/F— method B) filters.

Desirable
(1) Pressurized stirred filtration cell (e.g. Amicon) with 0.2 μm membrane or Nuclepore type filters (method A).
(2) Holder for wet filters (method B).
(3) Optional spectrophotometer features as in Section A2.4.

A2.5.3 Sampling
Sample waters as for Section A2.4.

A2.5.4 Procedure

Method A
(1) Concentrate the particulates from a measured volume of water by gentle filtration onto a 0.2 μm membrane filter in a standard filter assembly, or, preferably, using a stirred ultrafiltration cell.
(2) Using tweezers place the membrane filter residue-side down in a Petri dish of distilled water. Resuspend particulates by gentle rubbing with a glass rod (see Kirk 1980). Check to ensure that all visible material has been recovered (handle with care to ensure the membrane filter is not damaged).

(3) Wash all particulates into a small measuring flask and make up to volume with distilled water. The concentration factor, F, is the ratio of the volume filtered to the resuspension volume. F should be sufficient to ensure $D_p^*(440) > 0.01$ absorbance units. However, to avoid multiple scattering, the optical length, br should be <0.3, that is the scattering coefficient, b, should not exceed 30 m^{-1} in 10 mm cuvettes.

(4) Run blank and sample spectra as for total constituents (Section A2.4).

Method B

(1) Concentrate the particulates from a measured volume of water by gentle filtration onto a glass fibre GF/F filter in a standard filter assembly (e.g. Bricaud & Stramski 1990).

(2) Place the filter perpendicular to the sample beam of the spectrophotometer and as close as possible to the photomultiplier tube. Place a matched (unused, wetted) filter in the reference beam. A special holder can be constructed to hold the filters in position (e.g. Truper & Yentsch 1967).

(3) Run blank and sample spectra as for total constituents (Section A2.4).

A2.5.5 Reporting of results

Method A

The absorption coefficient of the particulates at the original concentration is calculated:

$$a_p(\lambda) = \frac{2.303}{Fr}\left[D_p^*(\lambda) - D_p^*(\Lambda)\frac{\Lambda}{\lambda}\right] \qquad (A2.7)$$

where $D_p^*(\lambda)$ is the measured absorbance of the concentrated particulates and F is the concentration factor. If the spectral trend of scattering is measured (Section A2.6), equation (A2.7) can be refined with a more accurate scattering correction.

Method B

The absorption coefficient of the particulates at the original concentration is calculated:

$$a_p(\lambda) = 2.303 D_p^*(\lambda)\frac{A_f}{V_f\beta(\lambda)} \qquad (A2.8)$$

where $D_p^*(\lambda)$ is the measured absorbance of the concentrated particulates, A_f is the filtration area of the filter assembly, V_f is the volume of filtered sample, and $\beta(\lambda)$ is the 'path amplification factor'. This last factor can be estimated as $1.63[D_p^*(\lambda)]^{-0.22}$ following Bricaud & Stramski (1990).

Finally, for both method A and B the total absorption coefficient of the water is calculated by adding the absorption of yellow substance and water itself to that of the particulates:

$$a(\lambda) = a_p(\lambda) + g_\lambda + a_w(\lambda) \qquad (A2.9)$$

A2.5.6 Notes

(1) The notes given in Section A2.4.5 apply equally here.

(2) A stirred filtration cell can be used to concentrate the particulates in method A. This has advantages over regular membrane filtration in that no resuspension step with inevitable loss of material is required. The water sample is simply filtered with continuous stirring until a sufficient degree of concentration has been achieved.

(3) A useful measure of resuspension efficiency in method A, at least in lake water, is the fraction of phytoplankton chlorophyll *a* remaining on the filter. Measurements of this tracer show that a recovery better than 90% can be routinely achieved for the phytoplankton (Davies-Colley, unpublished data), and, by inference, for all suspensoids present.

(4) Artefacts can arise with method B (and presumably method A also) owing to rapid decomposition of pigments in phytoplankton collected by filtration (Stramski 1990). Chemical fixation of the pigments captured on the filter (e.g. with paraformaldehyde) is the proposed remedy.

A2.6 ABSORPTION AND SCATTERING MEASUREMENT BY SPECTROPHOTOMETER. PART 4: SCATTERING MEASUREMENTS

A2.6.1 Principle

The measurement of the scattering spectrum is based on the principle that the difference in response of a spectrophotometer at two different scattered light detection efficiencies is entirely due to scattering. Thus the difference between the absorbance (diffuse attenuance), $D(\lambda)$, measured in the normal position in the spectrophotometer, and the (lower) absorbance $D^*(\lambda)$, measured on the same sample with increased scattered light detection efficiency, is a relative measure of scattering (Davies-Colley 1983; Davies-Colley *et al.* 1986).

A2.6.2 Checklist

(1) Equipment as for Sections A2.4, and A2.5 (method A).

(2) Nephelometer, preferably Hach 2100 or 2100A (see Section A2.7).

A2.6.3 Procedure

(1) Collect water samples and store as for Sections A2.3–A2.5 (process within 48 h).

(2) Measure turbidity with the nephelometer (see Section A2.7). Assume nephelometric turbidity in NTU is roughly numerically equal to the scattering coefficient (Kirk 1981b). Dilute or concentrate (see Section A2.5) the sample particulates as required to give a final turbidity between 5 and 30 NTU (for 10 mm cuvettes).

(3) Run absorbance spectra on the prepared samples from 740 to 340 nm as in Section A2.4. Repeat with the cuvette in the normal sample position of the spectrophotometer (absorbances will be higher owing to non-detection of scattered light).

A2.6.4 Reporting of results

Calculate the relative scattering coefficient:

$$\Delta\chi(\lambda) = \chi(\lambda) - \chi^*(\lambda) = \frac{2.303}{Fr}\left[D(\lambda) - D^*(\lambda)\right] \qquad (A2.10)$$

where D is apparent absorbance (diffuse attenuance) in the normal mode, D^* is that measured with increased detection efficiency of scattered light and F is the concentration factor (>1) or dilution factor (<1) ($F = 1$ for the unfractionated total water sample—Section A2.4).

A2.6.5 Notes

(1) Measuring scattering in the range $5\ m^{-1} < b < 30\ m^{-1}$ is a compromise between sensitivity (precision of measurements becomes too low for $b < 5\ m^{-1}$) and the desirability of avoiding significant multiple scattering which only exceeds 10% of total scattering in a 10 mm cuvette when $b > 30\ m^{-1}$.

(2) If the angular range of scattered light which is detected in the normal mode and in the second sample position can be evaluated, it is possible to make a rough estimate of the scattering coefficient. For a 'typical' angular dependence of scattering (e.g. that measured by Petzold (1972) in San Diego Harbour water— Fig. 2.3), a typical spectrophotometer collecting light scattered into a cone of 5° half-angle would detect roughly 50% of total scattered light. Use of an integrating sphere diffuser plate or turbid sample accessory increases scattered light detection efficiency to better than 90%. Therefore $\Delta\chi$ represents about 90% − 50% = 40% of total scattered light, and as a very rough approximation

$$b(\lambda) \sim \frac{100}{40}\Delta\chi(\lambda) = 2.5\,\Delta\chi(\lambda) \qquad (A2.11)$$

This approximation neglects multiple scattering (see Note (1)).

A2.7 NEPHELOMETRIC TURBIDITY MEASUREMENT

A2.7.1 Principle

Scattering, through a range of angles usually centred on 90°, is measured in arbitrary formazin turbidity units (NTU) with a nephelometer. The Hach 2100 or 2100A instrument is recommended, not because of any particular design virtues, but simply because this particular nephelometer has long been established in water resources work, and useful empirical correlations with a fundamental optical property of water, the scattering coefficient, have been established (e.g. Kirk 1981b).

Typical laboratory nephelometers, including the Hach 2100 and 2100A instruments, shine light from an incandescent lamp through the water sample contained in a glass cuvette, and a photomultiplier tube detects sideways-scattered light. Since the scattering function at 90°, $\beta(90)$, is only roughly correlated with the total scattering coefficient, 90° scattering measured in arbitrary units is only a rough index of total scattering. Instruments measuring scattering at forward angles (typically 15°) are to be preferred in principle, since $\beta(15)$ is typically better correlated with b than is $\beta(90)$ (Austin 1973; McCluney 1975).

Nephelometers vary widely in optical design (spectral power of light source and spectral sensitivity of detector, angular scattering range, optical geometry); thus nephelometric turbidity is a highly instrument-specific measurement, in spite of identical calibration to formazin.

A2.7.2 Checklist

Essential
(1) Clean sample bottles, plastic or glass.
(2) Nephelometer (preferably Hach 2100 or 2100A).
(3) Standard suspensions, typically rubber latex standards (secondary standards) calibrated to formazin (the primary standard).
(4) Glass cuvettes (cleaned with laboratory detergent).

Desirable
(1) High quality distilled water for making formazin primary standard.
(2) Membrane filtration unit for filtering distilled water used for zero checking.

A2.7.3 Procedure
(1) Collect water samples and store as for Sections A2.3–A2.6. Measure turbidity within 48 h.
(2) Calibrate instrument using secondary standards following manufacturer's handbook. Run a zero check with filtered, distilled water.
(3) Pour sample into glass cuvette after rinsing twice with sample water.
(4) Read turbidity in formazin units.
(5) Check instrument calibration frequently for drift.

A2.7.4 Reporting of results
Turbidity measurements with the Hach 2100 or 2100A instruments are reported directly as read on the instrument in nephelometric turbidity units (NTU). The readings cannot be converted to absolute scattering units (b in m^{-1} or $\beta(90)$ in m^{-1} sr^{-1}). However, by a happy coincidence nephelometric turbidity (in NTU) measured on the commonly used Hach 2100 and 2100A instruments is often numerically similar to the scattering coefficient, b (in m^{-1}) (Di Toro 1978; Kirk 1981b; Vant & Davies-Colley 1984).

A2.7.5 Notes
(1) Chilled samples should be warmed to room temperature prior to measurement to avoid condensation on the cuvette surface. Take care to avoid creating bubbles in the cuvette contents owing to dissolved gases coming out of solution on warming of chilled samples, or when mixing or transferring samples. Mixing is best accomplished by gently upending the sample container several times rather than overly vigorous shaking.
(2) Turbidity results should always be reported with a statement giving the instrument make and model number (e.g. Hach 2100A).
(3) Calibration of secondary standards should be checked at least twice yearly with formazin primary standard made up, following directions in standard methodological texts (e.g. APHA 1989).

A2.8 MEASUREMENT OF VISUAL WATER CLARITY BY SECCHI DISC

A2.8.1 Principle

The depth (vertical sighting range) at which a standard visual target (the white or black-and-white Secchi disc) can just be resolved by eye is recorded. This is a very convenient, direct, immediate, on-site, measure of visual water clarity which has a long tradition of use in marine and freshwater studies (Tyler 1968; Preisendorfer 1986).

The Secchi depth measurement of visual water clarity is often denigrated as 'rough' or 'semiquantitative'. In practice, it is probably more precise than turbidity measurement, for example, and is probably also more meaningful. The Secchi disc is now regarded as superseded by the black disc (Sections A2.9 and A2.10), but it is anticipated that some will wish to continue its use to maintain a historical time series.

A2.8.2 Checklist

Essential
 (1) Secchi disc(s) of appropriate size for the clarity range (see Note (4) in Section A2.8.5) painted matte white or in black-and-white quadrants.
 (2) Graduated line or mounting rod (turbid water).
 (3) Weight to hold graduated line vertical (the weight should be streamlined for work in currents).

Desirable
 (1) Underwater viewer (to reduce surface reflectance).

A2.8.3 Procedure
 (1) Proceed by boat to a suitable open water site.
 (2) Anchor boat in standing water or drift with the current in rivers or other flowing waters.
 (3) Lower the disc on the sunny side of boat (not the shaded side as is commonly, and wrongly, recommended).
 (4) Allow sufficient time (preferably 2 min) when looking at the disc near its extinction point for the eyes to adapt completely to the prevailing luminance level. Read the depth (z_1) at which the disc is judged to disappear.
 (5) Slowly raise disc and note the depth (z_2) of reappearance.
 (6) Record the Secchi depth as $z_{SD} = (z_1 + z_2)/2$.
 (7) A second observer should also read the Secchi depth to reduce the possibility of observer bias or an error in reading the graduated line. The two independent observations should agree within 10%.

A2.8.4 Reporting of results

The Secchi depth is reported as recorded, ideally with a note as to time of day, weather and sea conditions, and lighting conditions. Water colour (hue), observed through the underwater viewer, should also be recorded. The Secchi depth should not be regarded as giving reliable information about the penetration of diffuse light (irradiance attenuation) in the water body (Davies-Colley & Vant 1988).

A2.8.5 Notes

(1) Secchi depth measurements should normally be made as near to mid-day as possible since z_{SD} is slightly dependent on solar elevation and on adaptive illuminance. In practice avoid Secchi depth measurements at low solar altitudes, say below 20°.

(2) A record of cloud cover (10ths), time of day, and sea conditions at the time of observation may help with later interpretation.

(3) Occasionally, meaningful measurements can be made from a fixed structure (e.g. bridge or jetty) rather than from a boat, or by wading in very turbid waters. Care must be taken that the water through which the disc is viewed is not shadowed.

(4) A range of different Secchi discs should be used for different clarity waters to ensure that angular size of the disc remains constant in the range 2°–10°. Table A2.1 gives recommended disc size as a function of clarity (visual range).

(5) As a rule of thumb, water depth should be at least 50% greater than the Secchi depth so that the disc is viewed against the water background, not bottom-reflected light.

Table A2.1. Disc diameter as a function of visual range

Disc diameter	Visual range (2°–10° of arc)
20 mm	0.15–0.5 m
60 mm	o.5–1.5 m
200 mm	1.5–5 m (10 m maximum)
600 mm	5–15 m (30m maximum)

A 600 mm diameter disc is inconveniently large for some purposes. A 200 mm diameter disc can be used up to about 10 m maximum if the larger-sized disc is unavailable.

A2.9 MEASUREMENT OF VISUAL WATER CLARITY USING A BLACK DISC ('BLACK SPOT'). PART 1: *IN SITU* MEASUREMENT

A2.9.1 Principle

The maximum vertical and horizontal sighting ranges of an all-black disc in water are recorded. In shallow waters, particularly small rivers, only a horizontal sighting range can be observed, but this alone is a highly valuable measurement since it immediately yields an estimate of the beam attenuation coefficient of the water (Davies-Colley 1988).

The black disc observation has all the virtues of the Secchi disc observation ('low tech', direct and immediate) but in addition has important practical and theoretical advantages (Davies-Colley 1988). The black disc sighting range relates more closely to in-water sighting ranges of practical importance than does the depth of visibility of the high contrast Secchi disc, and can be measured in waters too shallow for Secchi depth observation. The black disc sighting range can be used to estimate important optical properties of the water: the beam attenuation coefficient, c, and irradiance attenuation coefficient, K (for illuminance).

The black disc is now regarded (e.g. Davies-Colley 1988) as having superseded the white (Secchi) disc for the measurement of visual water clarity. Use of the black disc for vertical measurements is identical to use of the Secchi disc.

A2.9.2 Checklist

Essential
(1) Black disc of appropriate size (Table A2.1) constructed identically to a Secchi disc but painted matt black.
(2) Graduated line or tape measure. (The 20 mm diameter disc used in very turbid water is conveniently mounted permanently on a graduated steel rod with a long axis parallel to the viewing path: see Fig. 3.12B.)
(3) Weight to hold line taut and vertical during vertical observations.
(4) Black-painted pole to hold disc during horizontal observations.
(5) Underwater viewer designed for vertical and horizontal observations in water (Fig. 3.12). Alternatively, observations can be made by snorkel divers.

Desirable
(1) Steel trough designed for off-site measurement of black disc visibility (Davies-Colley & Smith 1992) (see Section A2.10).

A2.9.3 Procedure

Method A—lake, reservoir, estuary or large ('optically deep') river
(1) Proceed by boat to a suitable open water site.
(2) Anchor boat in standing water or drift with the current in rivers or other flowing waters.
(3) Lower the weighted disc on the sunny side of the boat. Take vertical readings z_1 and z_2 as for the Secchi disc procedure. The vertical black disc reading is $z_{BD} = (z_1 + z_2)/2$.
(4) Attach the disc to the pole and have an assistant hold the disc out from the boat hull, perpendicular to the plane of the sun so as to avoid shadowing the path of sight. Observe the disc horizontally using the underwater viewer. The horizontal black disc reading is $y_{BD} = (y_1 + y_2)/2$ where y_1 is the horizontal disappearance distance and y_2 is the horizontal reappearance distance. In clear water a second boat may be needed to obtain sufficient path of sight in the water.
(5) A preferable, but less convenient, method is to have snorkel divers make both horizontal and vertical observations in the water. Observations by divers may be required in clear lakes where the sighting range exceeds the length of the boat.

Method B—small river or clearer large river
(1) Observer and assistant wade to >300 mm depth taking care to minimize bed sediment disturbance.
(2) Disturbed sediment is allowed to flush downstream before readings are taken.
(3) The assistant holds the black disc mounted on the pole out sideways and perpendicular to the plane of the sun so as to avoid shadowing the path of sight.

(4) The observer notes the horizontal black disc reading $y_{BD} = (y_1 + y_2)/2$ as above.

With both deep water and shallow water measurements, the observer and assistant should change places to provide two independent observations of black disc visibility. The two observations should agree within 10% when both observers have good, normal vision. Failing such agreement the observations should be repeated in an attempt to achieve a consensus.

A2.9.4 Reporting of results

The vertical and horizontal black disc ranges are reported as recorded with a note as to time of day, cloud cover (10ths) and sea conditions. Water colour (hue), observed as the underwater space light during the disc observations, should also be recorded (see Section A2.11).

The beam attenuation coefficient, c, can be estimated from the horizontal range, y_{BD}:

$$c = \frac{4.8}{y_{BD}} \qquad (A2.12)$$

The vertical black disc range (where measured) gives an estimate of $c + K$, where K is the illuminance attenuation coefficient (similar in magnitude to the irradiance attenuation coefficient):

$$c + K = \frac{4.8}{z_{BD}} \qquad (A2.13)$$

This permits estimation of the illuminance attenuation coefficient:

$$K = (c + K) - c = \frac{4.8}{z_{BD}} - \frac{4.8}{y_{BD}} \qquad (A2.14)$$

A2.9.5 Notes

(1) Measurements should not be made under poor lighting conditions. Definitive guidelines can not be given, but a guess is that illuminance should exceed 1 klx or about 1/100 of the illuminance of an open site on a bright clear day. In practice, avoid taking measurements near dawn or dusk.

(2) Vertical measurements should not be taken at solar altitudes lower than, say, 20°, because the illuminance attenuation coefficient, K, is slightly dependent on solar altitude.

(3) A range of different-sized black discs should be available for use in waters of different clarity, to ensure that angular size of the black disc stays in the range 2°–10°. Fig. 3.13 gives disc size as a function of clarity (visual range) as does Table A2.1.

(4) Avoid sighting over water paths with shadows, such as those cast on river waters by overhanging trees or high banks.

(5) The unrestricted path of sight in the water should exceed the sighting range of the black disc by 50%. As for the Secchi observation, this rule of thumb should ensure that the target is being seen against the water background, rather than against submerged surfaces.

A2.10 MEASUREMENT OF VISUAL WATER CLARITY USING A BLACK DISC ('BLACK SPOT'). PART 2: OFF-SITE MEASUREMENT IN A TROUGH

A2.10.1 Principle

The horizontal sighting range of an all-black disc in water can be measured, at least in relatively turbid waters, on water samples contained in a shallow trough (Davies-Colley & Smith 1992). The light field in the water, against which the black disc is observed as a silhouette, is set up by reflection of light from the trough sides and bottom. The off-site measurement should ideally be carried out within minutes of sample collection, say on the bank or shore of the water body, but later measurement might be feasible if flocculation of light-attenuating particles in the sample is not too rapid.

Off-site measurement of visual water clarity may be valuable when *in situ* measurement is precluded because of poor access or dangerous conditions, or because the lighting of the water body is poor (as is often the case with effluents accessed through a manhole). A more compelling reason for off-site measurement is when the visual clarity is very low (say, y_{BD} < 100 mm, as with rivers in flood or in effluents) when the clarity is better measured on a diluted water sample rather than observed directly.

A2.10.2 Checklist

Essential
 (1) Black disc.
 (2) Underwater viewer (Fig. 3.12).
 (3) Trough with reflective walls (e.g. uncoated galvanized sheet steel).
 (4) Sample container (volume depends on trough volume but 20 l may be required).
 (5) Tape measure or rule.

Desirable
 (1) Volumetric labware (5 and 1 l plastic measuring cylinders) for volumetric dilutions.
 (2) Tap water diluent of known high clarity.

A2.10.3 Procedure
 (1) Rinse sample container twice with sample water before collection. In standing waters, collect water samples from at least 100 mm below the water surface to avoid sampling any floatables or organics in the surface microlayer.
 (2) If necessary, dilute the sample volumetrically with tap water.
 (3) Set up the trough for black disc observations on the shore of the water body if possible, so as to minimize sample storage time. Orient the trough perpendicular to the plane of the sun. Level the trough and pour in the (mixed) sample, diluted if necessary.
 (4) Place the black disc in the trough, making sure that the distance behind the disc to the trough end is at least 50% of the sighting range. (This rule of thumb should avoid the possibility that end-reflections will distort the light field in the water.)
 (5) Take black disc readings as with *in situ* observations. Agitate the trough contents with a household spatula to prevent any settling of particles. Take great care not to contaminate the sample with dust.

A2.10.4 Reporting of results

The horizontal black disc ranges are reported as recorded on undiluted samples, with a note as to time of day and cloud cover (10ths) and general weather conditions.

For diluted samples, the *in situ* clarity is calculated from the mass balance on light attenuation:

$$c_{mix}V_{mix} = c_sV_s + c_{dil}V_{dil} \qquad (A2.15)$$

where V is volume and the subscripts denote the mixture (mix) of dilution water (dil) and sample (s). Solving for the unknown sample attenuation coefficient and sample visual clarity, with dilution factor, $F = V_{mix}/V_s$, we obtain:

$$c_s = c_{mix}F - c_{dil}(F-1) \qquad (A2.16a)$$

or, alternatively,

$$\frac{1}{y_{BDs}} = \frac{F}{y_{BDmix}} - \frac{F-1}{y_{BDdil}} \qquad (A2.16b)$$

A2.10.5 Note
(1) The notes given for Section A2.9 apply equally here. Measurements should not be made under poor lighting conditions, and care should be taken not to shadow the path of sight in the trough.

A2.11 COLOUR (HUE) OBSERVATIONS

A2.11.1 Principle
The hue of the water space light (i.e. the hue associated with the light field in the water) is viewed through an underwater viewer and matched to colour standards in the Munsell system (Davies-Colley *et al.* 1988; Davies-Colley & Close 1990) or described subjectively. The brightness and purity of the water colour can be noted but no firm recommendations for the specification of these attributes of colour can be given until the appropriate research is conducted. Note, though, that measurement of near-surface illuminace reflectance, $R(0-)$, quantifies the brightness of the colour of a water body (Section A2.1).

In shallow water the water colour can be viewed horizontally using the underwater viewer designed for black disc measurements as described in Sections A2.9 and A2.10 However, the colour of water viewed horizontally is somewhat different from that viewed vertically downwards (brighter and of lower purity, but usually only slightly different in hue), a fact well known to divers. Thus it is important to record the direction of viewing.

A2.11.2 Checklist
(1) Underwater viewer.
(2) Munsell book of colour or Munsell hue standards.

A2.11.3 Procedure
The assumption is that the colour observations will be made at the time the black disc equipment is deployed for the measurement of visual clarity.

(1) View the water colour horizontally or vertically, as appropriate, with the black disc viewer. Describe the perceived hue (see Table A2.2).
(2) Juxtapose the Munsell standards and the viewer and rapidly alternate viewing between the water and standard colour, while searching for a match.
(3) Record the best hue match available.

Table A2.2. Munsell hue numbers corresponding to hue descriptions

Hue	Munsell hue code	Munsell hue number (range)
Blue	5 B	65 (60–70)
Blue–green	5 BG	55 (50–60)
Green	5 G	45 (40–50)
Green–yellow	5 GY	35 (30–40)
Yellow	5 Y	25 (20–30)
Yellow–red (= 'orange')	5 YR	15 (10–20)
Red	5 R	5 (0–10)

A2.11.4 Reporting of results

Report the hue using the Munsell code (e.g. 7.5 GY is the 7.5th value of the green–yellow hue range). Also report the viewing direction (horizontal or vertical).

It will probably not be possible to match saturation and brightness—and if a match is obtained it may be spurious because of the viewing conditions. However, the hue match is meaningful. Davies-Colley *et al.* (1988) have shown that Munsell hue matches agree fairly well with Munsell hue correlations calculated from spectroradiometric observations.

If a Munsell hue standard set is not available, subjective description of the water colour should still be made as an adjunct to visual water clarity observations (e.g. Davies-Colley & Close 1990) (Sections A2.8–A2.10). The hue should be assigned to one of the categories in Table A2.2. A colour described informally as 'brown' for example is actually a low saturation orange (yellow–red).

A qualifier should be given to the hue as either 'bright' or 'dark' and either 'greyish' or 'pure'. Examples are 'A bright, greyish yellow–red' ('muddy brown') and 'A dark, pure blue–green' (colour of optically pure waters).

A2.11.5 Notes

(1) Non-spectral colours (purples) are not observed in natural waters except as a result of the growth of unusually pigmented organisms (e.g. *Euglena* spp.) or pollution by industrial dyes. Blues are only found in the very clearest natural waters and reds only in the most humic-stained waters.
(2) Hue description can be carried out in shallow water (e.g. small rivers) by viewing the water space light horizontally with the viewer used for the black disc observa-

tion (Sections A2.9 and A2.10). However, it must be realized that this hue may be very slightly different from the hue of zenith-directed light. It should be clearly recorded as 'Horizontal colour'.

(3) In optically deep water bodies, irradiance measurements (Section A2.1) can be used to calculate the near-surface reflectance coefficient, a useful index of the brightness of a water colour.

REFERENCES

APHA 1989: Standard methods for the examination of water and wastewater, 17th Ed. America Public Health Association, American Water Works Association, Water Pollution Control Federation, Washington, DC.

Anon 1984: Colour and turbidity of waters. 1981 tentative methods. Standing Committee of Analysts, Department of the Environment. HMSO, London. 25p.

Austin, R. W. 1973: Problems in measuring turbidity as a water quality parameter. USEPA seminar on methodology for monitoring the marine environment. Seattle, WA, October 16–18, 1973.

Bricaud, A.; Stramski, D. 1990: Spectral absorption coefficients of living phytoplankton and nonalgal biogeneous matter: a comparison between the Peru upwelling area and the Sargasso Sea. *Limnology and oceanography 35*: 562–582.

Bricaud, A.; Morel, A.; Prieur, L. 1983: Optical efficiency factors for some phytoplankters. *Limnology and oceanography 28*: 816–832.

Davies-Colley, R. J. 1983: Optical properties and reflectance spectra of three shallow lakes obtained from a spectrophotometric study. *New Zealand journal of marine and freshwater research 17*: 445–459.

Davies-Colley, R. J. 1988: Use of a black disc for measuring water clarity. *Limnology and oceanography 33*: 616–623.

Davies-Colley, R. J.; Close, M. E. 1990: Water colour and clarity of New Zealand rivers under baseflow conditions. *New Zealand journal of marine and freshwater research 24*: 357–365.

Davies-Colley, R. J. Smith, D. G. 1992: Offsite measurement of visual clarity of waters in a trough. *Water resources bulletin 28*: 1–7.

Davies-Colley, R. J.; Vant, W. N. 1987: Absorption of light by yellow substance in freshwater lakes. *Limnology and oceanography 32*: 416–425.

Davies-Colley, R. J.; Vant, W. N. 1988: Estimation of optical properties of water from Secchi disc depths. *Water resources bulletin 24*: 1329–1335.

Davies-Colley, R. J.; Vant, W. N.; Latimer, G. J. 1984: Optical characterisation of natural waters by PAR measurement under changeable light conditions. *New Zealand journal of marine and freshwater research 18*: 455–460.

Davies-Colley, R.J.; Pridmore, R. D.; Hewitt, J. 1986: Optical properties of some freshwater phytoplanktonic algae. *Hydrobiologia 133*: 165–178.

Davies-Colley, R. J.; Vant, W. N.; Wilcock, R. J. 1988: A comparison of lakewater colour as observed directly and as calculated from underwater spectral irradiance. *Water resources bulletin 24*: 11–18.

Di Toro, D. M. 1978: Optics of turbid estuarine waters: Approximations and applications. *Water research 12*: 1059–1068.

Kirk, J. T. O. 1976: Yellow substance (Gelbstoff) and its contribution to the attenuation of photosynthetically active radiation in some inland and coastal south-eastern Australian waters. *Australian journal of marine and freshwater research 27*: 61–71.

Kirk, J. T. O. 1977: Attenuation of light in natural waters. *Australian journal of marine and freshwater research 28*: 497–508.

Kirk, J. T. O. 1980: Spectral absorption properties of natural waters: contribution of the soluble and particulate fractions to light absorption in some inland waters of South-eastern Australia. *Australian journal of marine and freshwater research 31*: 287–296.

Kirk, J. T. O. 1981a: Monte-Carlo study of the nature of the underwater light field in, and the relationships between optical properties of, turbid yellow waters. *Australian journal of marine and freshwater research 32*: 517–532.

Kirk, J. T. O. 1981b: Estimation of the scattering coefficient of natural waters using underwater irradiance measurements. *Australian journal of marine and freshwater research 32*: 533–539.

Kirk, J. T. O. 1983: Light and photosynthesis in aquatic systems. Cambridge.

McCluney, W. R. 1975: Radiometry of water turbidity measurement. *Journal of the Water Pollution Control Federation 47*: 252–266.

Petzold, T. J. 1972: Volume scattering functions for selected ocean waters. Visibility Laboratory, Scripps Institution of Oceanography Technical Report, SIO Ref 72-78. 79 p.

Petzold, T. J.; Austin, R. W. 1968: An underwater transmissometer for ocean survey work. Scripps Institution of Oceanography Reference 68-9. 5 p.

Preisendorfer, R. W. 1986: Secchi disc science: visual optics of natural waters. *Limnology and oceanography 31*: 909–926.

Stramski, D. 1990: Artifacts in measuring absorption spectra of phytoplankton collected on a filter. *Limnology and oceanography 35*: 1804–1809.

Truper, H. G.; Yentsch, C. S. 1967: Use of glass fibre filters for the rapid preparation of in vivo absorption spectra of photosynthetic bacteria. *Journal of bacteriology 94*: 1255–1256.

Tyler, J. E. 1968: The Secchi disc. *Limnology and oceanography 13*: 1–6.

Vant, W. N.; Davies-Colley, R. J. 1984: Factors affecting clarity of New Zealand lakes. *New Zealand journal of marine and freshwater research 18*: 367–377.

Westlake, D. F. 1986: Measurement of underwater light. *In* The direct determination of biomass of aquatic macrophytes and measurement of underwater light. Standing Committee of Analysts, Department of the Environment. HMSO, London. 45 p.

Appendix 3: Discharge of light-attenuating effluents to a river: worked example calculations

PROBLEM 1: DISCHARGE OF CLAYS FROM A SETTLING POND

A settling pond effluent containing highly scattering clays is discharged to a river. The effluent water was originally abstracted from the river. The worst combination of conditions occurs at low river flow when the river water is relatively clear and, therefore, least able to assimilate light-attenuating materials, both because it is highly sensitive to change and because there is then little flow available for dilution. The data given in Table A3.1 and illustrated in Fig. A3.1A are for the 95 percentile low flow—which is chosen for the purposes of calculation. The spectral absorption and scattering data were obtained from spectrophotometric measurements. The effluent is weakly absorbing (cream coloured) but intensely scattering of light as shown in Table A3.1 and Fig. A3.1B. The river is to be protected for both recreational water uses and habitat for aquatic life.

- (a) Calculate the resulting visual clarity of the river once the effluent is fully mixed.
- (b) Calculate the resulting light penetration and colour once the effluent is fully mixed.
- (c) Define the assimilative capacity of the river for this effluent, i.e. what conditions on a consent to discharge would be appropriate (assuming the full river flow is available for mixing)?

Note that in general the whole river flow would not be regarded as available for effluent dilution before standards must be met.

Solution

- (a) The effect of the settling pond effluent on visual clarity of the river is easy to calculate since this does not require consideration of the spectral variation of the optical properties. We can calculate the beam attenuation coefficient, c, downstream, once the effluent is fully mixed, from the balance of attenuation cross-section flow:

Table A3.1. Optical properties of the river and effluents in the example calculations

---

River

At the 95 percentile flow (a fairly low flow) the river has the following character:

river flow, $Q = 26\ \mathrm{m^3\ s^{-1}}$
maximum river depth (pools) is 4 m
black disc clarity, $y_{BD} = 1.52\ \mathrm{m}$

Therefore beam attenuation coefficient, $c \sim 4.8/y_{BD} = 3.16\ \mathrm{m^{-1}}$.

$a(560) = 0.284\ \mathrm{m^{-1}}$, $b(560) = 2.87\ \mathrm{m^{-1}}$
$a(640) = 0.397\ \mathrm{m^{-1}}$, $b(640) = 2.61\ \mathrm{m^{-1}}$, from spectrophotometric measurements
$a_w(560) = 0.071\ \mathrm{m^{-1}}$, $a_w(640) = 0.329\ \mathrm{m^{-1}}$, from Smith & Baker (1981)
diffuse attenuation coefficient, $K(\mathrm{PAR}) = 0.42\ \mathrm{m^{-1}}$

Therefore euphotic depth, $z_{eu} = 4.6/K_d = 11\ \mathrm{m}$. (Note that this is a virtual euphotic depth, since maximum river depth is 4 m.)

reflectance coefficient, $R_0(\mathrm{PAR}) = 5.5\%$
hue is described as 'yellowish–green' (Munsell 7.5 GY or 37.5 Munsell hue units)

Fig. A3.1 gives a plot of the spectral absorption and scattering coefficients measured by spectrophotometry on the river water.

Problem 1: Settling pond effluent

effluent flow is up to 0.45 $\mathrm{m^3\ s^{-1}}$; this water was originally abstracted from the river upstream
black disc clarity of the effluent = 0.078 m (Measured indirectly on a diluted sample in a trough)

Therefore, $c_{eff}(550) = 62\ \mathrm{m^{-1}}$. Absorption is very low and can be neglected; therefore $b_{eff}(550) = c_{eff}(550) = 62\ \mathrm{m^{-1}}$. Fig. A3.1B gives a plot of the spectral scattering coefficient measured by spectrophotometry on the settling pond effluent.

Problem 2: Kraft effluent

effluent flow is 1.2 $\mathrm{m^3\ s^{-1}}$; this water was originally abstracted from the river upstream
$a_{eff}(560) = 9.39\ \mathrm{m^{-1}}$, $a_{eff}(640) = 3.40\ \mathrm{m^{-1}}$ (absorption coefficient of effluent constituents alone is 3.07) from spectrophotometry
black disc clarity = 0.27 m

That is, $c_{eff}(550) = 17.8\ \mathrm{m^{-1}}$. Scattering can be calculated using this c value together with the absorption coefficient at 550 nm: $b_{eff}(550) = c_{eff}(550) - a_{eff}(550) = 17.8 - 9.4 = 8.4\ \mathrm{m^{-1}}$. Fig. A3.1B gives a plot of the spectral absorption coefficient measured by spectrophotometry on the Kraft effluent. The absorption spectrum is approximately exponential in shape—resulting in a straight line plot on the logarithmic scale. The scattering is fairly low compared with absorption and is assumed constant at 8.4 $\mathrm{m^{-1}}$, independent of wavelength throughout the spectrum.

Directly measured values are underlined.

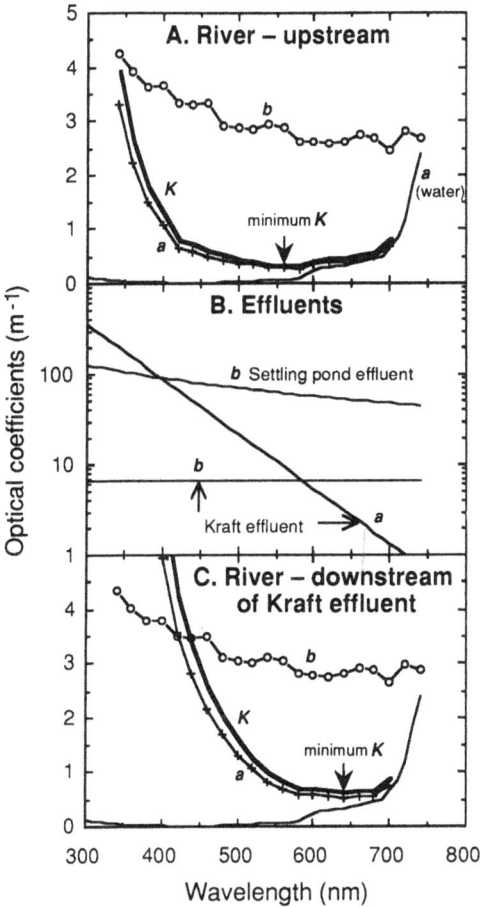

Fig. A3.1. Spectral absorption and scattering coefficients for the river and effluents in the worked examples. The spectra are derived from spectrophotometric measurements and published relationships. (A) Spectral trends of optical quantities characterizing the river upstream. b is the scattering coefficient, and a is the absorption coefficient. The absorption spectrum of pure water is also shown (this dominates absorption at wavelengths >600 nm). The irradiance attenuation coefficient, K, was calculated from Kirk's (1984a) equation for a solar altitude of 45°. (B) Spectral trends of optical quantities characterizing the two effluents. The scattering coefficient for the settling pond increases with falling wavelength, whereas the (much lower) scattering for the Kraft effluent is assumed to be independent of wavelength. The absorption spectrum for the Kraft effluent rises exponentially with decreasing wavelength (thus plotting as a straight line on a log scale of absorption). Absorption by the settling pond effluent is negligible. (C) Spectral trends of optical quantities characterizing the river downstream of the Kraft pulp mill. Symbols are as in (A). Notice how the wavelength of minimum light attenuation has shifted from about 560 to 640 nm.

$$c_d Q = c_u (Q - q) + c_{eff} q \qquad (A3.1)$$

where Q is river flow rate (upstream of the discharge but downstream of the point at which water is abstracted from the river), q is effluent flow rate, and the subscripts 'u' and 'd' refer to the effluent and the river up and downstream of the discharge. Note that this form of the 'mass balance' is slightly different from equation (5.2) because the water in the effluent was originally abstracted at flow rate q from the river.

Putting in values from Table A3.1 we obtain $c_d = 4.18$ m^{-1}. The proportional change in beam attenuation, $\Delta c/c_u = (4.18-3.16)/3.16 = 32\%$. Thus the change in c is greater than the guideline value of 25% corresponding to a maximum visual clarity reduction of 20%. The black disc clarity is predicted to drop from 1.52 m to $4.8/4.18 = 1.15$ m, a 24% reduction.

(b) Since the effluent is scattering rather than absorbing of light, it will have little influence on the spectral quality of light penetrating into the river water. The plot of absorption and scattering coefficients for the river in Fig. A3.1A shows that the light which penetrates deepest into the river is around 560 nm, and the corresponding hue is green–yellow. This is not expected to change as a result of the discharge because the effluent is only weakly absorbing.

The irradiance attenuation coefficient for PAR is usually only slightly greater than $K_d(\lambda)$ for the most penetrating waveband. Therefore, in this river water $K_d(\text{PAR}) \sim K_d(560)$ and we can examine the proportional change in $K_d(\text{PAR})$ by calculating the effect on $K_d(560)$ using Kirk's (1984a) equation (for a sun angle of 45°):

$$K_d = 1.181\left(a^2 + 0.0170ab\right)^{1/2} \qquad (A3.2)$$

We will henceforth drop the subscript d to avoid potential confusion with that denoting 'downriver'. Fig. A3.1A shows the calculated $K(\lambda)$ spectrum for the river. With $a(560) = 0.284$ m^{-1} and $b(560) = 2.87$ m^{-1} from the data given in Table A3.1 we obtain $K(560) = 0.36$ m^{-1} for the river upstream. This appears reasonably consistent with the measured value of $K(\text{PAR})$, being slightly higher at 0.42 m^{-1}.

To calculate the change in $K(560)$ we first need to calculate the change in $b(560)$ from a mass balance on b identical in form to that for c (equation (A3.1)). Using the data in Table A3.1 this yields $b_d(560) = 3.89$ m^{-1}, a 36% increase due to the effluent. Absorption can be assumed to be unchanged by the weakly absorbing settling pond effluent, so $a_d(560)$ remains equal to the upstream value of 0.284 m^{-1}. The corresponding value of K, using equation (A3.2), is 0.373 m^{-1}, and therefore K is only increased by 2.8% which is well within the 11% increase corresponding to the guideline maximum reduction in euphotic depth of 10% (Sections 5.3.3, 5.3.5). This calculation demonstrates how much more strongly K is dependent on absorption than on scattering of light, such that turbid, but weakly coloured, effluents affect light penetration only slightly.

The maximum river depth (4 m) is less than half the (virtual) euphotic depth ($z_{eu} = 11$ m), so we need to check that change in lighting at the riverbed is less than 20%. This will certainly be the case since the '20% maximum reduction in light' guideline (Section 5.3.5) is less restrictive than the '10% maximum change in euphotic depth' guideline at depths shallower than $0.5z_{eu}$.

The most penetrating light is also the light backscattered most strongly so we can examine the change in brightness in relation to $R(0-,560)$, i.e. the reflectance coefficient at zero depth for 560 nm light. Assuming $R(0-) \sim 0.0063b/a$ we obtain

$R(0-,560) = 0.0637$. This is in fair agreement with the measured $R(0-,\text{PAR})$ value of 0.051 (Table A3.1). Downstream of the discharge $R(0-,560) = 0.0863$, a 35% increase which is within the guideline of 50% change (Section 5.3.5). This illustrates what is probably a near-universal rule: addition of light-attenuating constituents to a water is unlikely to change brightness significantly (i.e. more than the guideline) unless other attributes of optical water quality, particularly visual clarity, exceed their guideline values.

(c) Since only the guideline for visual clarity is exceeded (slightly) by the effluent discharge we can calculate the assimilative capacity of the river in terms of visual clarity alone. The guideline condition is that visual clarity should not be reduced by more than 20%, which can be stated

$$y_{BDd} > 0.8 y_{BDu} \text{ or, equivalently,} c_d < 1.25 c_u \quad (A3.3)$$

Combining this condition with the mass balance equation (A3.1) defines the assimilative capacity in terms of the effluent mass flow:

$$c_{\text{eff}} q < c_u Q(q/Q + 0.25) \quad (A3.4)$$

That is, the maximum attenuation cross-section flow of the effluent is proportional to that in the river ($c_u Q$) and also depends on the ratio of effluent to river flow. Equation (A3.4) is general but when, as is typical, q/Q is very small, $q \ll Q$ (in practice $q/Q < 0.025$ to avoid an error greater than 10%) we have simply $c_{\text{eff}} q < 0.25 c_u Q$. That is, the attenuation cross-section flow of the effluent must not exceed one-quarter that of the river. This is a very useful simplification for rough calculations with relatively small discharges of turbid effluents.

For the river and effluent under consideration, at the 95 percentile river flow, $c_{\text{eff}} q$, must be <22.0 m^2 s^{-1} (note the units: optical 'area' per unit time). This could be achieved if the effluent quality were improved (e.g. by increasing pond size and residence time) so as to reduce attenuation coefficient in the pond effluent to <22.0 m^2 s^{-1}/(0.45 m^3 s^{-1}) = 49 m^{-1} (corresponding to an increase in visual clarity of the effluent from 78 mm to about 98 mm). Alternatively, if the effluent quality were unchanged ($c_{\text{eff}} = 62$ m^{-1}), effluent flow would need to be reduced from 0.45 to about 0.35 m^3 s^{-1}.

PROBLEM 2: DISCHARGE OF KRAFT PULP MILL EFFLUENT

A Kraft pulp mill discharges to the river in Problem 1. The optical quality of the bleached Kraft effluent is given in Table A3.1 and illustrated in Fig. A3.1B.

(a) Calculate the resulting change in visual clarity of the river caused by this effluent alone.

(b) Calculate the resulting change in light penetration.

(c) Calculate the change in colour.

Note that if the settling pond were using up all the assimilative capacity of the river in terms of visual clarity, the Kraft mill effluent could not be permitted to degrade the visual clarity further. Otherwise, in a reach of a river with multiple discharges the overall cumulative degradation of optical water quality would be very great. In practice the assimilative capacity of the river would normally be shared between the settling pond, the Kraft mill, and any other dischargers.

Solution

(a) The bleached Kraft effluent is strongly light absorbing but rather weakly light scattering. Spectrally selective light absorption will shift the wavelength of maximum light penetration in the river from about 560 nm to longer wavelengths. Thus it is not possible in this case (in contrast to Problem 1) to examine colour and clarity effects using single-wavelength calculations—with one exception. The visual clarity effect can still be simply calculated.

The beam attenuation coefficient of the Kraft effluent calculated from the effluent clarity is 17.8 m^{-1} (Table A3.1). From equation (A3.1) we can calculate the resulting beam attenuation in the river downstream as 3.84 m^{-1} (a 21% increase) and the corresponding visual clarity is 1.25 m (an 18% decrease). Evidently this discharge on its own does not quite breach the guideline for visual clarity. However, the strongly light-absorbing nature of the effluent suggests that changes in light penetration or hue could be outside guideline values.

(b) Fig. A3.1C shows that the strong blue light absorption by the Kraft effluent shifts the wavelength of maximum light penetration to about 640 nm. Based on the fact that $K(PAR)$ is only slightly larger than K for the wavelength of maximum light penetration, we can compare $K(640)$ in the river downstream with $K(560)$ upstream (= 0.363 m^{-1}) as a rough guide to change in total light penetration. Using the absorption and scattering coefficients at 640 nm that are given in Table A3.1 the resulting absorption and scattering coefficients of the river downstream can be calculated from the mass balance. The scattering coefficient, $b_d(640)$, is easily calculated from the mass balance as 2.88 m^{-1}.

Calculation of the change in absorption coefficient is a little more complicated because the measured absorption of the effluent is that of the wastewater constituents alone. At 640 nm in the red part of the spectrum, the absorption by pure water is not negligible, even in a highly light-absorbing effluent. So we have to calculate the total absorption of light in the effluent by adding the absorption of pure water to that of the effluent constituents, that is, $a = a_c + a_w = 3.07 + 0.33 = 3.40$ m^{-1}, using values in Table A3.1. Absorption downstream (at 640 nm) is calculated: $a_d = [a_u(Q - q) + a_{eff} q]/Q = [0.397(26 - 1.2) + 3.4 \times 1.2]/26 = 0.535$ m^{-1}. Using these values for $a_d(640)$ and $b_d(640)$ in equation (A3.2) then gives $K(640) = 0.660$ m^{-1}, suggesting a marked reduction in light penetration (roughly 45% change in virtual euphotic depth) which is well outside the guideline value of 10% (Sections 5.3.3 and 5.3.5).

Again, since the maximum river depth (4 m) is less than half the (virtual) natural euphotic depth (z_{eu} = 11 m) the appropriate guideline is that change in lighting at the riverbed be less than 20%. Upstream the ratio of irradiance at 560 nm at the bed to that at the water surface is $\exp(-Kz) = \exp(-0.363 \times 4) = 0.23$ (bed lighting is 23% of surface lighting). Downstream of the Kraft mill discharge the ratio at 640 nm is exp(-0.660×4) = 0.07, that is, bed lighting is roughly ($0.23 - 0.07$) $\times$ 100/0.23 = 70% lower than the upstream level, well outside the 20% guideline (Section 5.3.5).

Many river plants are strongly light limited owing to self-shading and are thus sensitive to any light reduction. Thus the severe reduction in light penetration induced by the Kraft discharge may have far-reaching ecological effects and is not acceptable.

A more accurate analysis of the reduction in light penetration requires modelling of the photosynthetically available radiation (PAR) in the water column as a function of depth. Kirk (1984b) has shown how this can be done using a simple program on a personal computer. Briefly, the method involves calculating the spectral irradiance at 10 nm intervals across the PAR band (400–700 nm) for about ten successive depths spaced at depth interval, Δz, through the euphotic zone: $E(\lambda, z) = E(\lambda, z + \Delta z) \exp(-K \Delta z)$. The quantum irradiance in the PAR band, $E(PAR, z)$ is then calculated at each depth by integrating the spectral irradiance, and, finally, $K(PAR)$ is estimated by linear regression of ln $E(PAR)$ versus depth, z. The results are very weakly dependent on the incident spectral irradiance assumed for the purpose of calculation: Kirk (1984b) used the spectrum reported by Bird *et al.* (1982) (Fig. A1.1).

Application of Kirk's method to the river under consideration gives $K(PAR)$ = 0.46 m^{-1} upstream (a little higher than the measured value of 0.42 m^{-1}) compared with 0.76 m^{-1} downstream of the Kraft effluent discharge, confirming the above preliminary analysis using the wavelength of maximum light penetration.

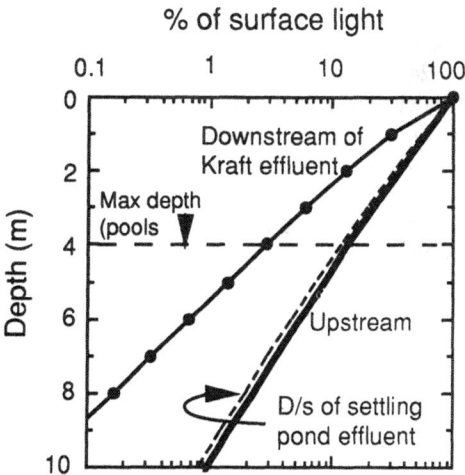

Fig. A3.2. Depth profiles of irradiance (PAR) in the river showing approximately exponential attenuation. PAR was calculated at 1 m depth intervals from the spectral attenuation coefficient by the method of Kirk (1984b) assuming that the ambient irradiance (sunlight) had the spectral distribution given by Bird *et al.* (1982). The departure of the profiles from linearity near the water surface arises because the colour of the light changes rapidly with depth from (nearly) white sunlight to the characteristic water colour. This colour change, and the resulting curvilinearity, is most pronounced downstream of the Kraft mill.

Fig. A3.2 shows the PAR profiles calculated by Kirk's method for the river upstream, downstream of the settling pond discharge, and downstream of the Kraft mill (each curve is for one or the other effluent in isolation). These simulated irradiance profiles on a semilog scale deviate slightly from linearity near the river water surface. This non-linearity, which is especially marked for the profile downstream of the Kraft mill, results from the rapid change in colour of the light entering the water (initially white light from the sun) due to selective absorption. The ratio of the PAR at the bed in the deeper pools of the river (at 4 m depth) to the surface PAR would be about 14% upstream, about 12% downstream of the settling pond, and only 3% downstream of the Kraft mill (Fig. A3.2).

(c) The shift in wavelength of maximally penetrating light is associated with a shift in hue, probably to orange, which would be expected to be outside the guideline hue shift of 10 Munsell units. To demonstrate this would require so-called 'chromaticity analysis' of the calculated reflectance spectrum for the water (e.g. Davies-Colley *et al.* 1988). Chromaticity analysis would also permit the brightness change to be calculated in terms of illuminance reflectance. Such sophisticated analysis is a specialist task and would only be warranted in the (comparatively rare) situation where there was a problem with hue when light penetration and visual clarity guidelines were met.

REFERENCES

Bird, R. E.; Hulstrom, R. L.; Kliman, A. W.; Eldering, H. G. 1982: Solar spectral measurements in the terrestial environment. *Applied optics 21*: 1430–1436.

Davies-Colley, R. J.; Vant, W. N.; Wilcock, R. J. 1988: A comparison of lake water colour as observed directly and as calculated from underwater spectral irradiance. *Water resources bulletin* 24: 11–18.

Kirk, J. T. O. 1984a: Dependence of relationship between inherent and apparent optical properties of water on solar altitude. *Limnology and oceanography 29*: 350–356.

Kirk, J. T. O. 1984b: Attenuation of solar radiation in scattering–absorbing waters: a simplified procedure for its calculation. *Applied optics 23*: 3737–3739.

Smith, R. C.; Baker, K. S. 1981: Optical properties of the clearest natural waters (200–800 nm). *Applied optics 20*: 177–184.

Index

www.ingramcontent.com/pod-product-compliance
Lightning Source LLC
Chambersburg PA
CBHW060809220326
41598CB00022B/2577